Gerhard Pfaff (Ed.)
**Encyclopedia of Color, Dyes, Pigments**

## Also of Interest

*Encyclopedia of Color, Dyes, Pigments.*
*Volume 2: Color Measurement - Metal Effect Pigments*
Gerhard Pfaff (Ed.), 2021
ISBN 978-3-11-058684-8, e-ISBN (PDF) 978-3-11-058710-4,
e-ISBN (EPUB) 978-3-11-058694-7

*Encyclopedia of Color, Dyes, Pigments.*
*Volume 3: Mixed Metal Oxide Pigments - Zinc Sulfide Pigments*
Gerhard Pfaff (Ed.), 2021
ISBN 978-3-11-058686-2, e-ISBN (PDF) 978-3-11-058712-8,
e-ISBN (EPUB) 978-3-11-058688-6

*Inorganic Pigments*
Gerhard Pfaff, to be published in 2024
ISBN 978-3-11-074391-3, e-ISBN (PDF) 978-3-11-074392-0,
e-ISBN (EPUB) 978-3-11-074399-9

*Applied Inorganic Chemistry.*
*Volume 1: From Construction to Photofunctional Materials*
Rainer Pöttgen, Thomas Jüstel, Cristian A. Strassert (Eds.),
to be published in 2022
ISBN 978-3-11-073814-8, e-ISBN (PDF) 978-3-11-073314-3,
e-ISBN (EPUB) 978-3-11-073332-7

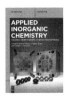

*Applied Inorganic Chemistry.*
*Volume 2: From Magnetic to Bioactive Materials*
Rainer Pöttgen, Thomas Jüstel, Cristian A. Strassert (Eds.),
to be published in 2022
ISBN 978-3-11-073837-7, e-ISBN (PDF) 978-3-11-073347-1,
e-ISBN (EPUB) 978-3-11-073357-0

# Encyclopedia of Color, Dyes, Pigments

Volume 1: Antraquinonoid Pigments – Color Fundamentals

Edited by
Gerhard Pfaff

**DE GRUYTER**

**Editor**
Prof. Dr. Gerhard Pfaff
Allee der Kosmonauten 69
12681 Berlin
Germany
pfaff.pigmente@gmx.de

ISBN 978-3-11-058588-9
e-ISBN (PDF) 978-3-11-058807-1
e-ISBN (EPUB) 978-3-11-058663-3

**Library of Congress Number:** 2021948063

**Bibliographic information published by the Deutsche Nationalbibliothek**
The Deutsche Nationalbibliothek lists this publication in the Deutsche Nationalbibliografie;
detailed bibliographic data are available on the Internet at http://dnb.dnb.de.

© 2022 Walter de Gruyter GmbH, Berlin/Boston
Cover image: Gettyimages/Sagittarius Pro
Typesetting: Integra Software Services Pvt. Ltd.
Printing and binding: CPI books GmbH, Leck

www.degruyter.com

MIX
Papier aus verantwor-
tungsvollen Quellen
FSC
www.fsc.org    FSC® C083411

# Preface

The three-volume "Encyclopedia of Color, Dyes, Pigments" represents this attempt to summarize the current expertise in the fields stated in the title of the book. The main objective is to present the information that is available today encompassed by these three terms in a scientifically and technically correct, high-level and up-to-date manner. All aspects from theory to practical application are covered.

As the title suggests, all major classes of dyes and pigments are covered in detail. Emphasis is given so that the reader obtains an overview of the basic principles, the synthesis possibilities, the production, the chemical and physical properties as well as the technical application of these colorants. Separate chapters are provided for the main areas of application of dyes and pigments, for color fundamentals and color measurement as well as for historical pigments, dyes and binders. The encyclopedia is addressed to color specialists in industry and academia as well as to dye and pigment users in the applications areas of coatings, paints, cosmetics, plastics, printing inks, ceramics, and building materials. In order to make orientation as easy as possible for the interested reader, the topics covered in the book are arranged in alphabetic order.

Color has always played an important role in the lives of humans and animals. Dyes and pigments were therefore important substances early on in the development of mankind for expressing oneself through color, white or black and for shaping life. Colors and their functions have always been fascinating, playing a major role in the human psyche, and are of great importance in the design of a wide variety of surfaces. While natural colorants were initially used by people for thousands of years, the development of modern natural sciences, especially chemistry, led to the introduction of synthetically produced dyes and pigments in the lives of people. These play a predominant role among today's colorants.

The treatment of the individual topics differs in many cases from that in other reference books. New ways of presentation are chosen for different classes of dyes and pigments, but also for the application of colorants in various systems. The encyclopedia is thus up to date, especially since the latest findings from the field have been included. The authors involved are very familiar with the contents of their chapters, most of them having researched and worked on them for many years.

The objective of the Encyclopedia of Color, Dyes and Pigments is to provide a comprehensive overview of the state of knowledge about dyes and pigments, while at the same time identifying the connections to the relevant coloristic and application fundamentals. Special attention was paid to developing a clear structure of the approximately 70 chapters to provide access to the desired information quickly and without a long search. Figures and tables inserted illustrate fundamental aspects, dye and pigment structures, manufacturing details as well as application examples. At the end of each chapter, references to further reading are given. These include a wide range of journal articles, reference books and patents.

https://doi.org/10.1515/9783110588071-202

The chapters to follow are written by some of the most knowledgeable authors in the subject areas covered in the encyclopedia. Most of them were and are involved in innovative developments and practical applications of colorants. They are authors of a variety of publications, patents, presentations and lectures in the respective areas. Their insights will certainly prove to be valuable to the reader and their contributions to the Encyclopedia of Color, Dyes, Pigments are greatly appreciated.

At this point, I would like to take the opportunity to express my sincere thanks to all the authors involved in this book project. Without their support and cooperation, the completion of the encyclopedia would not have been possible. Finally, a big thank you goes to Karin Sora from de Gruyter, who has supported and accompanied the project from the early beginning and to Vivien Schubert and Esther Markus from the same publisher, who have done an extraordinarily good job in the completion of the encyclopedia.

Berlin, December 2021                                                    Gerhard Pfaff

# Contents

# Volume 2

## Volume 3

# List of contributors

**Adrian Abel**
DCC Europe
Rossendale
Lancashire
United Kingdom
adrianfixby@gmail.com

**Michael R. Bartelt**
Business Unit Effect Pigments
Schlenk Metallic Pigments GmbH
Barnsdorfer
Hauptstr. 5, Roth
91154 Germany
michael.bartelt@schlenk.de

**Robert Christie**
School of Textiles & Design
Heriot-Watt University
Scottish Borders Campus
Galashiels TD1 3HF, United Kingdom
rmchristie66@gmail.com

**Magdalene Gärtner**
Maltechnik
Institut für Maltechnik
Muenstergasse
3, Schwaebisch Gmuend
73525 Germany
magdalene.gaertner@yahoo.de

**Uwe Hempelmann**
Technical Operations, Lanxess Deutschland
GmbH
Rheinuferstraße 7-9, Krefeld
Nordrhein-Westfalen
47812 Germany
uwe.hempelmann@-lanxess.com

**Heinz Langhals**
Department of Chemistry
Ludwig-Maximilians-Universität
Butenandtstr. 13,
Munich D-81377
Germany
langhals@lrz.uni-muenchen.de

**Ghita Lanzendörfer-Yu**
Owner
DejaYu Kosmetikblog
Fuchsgrube 36
Mülheim
45478 Germany
info@dejayu.de

**Frank Maile**
Business Unit Effect Pigments
Schlenk Metallic Pigments GmbH
Barnsdorfer
Hauptstr. 5, Roth
91154 Germany
frank.maile@schlenk.de

**Heinz Mustroph**
Former FEW Chemicals GmbH
Technikumstraße 1,
Bitterfeld-Wolfen
06756 Germany
mustroph@few.de

**Gerhard Pfaff**
Former Merck KGaA
Allee der Kosmonauten 69, Berlin
12681 Germany
pfaff.pigmente@gmx.de

https://doi.org/10.1515/9783110588071-204

**Thomas Rathschlag**
Weilburger Graphics GmbH
Am Rosenbühl 5,
Gerhardshofen 91466
Germany
t.rathschlag@weilburger-graphics.de

**Werner Rudolf Cramer**
Optical research
Cramer Lackdesign
Hafenweg
22, Muenster 48155
Germany
info@wrcramer.de

**Andrew Town**
Lambson Ltd., Clifford House
York Road, Wetherby
United Kingdom of Great Britain and Northern
Ireland
andrew.towns@lambson.com

Robert Christie and Adrian Abel

# 1 Anthraquinonoid pigments

**Abstract:** Colorants based on the anthraquinone structure are categorized as a sub-class of carbonyl colorants. Anthraquinone textile dyes rank second in importance to azo dyes, especially within the vat dye application class. Vat dyes became of interest to the pigment industry because of their insolubility. This insolubility and generally excellent fastness properties inspired investigations into the selection of suitable established anthraquinonoid vat dyes for use as pigments after conversion to a physical form that is appropriate for their applications. Originally this proved difficult, but was eventually achieved following the development of appropriate conditioning after treatment processes. The structural chemistry of the various types of anthraquinonoid pigments in relation to their technical and coloristic performance is discussed. The chapter concludes with an illustrated description of the main synthetic routes and finally with a description of the principal applications of the individual commercial products. Anthraquinonoid pigments are generally regarded as high-performance products, suitable for highly demanding applications, although they tend to be expensive.

**Keywords:** anthraquinonoid, madder, alizarin, purpurin, quinizarin, vat dyes, vat pigments, aminoanthraquinones, heterocyclic, polycarbocyclic, anthrapyrimidine, indanthrone, flavanthrone, pyranthrone, anthanthrone, isoviolanthrone

## 1.1 Fundamentals

Colorants based on the anthraquinone structure, categorized as a sub-class of carbonyl colorants, hold a special place in the field of color chemistry. The parent unsubstituted anthraquinone molecule (**1a**) is illustrated in Figure 1.1, with reference to Table 1.1. Anthraquinone textile dyes rank second in importance to azo dyes. Many textile dyes are anthraquinones carrying a range of substituents. However, the important commercial anthraquinonoid pigments are invariably large polycyclic molecular structures, a feature that is primarily responsible for their high levels of technical performance, including fastness to light, weather, solvents, chemicals, and heat [1–3].

This article has previously been published in the journal Physical Sciences Reviews. Please cite as: R. Christie, A. Abel, Anthraquinonoid Pigments *Physical Sciences Reviews* [Online] 2021, 6. DOI: 10.1515/psr-2020-0146

https://doi.org/10.1515/9783110588071-001

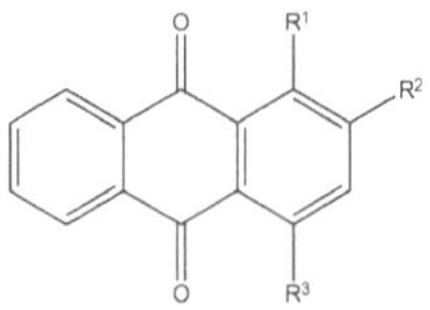

Figure 1.1: Structures of anthraquinones (1a-1d).

Table 1.1: Substituent pattern in some simple anthraquinone molecules.

| Compound | $R^1$ | $R^2$ | $R^3$ |
|---|---|---|---|
| 1a | H | H | H |
| 1b | OH | OH | H |
| 1c | OH | OH | OH |
| 1d | OH | H | OH |

## 1.2 History

Madder, one of the earliest natural dyes, is obtained from the root of Rubia tinctorum, also known as dyer's madder. It is an herbaceous perennial plant related to the Rubiaceae family, which includes coffee. The roots are harvested after two years, and an inner layer provides the best quality dye. The dye is applied to the cloth with alum (a hydrated sulfate salt of aluminum), which acts as a mordant that fixes the dye to the cloth. It is claimed that the dye was being used in the Indian sub-continent as early as 2300 BC, based on the discovery of a piece of cotton colored with madder in Mohenjo-daro, an archaeological site in the Indus Valley, Pakistan [4]. This dye was later traded around the world and was of such importance that it was the cause of trade wars between various European countries and the colonized lands in the Americas. It was one of Sir Isaac Newton's "colour spectrum", that he postulated in 1672. The main active components of this dye are alizarin (1b) and purpurin (1c), hydroxy derivatives of anthraquinone, as illustrated in Figure 1.1 with reference to Table 1.1. The dye could be converted to an insoluble pigment, madder lake, by precipitating (laking) with alum. This lake pigment has low lightfastness and is remembered mainly as a component of mixtures used to paint miniatures, popular in the fifteenth and sixteenth centuries, especially in the Flemish region of Belgium. German chemists, Graebe and Liebermann discovered a synthetic route to alizarin in 1868 [5], around the same time as Perkin, who had discovered Mauveine, devised an alternative process that he patented in the UK. A year later, Perkin developed a more practical route, which he adopted at his Greenford manufacturing plant, London and quickly brought it into production. Soon after these events, the cultivation of madder root virtually ceased. A pigment that was originally made from natural madder, marketed as CI Pigment Red 83, is still made but using synthetic alizarin. This calcium lake, strictly a metal salt pigment, was discovered in 1826 by Robiquet and Colin. Older versions of the Colour Index refer to it as a metal complex. The main current use of this pigment, still referred to as madder lake, is in the coloration of soap, cosmetics, and artists' colors. It has very poor fastness to solvents and is not fast to light. Confusingly, this product is often referred to as CI Pigment Red 83:1 (especially in artists' colors), but there is no such product listed in the Colour Index. Another metal salt pigment, made from quinizarin (**1d**), was discovered by Bayer and introduced as Helio Fast Rubine 4BL, either as a bright violet sodium salt (CI Pigment Violet 5) or, more often, as an aluminum salt (CI Pigment Violet 5:1), which has a bright reddish violet hue. Despite its poor tinctorial strength, low solvent resistance and low lightfastness in reductions, the aluminum salt was still being used for general industrial paints until relatively recently, because it offered an inexpensive way to obtain violet colors.

In terms of volume, anthraquinones were second only to azo colorants until the discovery of copper phthalocyanine. Many anthraquinone textile dyes provided the highest quality dyeings with respect to lightfastness and washfastness. Anthraquinones have been developed as acid, disperse, mordant, reactive, and vat dyes. Vat dyes were the application class of most interest to the pigment industry because of their insolubility. They constitute a group of insoluble colorants that are applied to cellulosic fibers (e. g., cotton) via a water-soluble (leuco) form that is obtained by reduction with sodium dithionite in alkaline solution. They are then oxidized to regenerate the pigment as insoluble particles trapped within the fibers. Vat dyes set new standards for washfastness in textiles, to such an extent that the German dye industry allowed vat dyed articles to carry a special label, the Indanthren label, assuring purchasers of goods that had been dyed in this way that they would remain fast during washing. Application of the commercial forms of the early vat dyes as pigments was not straightforward, because of difficulties in wetting out the particle surfaces and in producing fine particle dispersions, and consequently they lacked tinctorial strength and brightness. However, the insolubility and very good fastness properties of vat dyes inspired investigations into the selection of suitable established products for use as pigments after conversion to a physical form that is appropriate for their applications. Lessons learned, especially from experience of the conditioning aftertreatment processes developed for copper phthalocyanine pigments, were applied to many of these vat dyes, and a few became important as pigments, generally referred to as vat pigments. These products offered the technical and coloristic advantages, notably high levels of fastness performance and bright colors, that were necessary to justify the high price required for commercial viability, which resulted from the expensive synthesis processes together with additional costs associated with the conditioning. They were aimed at applications where the high cost could be justified, such as automotive finishes and the coloration of plastics. However, as demand for high-performance pigments increased as the twentieth century progressed, industry responded with the development of pigments based on new chromophores which ultimately undercut the anthraquinones in price. Consequently, most original anthraquinonoid pigments have declined in importance, with many withdrawn, and only a few remain as commercial products. The use of anthraquinonoid vat dyes for textiles has also declined over the years, mainly because very good fastness to washing can be obtained using alternative dye application classes, especially reactive dyes, which offer better economics.

## 1.3 Structures and properties

Quinones, defined as cyclohexadienediones, contain a 6-membered ring with two ketone carbonyl groups and two double bonds as the essential structural arrangement. Their molecular structures frequently give rise to highly colored species. Anthraquinones, as illustrated by compounds (**1a-1d**) (Figure 1.1), contain a linear arrangement of three fused six-membered rings with two outer aromatic rings and two carbonyl groups in the middle ring. In the chemistry of textile dyes, the anthraquinone chemical class is second in importance to azo dyes. However, it is much less important in organic pigments. One reason is that the role of anthraquinones in many textile dye application classes is to provide lightfast blue colors, to complement azo dyes in yellow, orange and red colors, while this position is occupied by copper phthalocyanines in organic pigments. As discussed in the previous section, anthraquinonoids dominate the vat dye textile application class. Around twenty of the hundreds of known anthraquinonoid vat dyes have been converted to pigment use, but only a few have current industrial significance. These pigments essentially cover the entire shade range and offer high levels of performance suitable for demanding paint and plastics applications. However, their high cost limits their use. Anthraquinonoid pigments may be separated into three main categories based on their polycyclic chemical structures: aminoanthraquinone derivatives, heterocyclic derivatives and polycarbocyclic derivatives.

### 1.3.1 Aminoanthraquinone derivatives

In this category, CI Pigment Red 177 (**2**) is the most important product (Figure 1.2). Other pigments in this category that have been important in the past, such as CI Pigments Yellow 123 and 147, no longer have significant commercial use. X-ray structural analysis has revealed that, in the crystalline solid state, CI Pigment Red 177 (**2**) exhibits a twisting of the two anthraquinone units by 75° relative to one another, which is reported to provide the optimum geometry for intramolecular hydrogen bond formation [6].

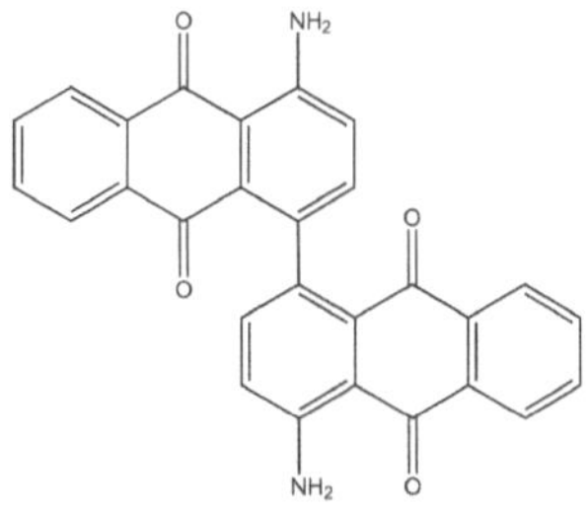

**Figure 1.2:** Structure of the aminoanthraquinone Pigment, CI pigment Red 177 (**2**).


### 1.3.2 Heterocyclic derivatives

Polycyclic anthraquinonoid pigments containing heterocyclic ring systems, illustrated structurally in Figure 1.3, are based on some of the longest-established vat dyes. CI Pigment Yellow 108 (**3**) is based on the anthrapyrimidine heterocyclic system. Indanthrone (**4**), CI Pigment Blue 60, is a long-established pigment that is highly lightfast and weatherfast and suitable for use in automotive finishes, especially in combination with metallic pigments. A dichloroindanthrone, CI Pigment Blue 64, has had some market impact, but there are no current manufacturers. Indanthrone exists in four polymorphic forms. The α-form is the most stable and provides greenish blue hues. The β- and γ- forms provide reddish shades. The δ- form is not of coloristic interest. Flavanthrone yellow (**5**), CI Pigment Yellow 24, has in the past been a highly important vat pigment although its use has declined significantly. α-Indanthrone

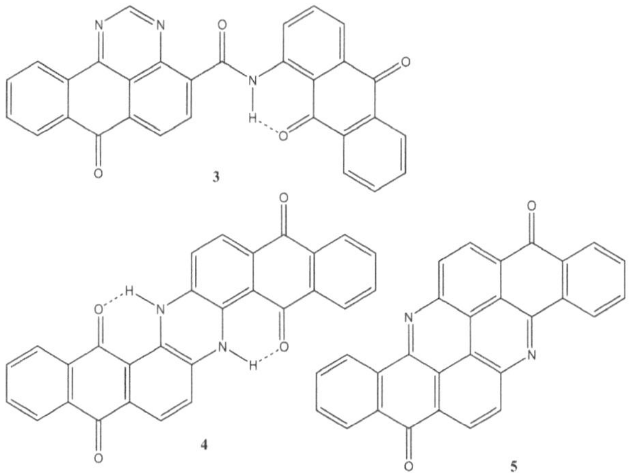

**Figure 1.3:** Structures of heterocyclic anthraquinonoid pigments.

and flavanthrone have similar crystal structures with their planar molecules arranged in stacks in a herringbone arrangement, as is commonly encountered in organic pigments [7].

### 1.3.3 Polycarbocyclic derivatives

The prominent examples of polycarbocyclic anthraquinonoid pigments are based on the pyranthrone (**6**), anthanthrone (**7**) and isoviolanthrone (**8**) ring systems, as illustrated in Figure 1.4. The similarity in the molecular structures of flavanthrone (**5**) and pyranthrone (**6**) is notable, with the heterocyclic N-atoms in compound (**5**) replaced by CH in compound (**6**). The parent pyranthrone (**6**), CI Pigment Orange 40, has been discontinued. However, CI Pigment Orange 51, a dichloro derivative, CI Pigment Red 216, a tribromo derivative, and CI Pigment Red 226, a dichlorodibromo derivative, find some use in industrial paints. 4,10-Dibromoanthanthrone, CI Pigment Red 168, is the most important derivative of the heterocyclic system (**7**), while a dichloro derivative, CI Pigment Violet 31, is the most prominent example of a commercial pigment based on the isoviolanthrone system (**8**). The halogenated pigments are planar molecules and exhibit similar crystal structures, arranged in stacks that are slightly inclined to the stacking direction, an arrangement that ensures that spaces are perfectly filled.

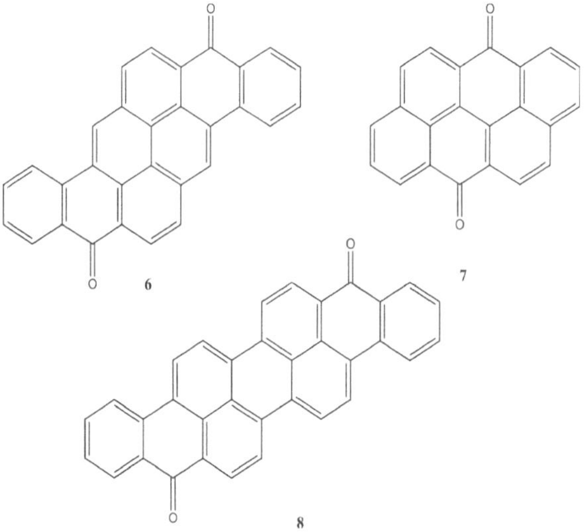

**Figure 1.4:** Structures of Polycarbocyclic Anthraquinonoid Pigments.

## 1.4 Synthesis and manufacture

The synthesis of anthraquinone dyes is well documented [1–3, 8, 9]. Although the chemistry of the syntheses is for the most part long established, some mechanistic detail of individual reactions remains speculative. The synthetic sequences most commonly start with an appropriate substituted anthraquinone, with cyclization reactions subsequently used to construct the polycyclic systems. In a few cases, the anthraquinone ring system is generated during the synthesis by cyclization reactions. A common feature of these pigments is the prevalence of halogenated (chloro and bromo) derivatives. These products may be obtained by halogenation of an intermediate in the process followed by cyclization or by direct halogenation of the parent anthraquinonoid system. In most cases, the pigments are produced initially in a large particle size form. They are then reduced to pigmentary particle size by processes such as acid pasting, which involves reprecipitation from solution in concentrated sulfuric acid, or conversion by reduction in alkali to the water-soluble leuco forms followed by re-oxidation to the insoluble pigment, adapted from the processes used in the vat dyeing of textiles.

The synthetic route to pigment (2) from 1-amino-4-bromoanthraquinone-2-sulfonic acid, known as bromamine acid (9), a frequently used textile dye intermediate, is outlined in Figure 1.5. When treated with copper powder in dilute sulfuric acid (an Ullmann reaction), compound (9) undergoes dimerization leading to disulfonate intermediate (10) as its sodium salt. This intermediate is then treated with 80% sulfuric acid at around 140°C to remove the sulfonate groups and thus form pigment (2).

Figure 1.5: Synthesis of the aminoanthraquinone pigment, CI Pigment Red 177 (2).

CI Pigment Yellow 108 (3) is synthesized by condensation of anthrapyrimidine derivative (11) with 1-aminoanthraquinone (12) in the presence of thionyl chloride at 140–160°C in a high boiling aromatic solvent, as illustrated in Figure 1.6.

While indanthrone (4) may be prepared in various ways, the principal industrial method involves oxidative dimerization of 2-aminoanthraquinone (13) by fusion with sodium/potassium hydroxide at around 220°C in the presence of an oxidant such as

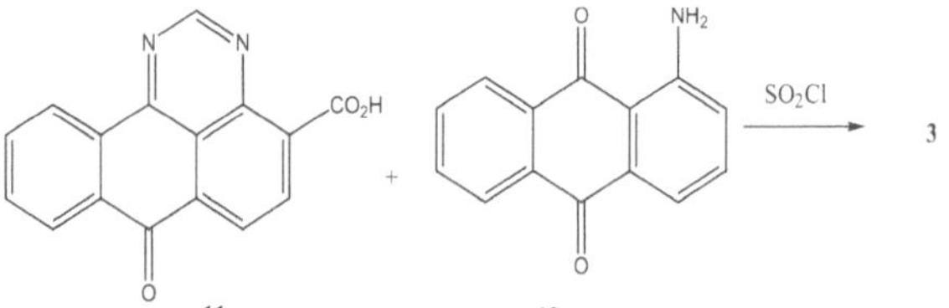

**Figure 1.6:** Synthesis of CI Pigment Yellow 108 (**3**).

sodium nitrate (Figure 1.7). Curiously, the same product can be prepared from 1-aminoanthraquinone (**12**) in a similar way, although the 2-isomer is the usual industrial starting material. The pigmentary form may be produced by conversion to its leuco form by reduction with an aqueous alkaline solution of sodium dithionite, followed by oxidation to the insoluble pigment.

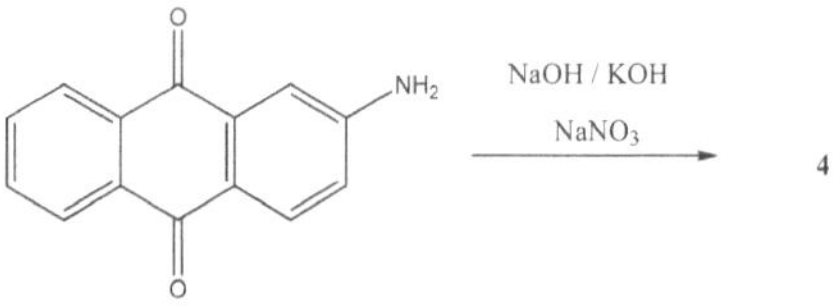

**Figure 1.7:** Synthesis of indanthrone (**4**).

The industrial manufacture of flavanthrone (**5**) (Figure 1.8) initially involves a condensation reaction of 1-chloro-2-aminoanthraquinone (**14**) with phthalic anhydride (PA) to form intermediate (**15**), which subsequently undergoes an Ullman reaction with copper powder in trichlorobenzene leading to the dimerized product (**16**). This intermediate is then treated with boiling dilute aqueous sodium hydroxide, undergoing cyclization with cleavage of the phthalimide groups to yield the pigment (**5**). The pigmentary form may be obtained by reprecipitation from a solution in concentrated sulfuric acid or via the leuco form as described for the synthesis of indanthrone (**4**). Solvent treatment processes may also be used.

In the synthesis of pyranthrone (**6**) (Figure 1.9), 1-chloro-2-methylanthraquinone (**17**) is treated with copper powder, pyridine, and sodium carbonate in an organic solvent at around 180°C to form 2,2′-dimethyl-1,1′-dianthraquinonyl (**18**), which then undergoes cyclization by refluxing with an alcoholic solution of sodium hydroxide to afford the leuco form, which is then oxidized with air to provide pyranthrone (**6**). The halogenated derivatives that are used as pigments are prepared either by halogenation of intermediate (**18**) followed by cyclization, or by halogenation of the parent compound (**6**). Conditioning of the pigments involves procedures broadly similar to those outlined for indanthrone and flavanthrone.

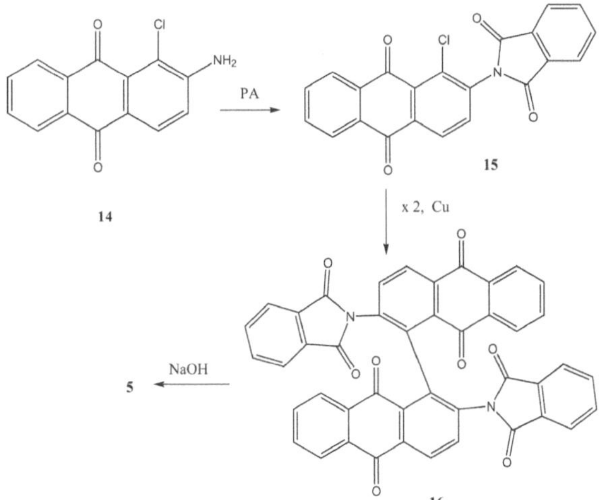

**Figure 1.8:** Synthesis of flavanthrone (**5**).

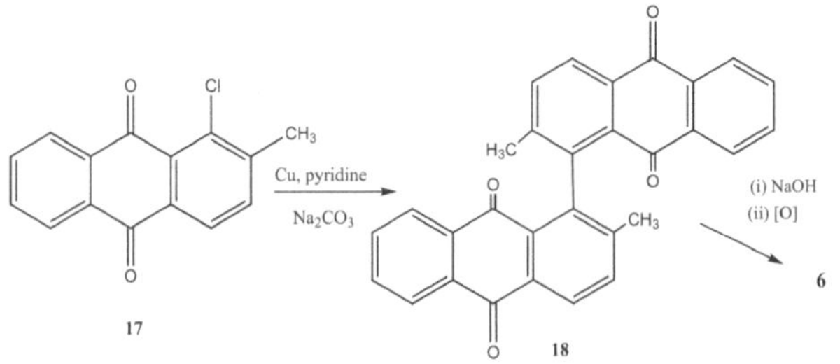

**Figure 1.9:** Synthesis of pyranthrone (**6**).

The synthesis of anthanthrone (**7**) (Figure 1.10) involves the diazotization of 1-aminonaphthalene-8-carboxylic acid (**19**) followed by treatment of the diazonium salt solution with copper powder at the boil to form 1,1-dinaphthyl-8,8′-dicarboxylic acid (**20**). Compound (**20**) undergoes intramolecular Friedel Crafts acylation in concentrated sulfuric acid, cyclizing to give anthanthrone (**7**). 4,8-Dibromoanthranthrone (CI Pigment Red 168), the most important commercial product of this structural type, is formed by bromination of the parent compound, with or without its isolation from the synthetic sequence.

Isoviolanthrone (**8**) is obtained by boiling 3,3′-dibenzanthronyl sulfide (**21**) with alcoholic potassium hydroxide (Figure 1.11). Chlorinated and brominated isoviolanthrones are obtained by direct halogenation of parent compound (**10**).

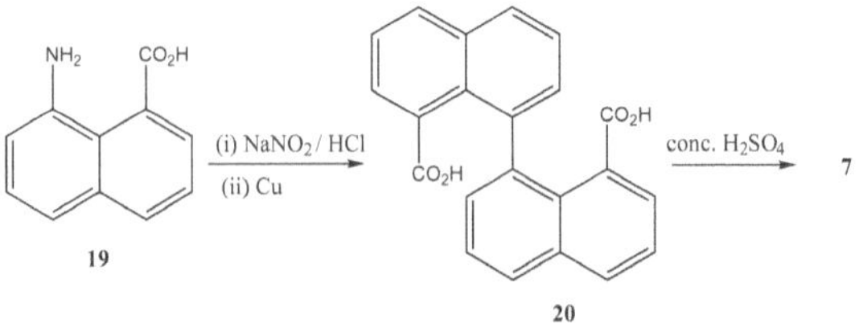

**Figure 1.10:** Synthesis of anthanthrone (**7**).

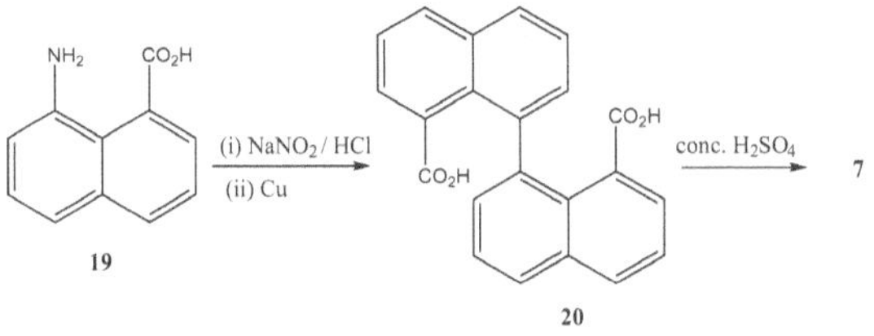

**Figure 1.11:** Synthesis of isoviolanthrone.

## 1.5 Applications

The use of some important classical anthraquinonoid pigments has declined substantially over the years and many are no longer offered by traditional pigment manufacturers, although some may still be available from manufacturers in Asia. Applications of the few products that remain as important commercial pigments are described.

### 1.5.1 CI Pigment Yellow 24 (5)

This product, known as flavanthrone yellow, and also as CI Vat Yellow 1, is offered as a bright reddish yellow version of the pigment, which is quite opaque, and also as a transparent, high strength version. The transparent form gained acceptance for use in automotive metallic finishes. It was discovered as a vat dye in 1901 by Bohn, and then patented and marketed by BASF. The pigment tends to darken when exposed to light providing the unusual property of higher rated lightfastness in pale shades than in strong shades, which is the opposite performance compared with most other pigments, although the phenomenon is more common with anthraquinononoid pigments. In these cases, the darkening with exposure masks a degree of fading in pale reductions. In plastics, the pigment is stable up to 230°C in polyolefins. Above that temperature, it starts to dissolve and gain tinctorial strength, but remains locked in the polymer without migrating on cooling. Its high color strength and excellent lightfastness allowed its use in the spin coloration of fibers for exterior use. Its use in plastics never matched its use in paint. Its importance has declined substantially over the years.

### 1.5.2 CI Pigment Yellow 108 (3)

This pigment, known as anthrapyrimidine yellow and CI Vat Yellow 20, was introduced to the market by BASF, and offered a much greener shade than flavanthrone (**5**). While lacking brightness of color, it was for many years considered the ultimate yellow pigment for fastness to weathering and was extensively used in automotive finishes, including metallics. It darkened in full shades, as with flavanthrone, and so it found some use in pale shades of decorative paints. It is no longer seriously marketed either as a pigment or as a vat dye.

### 1.5.3 CI Pigment Yellow 147

This pigment provided a mid-yellow shade with only limited use in the coloration of plastics, where it offered good heat stability. It was one of many anthraquinononoid pigments developed by Ciba in the late 1950s to early 60s, and one of the few that was successfully introduced to the market [10]. The pigment still has some applications and Orient Chemicals, who are very active in non-impact printing are now recommending it as a dispersion for inkjet applications.

### 1.5.4 CI Pigment Orange 40 (6)

This pigment, known as pyranthrone and as CI Vat Orange 9, was discovered by Scholl in 1905 and introduced by BASF into their Indanthren range. As a pigment it had limited application in the spin coloration of viscose, where it offered very good lightfastness. It was withdrawn as a powder pigment many years ago.

### 1.5.5 CI Pigment Orange 51

This pigment is a dichloropyranthrone, which is redder and brighter than the parent compound CI Pigment Orange 40 (**6**). It has been recommended for the coloration of paints, where its high lightfastness, excellent heat stability, and good solvent fastness meant that it was considered for automotive finishes, if over-coated with a protective clear finish, to enhance its fastness to weather.

### 1.5.6 CI Pigment Red 177 (2)

This pigment does not appear to have been used as a vat dye and was introduced directly into the Ciba Cromophtal pigment range. There are two alternative versions of the pigment offering either a transparent bluish red or a more opaque yellower shade. The brilliant bluish red shade product is a useful pigment for paints and plastics. In paints, it was a favorite choice for blending with molybdate orange to meet the requirements for automotive original finishes. As these applications were replaced with lead free finishes, the more transparent grade dominated. In spite of its good fastness to weathering, it does not quite meet the requirements of modern automotive original finishes but finds use in general industrial and vehicle repair finishes. In plastics it has some use in polyolefins, in which it is stable to 300°C and does not affect the dimensional stability when used in injection moldings. It can be used for spin coloration of polypropylene. In PVC it offers a bright shade but is limited by its slight solubility in plasticizers.

### 1.5.7 CI Pigment Red 168

This pigment is a dibromoanthanthrone and is also CI Vat Orange 3. The pigment offers a brilliant orange-scarlet shade. It was discovered in 1913 by Kalb of Cassella AG, Frankfurt, which became part of Hoechst. It is now an important member of the Clariant pigment range. Its main use is in the coloration of paints, extending also to coil coatings and specialty printing inks. As with most anthraquinonoid vat pigments, it can be prepared in either transparent or more opaque forms. In paints, it offers excellent lightfastness in all but very pale shades, and even then, is rated as very good to excellent. It is widely used for automotive original finishes, with the transparent form appropriate for metallic and pearlescent finishes, where it can provide useful bronze shades with an interesting "flop" tone. At the other end of the scale, it can be used in decorative (architectural) finishes in very pale shades, where its lack of tinctorial strength (about half that of the perinone pigment, CI Pigment Orange 43) is of little consequence. It has heat stability up to 180°C, although it bleeds slightly when overcoated with an alkyd melamine based stoving paint at as low as 120°C, the bleed increasing at higher temperatures. Its low tinctorial strength and high price means that it does not find application in printing inks or plastics.

### 1.5.8 CI Pigment Violet 31

This pigment is a dichloroisoviolanthrone and is also known as CI Vat Violet 1.

It is currently not widely used as a pigment but is still important as a vat dye. It is a reddish shade violet and has been used for the spin coloration of viscose, where its lightfastness is very good to excellent. There are several sources of the powder pigment in China.

### 1.5.9 CI Pigment Blue 60 (6)

This long-established pigment is known as indanthrone and also as CI Vat Blue 4 [11]. The pigment still creates much industrial interest despite its high cost compared with other blue pigments such as the copper phthalocyanines. Indanthrone was discovered as a vat dye by Bohn, who also discovered flavanthrone in 1901. In 2009, when BASF acquired the Ciba pigment range, the Federal Trade Commission approved the deal on condition that Ciba divested itself of indanthrone (and bismuth vanadate) as both companies had a such a high market share. The manufacturing plant in the Netherlands was eventually acquired in 2010 by Dominion Colour Corporation (DCC), now Dominion Colour Lansco (DCL). Indanthrone is much redder than most phthalocyanine pigments and is somewhat duller. However, it is similar in hue to the ε- crystal modification of copper phthalocyanine but is somewhat duller. Compared with most pigments it might be considered to have high tinctorial strength, but it is weaker than copper phthalocyanine, and is more transparent. With these negative features, it may be surprising that such a pigment still finds significant use. In paints, interest is due to the excellent fastness to weathering, which surpasses copper phthalocyanine particularly in deep blue automotive original equipment and in metallic or pearlescent finishes. It is superior to the α- modification of copper phthalocyanine tinted with a redder pigment such as the dioxazine, CI Pigment Violet 23. It also matches copper phthalocyanine pigments for resistance to most solvents used in paints. In plastics, it maintains its excellent lightfastness in both rigid and plasticized PVC, although its superiority to copper phthalocyanine does not extend to PVC. It is just short of excellent in terms of bleeding in plasticizers. In polyolefins, it has very high heat stability, achieving 300°C at all but the lowest concentrations, when it decreases to 280°C. However, the presence of titanium dioxide severely reduces its heat stability, limiting its use in reductions. It is used in polypropylene fibers and can be used in injection moldings where it has little effect on dimensional stability and hence does not cause warping. It is recommended for both solvent- and water-based inkjet applications.

## References

1.  Hunger K, Schmidt MU. Industrial organic pigments, 4th ed, Weinheim: Wiley-VCH Verlag GmbH, Ch 3 2019.
2.  Zollinger H. Color chemistry. Weinheim: Wiley VCH, Ch 8 2003.
3.  Christie RM. Colour chemistry, 2nd ed, London: RSC, Ch 4 2015.
4.  Bhardwaj H, Kamal K. Indian dyes and dyeing industry during the eighteenth-nineteenth century. Indian J Hist Sci. 1982;17:70–81.

5. Graebe C, Liebermann C. Ueber Alizarin und Anthracen. Berichte der Deutschen Chemischen Gesellschaft. 1868;1:41–56.
6. Ogawa K, Scheel HJ, Laves F. Kristallstrukturen organischer Pigmentfarbstoffe. II. 4,4′-Diamino-1,1′-dianthrachinonyl $C_{28}H_{16}N_2O_4$. Naturwissenschaften. 1966;53:700–1.
7. Thetford D, Cherryman J, Chorlton AP, Docherty R. Intermolecular interactions of flavanthrone and indanthrone pigments. Dyes Pigm. 2005;67:139–44.
8. Hallas G. In colorants and auxiliaries. In: Shore J, editor. Bradford: society of dyers and colourists. Ch 6 1990.
9. Gordon PF, Gregory P. Organic chemistry in colour. Berlin: Springer-Verlag, Ch 2 1983.
10. Gaertner H. Modern chemistry of organic pigments. J Oil Colour Chem Assoc. 1963;46:23–5.
11. Bradley W, Leete E. Chemistry of indanthrone. Part 1. The mode of formation of indanthrone from 2-aminoanthraquinone and potassium hydroxide. J Chem Soc. 1951:2129.

Heinz Mustroph

# 2 Apocyanine dyes

**Abstract:** Cyanine dyes without a methine group or polymethine chain between their two terminal unsaturated heterocycles are called apocyanines. Such colorants which are yellow and red received the labels xantho-apocyanine and erythro-apocyanine dyes, respectively. Nevertheless, from time to time in the literature, the term "apocyanine dyes" is wrongly applied to various *N*-bridgehead heterocycles. In addition, the phenylogous zeromethine hemicyanine dyes are often incorrectly referred to with the term "apocyanine dyes". This chapter clarifies the definition of apocyanine, describes how confusion over the term has come about, and points the ways in which it is sometimes incorrectly employed.

**Keywords:** apocyanine dyes, cyanine dyes, erythro-apocyanine, hemicyanine dyes, xantho-apocyanine

In a patent, filed in 1903, the Farbwerke Hoechst describes the synthesis of two isomeric red and yellow dyes by boiling a highly concentrated methanolic solution of quinoline alkyl halides with caustic soda [1].

Cyanine dyes comprise two terminal nitrogen atoms which form part of separate unsaturated heterocyclic rings that are in conjugation with a polymethine chain. Adolf Kaufmann and Paul Strübin assumed that these new dyes formed in the same reaction as the cyanines and isocyanines, but due to the lack of a methine group or polymethine chain between the two rings they called them *apocyanines* (*απο* = lack). They termed the yellow dye *erythro-apocyanine* and the red *xantho-apocyanine* and assigned formula **1** to the erythro-apocyanine and formula **2** to the xantho-apocyanine [2].

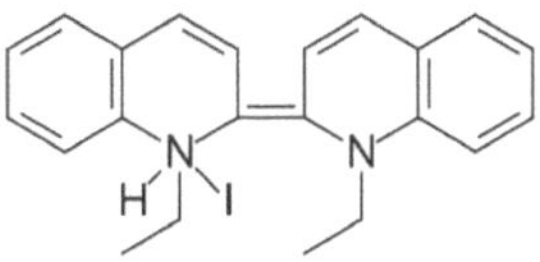

**1** 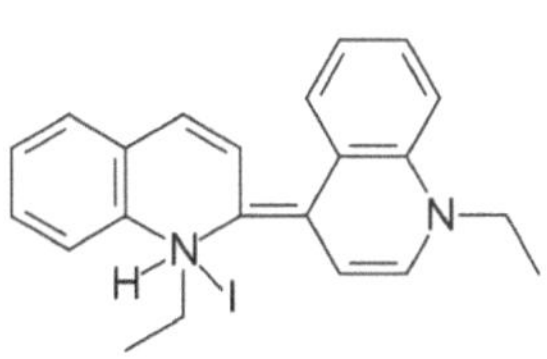        **2**

This article has previously been published in the journal Physical Sciences Reviews. Please cite as: H. Mustroph, Apocyanine Dyes *Physical Sciences Reviews* [Online] 2021, 6. DOI: 10.1515/psr-2020-0147

https://doi.org/10.1515/9783110588071-002

Without any experimental support Walter König suggested the alternative formulae **3** and **4**, respectively [3]. He had this conviction owing to the knowledge that in cyanine dyes there is always an odd number of methine groups between the two nitrogen atoms. Therefore, he assumed that the ring was linked via the 3-position of the first quinoline ring. In addition, he believed the compound with three methine groups in the conjugated chain is the yellow dye, the xantho-apocyanine **3**, and the one with the longer chain of five methine groups is the red dye, the erythro-apocyanine **4**, vice versa to the original trivial names that Kaufmann and Strübin had given them [2].

**3**                                        **4**

William Hobson Mills and Henry Geoffrey Ordish supported König´s view by experimentally demonstrating that erthryo-cyanine's structure is **4** [4]. Without further experimental verification, the structure **3** was also assumed to be correct in the following years, despite the chemical and physical properties of xantho-apocyanines differing significantly from those of the erythro-apocyanines.

That was the state-of-the-art up to 1961. This year, Fritz Kröhnke et al. determined the chemical structure of xantho-apocyanine and reported that the electronic absorption spectra in MeOH of both differ considerably with $\lambda_{max}$ = 516 nm ($\varepsilon$ = 34 700 $M^{-1}$ $cm^{-1}$) for erythro-apocyanine and $\lambda_{max}$ = 461 nm ($\varepsilon$ = 19 500 $M^{-1}$ $cm^{-1}$) for xantho-apocyanine [5]. By means of extensive experimental work, they were able to show that coupling of $-CH_2-$ in the $-N(+)-CH_2-R$ group of the first heterocyclic ring with the second heterocyclic ring happens, giving structure **5** [5].

**5**

This result was later confirmed [6], but without referring to the results of Kröhnke et al. [5].

With this structural elucidation it was clear that "xantho-apocyanine" is a misnomer. The "xantho-apocyanines" are not true apocyanines because in addition to the direct bond between the two quinoline rings they contain a substituted methine bridge. Nevertheless, from time to time in the literature, the term "apocyanine dyes" is wrongly applied to various *N*-bridgehead heterocycles, e. g. indolo[3,2-a]carbazoles [7] and pyrazolo[5,4-b]pyrido[2,1-c]pyrimidines [8].

The *phenylogous zeromethine hemicyanine dyes* are often referred to with the term "apocyanine dyes". However, they are neither cyanine dyes nor apocyanine dyes. Instead they are hemicyanine dyes, where one terminal component is a heterocyclic ring with a nitrogen atom in conjugation with the polymethine chain, as it is typically in cyanine dyes, and the second is obtained from a terminal nitrogen atom which is not part of a heterocyclic ring.

# References

1.  DE 154448. 20 Sep 1903.
2.  Kaufmann A, Strübin P. Über Chinolin-Farbstoffe. (I. Mitteilung: Die Apocyanine). Ber Dtsch Chem Ges. 1911;44:690.
3.  König W. Über die Konstitution der Pinacyanole, ein Beitrag zur Chemie der Chinocyanine. Ber Dtsch Chem Ges. 1922;55:3293.
4.  Mills WH, Ordish HG. The cyanine dyes. Part X. The constitution of the apocyanines. J Chem Soc. 1928;89.
5.  Kröhnke F, Dickhäuser H, Vogt I. Zur Konstitution der sogenannten Xantho-Apocyanine. Liebigs Ann Chem. 1961;644:93.
6.  Afarinkia K, Ansari M-R, Bird CW, Gyambibi I. A reinvestigation of the structure of the erythro and xanthoapocyanine: some unusual aspects of quinoline chemistry. Tetrahedron Lett. 1996;37:4801.
7.  Niesobski P, Nau J, May L, Moubsit A-E, Müller TJ. A mild and sequentially Pd/Cu-catalyzed domino synthesis of acidochromic Indolo[3,2-a]carbazoles – Free bases of apocyanine dyes. Dyes Pigm. 2020;173:107890.
8.  Koraiem AI, Abdellah IM. Synthesis and photophysical characterization of highly stable cyanine dyes based on pyrazolo[5,4-b]pyrido[2,1-c]pyrimidine and pyrazolo[5,4-b]pyrido[2,1-d][1,3,4]triazepine. J Appl Chem. 2018;7:821.

Robert Christie
# 3 Azo (Hydrazone) pigments: general principles

**Abstract:** This paper presents an overview of the general chemical principles under-lying the structures, synthesis and technical performance of azo pigments, the dom-inant chemical class of industrial organic pigments in the yellow, orange, and red shade areas, both numerically and in terms of tonnage manufactured. A description of the most significant historical features in this group of pigments is provided, starting from the discovery of the chemistry on which azo colorants are based by Griess in the mid-nineteenth century, through the commercial introduction of the most important classical azo pigments in the early twentieth century, including products known as the Hansa Yellows, β-naphthol reds, including metal salt pig-ments, and the diarylide yellows and oranges, to the development in the 1950s and 1960s of two classes of azo pigments that exhibit high performance, disazo conden-sation pigments and benzimidazolone-based azo pigments. A feature that compli-cates the description of the chemical structures of azo pigments is that they exist in the solid state as the ketohydrazone rather than the hydroxyazo form, in which they have been traditionally been illustrated. Numerous structural studies conducted over the years on an extensive range of azo pigments have demonstrated this feature. In this text, they are referred to throughout as azo (hydrazone) pigments. Since a com-mon synthetic procedure is used in the manufacture of virtually all azo (hydrazone) pigments, this is discussed in some detail, including practical aspects. The procedure brings together two organic components as the fundamental starting materials, a diazo component and a coupling component. An important reason for the dominance of azo (hydrazone) pigments is that they are highly cost-effective. The syntheses gen-erally involve low cost, commodity organic starting materials and are carried out in water as the reaction solvent, which offers obvious economic and environmental ad-vantages. The versatility of the approach means that an immense number of products may be prepared, so that they have been adapted structurally to meet the require-ments of many applications. On an industrial scale, the processes are straightfor-ward, making use of simple, multi-purpose chemical plant. Azo pigments may be produced in virtually quantitative yields and the processes are carried out at or below ambient temperatures, thus presenting low energy requirements. Finally, provided that careful control of the reaction conditions is maintained, azo pigments may be prepared directly by an aqueous precipitation process that can optimise physical form, with control of particle size distribution, crystalline structure, and surface

This article has previously been published in the journal Physical Sciences Reviews. Please cite as: R. Christie, Azo (Hydrazone) Pigments: General Principles *Physical Sciences Reviews* [Online] 2021, 6. DOI: 10.1515/psr-2020-0148

https://doi.org/10.1515/9783110588071-003

character. The applications of azo pigments are outlined, with more detail reserved for subsequent papers on individual products.

**Keywords:** azo, pigment, hydrazone, monoazo, disazo, diazotization, azo coupling, diazo component, coupling component

## 3.1 Fundamentals

Azo pigments form the dominant chemical class of commercial organic pigments in the yellow, orange, and red shade areas, both numerically and in terms of quantities manufactured. They are complemented commercially in the blue and green shade areas predominantly by copper phthalocyanine pigments. Azo pigments are mostly contained within the classical organic pigment category. However, there are also a few groups of azo pigments that are categorised as high performance. Azo pigments can also be classified chemically according to the number of azo (hydrazone) groups present, as either monoazo (monohydrazone) or disazo (bishydrazone) structures [1–4].

## 3.2 History

A highly significant historical discovery in the development of organic pigments is attributable to the German chemist Griess, whose work provided the foundation for the chemistry of azo dyes and pigments. In 1858, he demonstrated that the treatment of a primary aromatic amine with nitrous acid gave rise to an unstable salt (a diazonium salt) that could be used to prepare highly coloured compounds, which became known as azo dyes [5, 6]. In 1863, Martius discovered a brown dye which he named Bismarck Brown after the German Chancellor, Otto von Bismarck. This was the first azo dye, designated as CI Basic Brown 1. It is still used today in histology for staining tissues. Martius, with Bartholdy, founded Gesellschaft für Anilinefabrikation mbH in 1873, the forerunner of AGFA (Aktiengesellschaft für Anilinfabrikation), which was to become better known for its photographic products. The chemistry was initially applied industrially to the development of water-soluble azo dyes. Initially, the azo dyes were obtained by treatment of a primary aromatic amine with a half equivalent of nitrous acid, so that diazotisation and azo coupling was involved with the amine acting as both diazo and coupling component. Bismarck Brown was an example of such a product. However, the range of azo dyes expanded rapidly after it was discovered that separate diazo and coupling components could be used. The development of a group of yellow, orange and red monoazo pigments took place around the turn of the twentieth century, including products such as the Hansa Yellows and β-naphthol reds, many of which still enjoy commercial importance today amongst the classical azo pigments. Diarylide yellow and orange pigments, which are disazo pigments, reached commercial significance around the 1930s, although they had been discovered much earlier. The subsequent advances in synthetic chemistry were important in developing the current range of high performance pigments, and led to the introduction of two classes of azo pigments exhibiting high performance in the 1950s and 1960s, namely disazo condensation pigments and benzimidazolone-based azo pigments.

## 3.3 Structures and properties

Azo compounds are described structurally as compounds containing one or more azo groups (-N = N-) linked to two carbon atoms. In the case of azo pigments, the carbon atoms are often, though not exclusively, part of aromatic ring systems. The simplest aromatic azo compound is azobenzene (Figure 3.1).

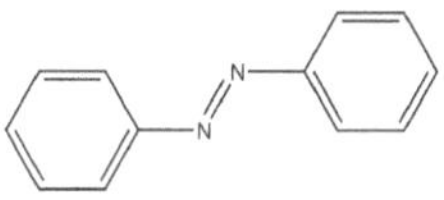

**Figure 3.1:** The structure of azobenzene.

The feature that complicates accurate description of the structures of those azo compounds which contain a hydroxy group at a position *ortho* to the azo group, and virtually all azo pigments fall into this category, is the possibility of tautomerism involving hydroxyazo (a) and ketohydrazone (b) forms, as illustrated in Figure 3.2. In fact, structural studies conducted on an extensive range of azo pigments have demonstrated that, in the solid state, the pigments exist exclusively in the ketohydrazone form [4, 7, 8]. A factor that contributes to explaining the predominance of the ketohydrazone isomer is that, in this form, intramolecular hydrogen-bonding is significantly stronger than in the hydroxyazo form, because of the higher bond polarities in the ketohydrazone system. Most older texts dealing with organic pigments followed the conventional, if strictly inaccurate, approach to illustrate the structures of azo pigments as if they existed in the azo form. Most current texts illustrate the structures as hydrazones although frequently continuing to refer to them as azo pigments. It seems unlikely that the terminology "azo" pigment will be completely replaced by "hydrazone" pigment because of entrenchment of the former term in common usage. However, a recent edition of the extremely important reference work, "Industrial Organic Pigments", describes them as "Hydrazone Pigments (formerly known as Azo Pigments)" [4].

## 3.4 Synthesis and manufacture

An important reason for the dominant position of pigments of the azo chemical class in the range of commercial industrial organic pigments is that they are highly

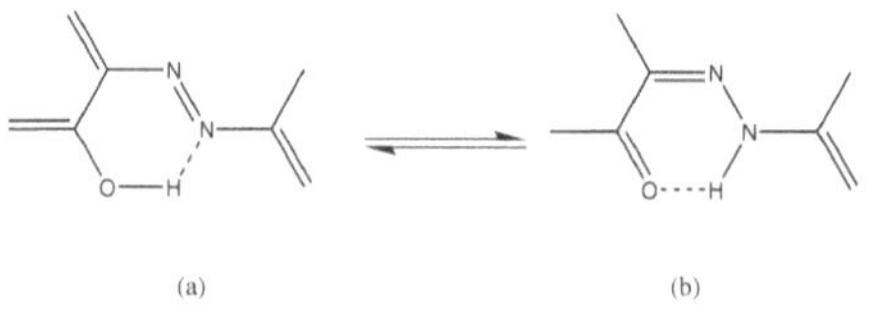

**Figure 3.2:** Hydroxyazo (a)/ketohydrazone (b) tautomerism.

cost-effective. The reasons for their relatively low cost become clear by examining the nature of the materials and processes used in their manufacture. The classical azo pigments are manufactured by variations of their common synthetic route [1–4, 9, 10]. The process leading to monoazo (monohydrazone) pigments brings together two organic components as the fundamental starting materials, a *diazo* component and a *coupling* component, in a two-stage reaction sequence known as diazotization and azo coupling. For disazo (bishydrazone) pigments, which are generally symmetrical, a bisdiazo (tetrazo) component is used. The syntheses generally involve low cost, readily available, commodity organic starting materials and are carried out in water, which offers the obvious economic and environmental advantages over other reaction solvents. The versatility of the chemistry involved in the synthetic sequence means that an immense number of products may be prepared. This explains why azo pigments have been adapted structurally to meet the requirements of many applications. On an industrial scale, the processes are straightforward, making use of simple, multi-purpose chemical plant. Azo pigments may be produced in virtually quantitative yields and the processes are carried out at or below ambient temperatures, thus presenting low energy requirements. Finally, provided that careful control of the reaction conditions is maintained, azo pigments may be prepared directly by an aqueous precipitation process that can optimise physical form, with control of particle size distribution, crystalline form and surface character, often achieved by the inclusion of additives such as surfactants or resins. It is also important to control the synthesis conditions carefully in order minimise side reactions that can lead to formation of impurities, which can impair product quality, especially colouristics, and present environmental and toxicological hazards.

$$ArNH_2 + NaNO_2 + 2\ HCl \rightarrow ArN_2^+\ Cl^- + H_2O \tag{3.1}$$

Diazotisation involves the treatment of a primary aromatic amine ($ArNH_2$), the diazo component, with nitrous acid to form a diazonium salt ($ArN_2^+Cl^-$), as described by eq. (3.1). Nitrous acid, $HNO_2$, is an unstable inorganic substance that decomposes relatively easily forming oxides of nitrogen. It is therefore usually generated *in situ* by treating sodium nitrite, a stable species, with a strong acid. The mineral acid of choice for most diazotisations is hydrochloric acid. A reason for this is that the chloride ion can exert a catalytic effect on the reaction. Most primary aromatic amines undergo diazotisation with little interference from the presence of other substituent groups, although the nature of substituents may influence the required reaction conditions.

With careful control of conditions, diazotisation usually proceeds smoothly and essentially quantitatively. Control of the degree of acidity is particularly important in ensuring smooth reaction. As illustrated by the reaction eq. (3.1), reaction stoichiometry requires the use of two moles of acid per mole of amine. However, for several reasons, a greater excess of acid is generally used. The first stage in the mechanism of diazotisation usually involves N-nitrosation of the free amine [9]. One reason for the use of highly acidic conditions is that this favours the generation from nitrous acid of the reactive nitrosating species that are responsible for the reaction. A second reason is that acidic conditions suppress the formation of side-products which may be formed by N-coupling reactions between the diazonium salts and the aromatic amines from which they are formed. A practical reason for the acidic conditions is to convert the insoluble free amine ($ArNH_2$) to its water-soluble protonated form ($ArNH_3^+Cl^-$). However, too strongly acidic conditions are avoided so that the position of the equilibrium is not too far in favour of the protonated amine and allows a reasonable equilibrium concentration of the free amine, which under most conditions is the reactive species. There is thus an optimum acidity level for the diazotisation of a particular aromatic amine, which depends on the basicity of the amine in question. In the case of aniline derivatives, electron-withdrawing groups, such as the nitro group, reduce the basicity of the amino group. Thus, for example, the diazotisation of nitroanilines requires much more acidic conditions than aniline itself. Very weakly basic amines, such as 2,4-dinitroaniline, require extremely acidic conditions. They are usually diazotised using a solution of sodium nitrite in concentrated sulphuric acid, which forms nitrosyl sulphuric acid ($NO^+HSO_4^-$), a powerful nitrosating agent. It is critically important to maintain control of the temperature of the reaction medium in diazotisation reactions. The reactions are commonly carried out in the temperature range 0–5 °C, necessitating the use of ice cooling. Efficient cooling is essential, not least because the reactions are exothermic. Another reason for keeping the temperature low is that higher temperatures promote the decomposition of nitrous acid giving rise to the formation of oxides of nitrogen and potentially undesirable by-products. A further consideration is the instability of diazonium salts, which decompose readily with the evolution of nitrogen.

The diazo components used in monoazo pigment synthesis are usually aniline derivatives containing an appropriate substituent pattern. Normally, such amines are diazotised using a *direct* method which involves the addition of sodium nitrite solution to an acidic aqueous solution of the amine. Aromatic amines that also contain sulphonic acid groups are commonly used in the synthesis of metal salt azo pigments (lakes/toners). Since such amines often dissolve with difficulty in aqueous acid, they are commonly diazotised using an *indirect* method, which involves dissolving the component in aqueous alkali as the water-soluble sodium salt of the sulphonic acid, adding the appropriate quantity of sodium nitrite and then adding this combined solution with efficient cooling and stirring to the acid.

$$NH_2SO_3H + HNO_2 \rightarrow N_2 + H_2SO_4 + H_2O \qquad (3.2)$$

The quantity of sodium nitrite used in diazotisation is usually the equimolar amount required by reaction stoichiometry, or as a very slight excess. A large excess of nitrite is undesirable because of the instability of nitrous acid and because high concentrations can promote diazonium salt decomposition. With direct diazotisation, sodium nitrite solution is usually added at a controlled rate such that slight excess is maintained throughout the reaction. In practice, this can be monitored easily by the characteristic blue colour given by nitrous acid with starch/potassium iodide paper. When diazotisation is judged to be complete, any remaining nitrous acid excess is destroyed prior to azo coupling to avoid side-products, for example due to C-nitrosation of the coupling component. This can be achieved by addition of sulphamic acid, which is described by eq. (3.2).

$$ArN_2{}^+ + Ar'-H \rightarrow Ar-N=N-Ar' \qquad (3.3)$$

Azo coupling is an example of aromatic electrophilic substitution in which the electrophile is the diazonium cation, $ArN_2^+$. The process is described by eq. (3.3). Electrophilic substitution reactions, of which nitration, sulphonation and halogenation are among the best-known examples, are the most frequently encountered group of reactions undergone by aromatic systems. However, the diazonium cation is a relatively weak electrophile and will therefore only react with aromatic systems that are highly activated to electrophilic attack by the presence of strongly electron-releasing groups. The most common strongly electron-releasing group is the hydroxy group. This in turn means that the most common aromatic compounds which undergo azo coupling, referred to as *coupling components*, are phenols. There is a further type of coupling component, commonly a β-ketoacid derivative, in which coupling takes place at a reactive methylene group, a $CH_2$ group that is adjacent to two electron with drawing groups, in azo pigments either $C=O$ or $C=N$.

As illustrated in Figure 3.3, the most common range of coupling components used in the synthesis of azo pigments include reactive methylene compounds, notably the acetoacetanilides (**1**), heterocyclic derivatives, notably 3-methyl-1-aryl-5-pyrazolones (**2**) and phenols, especially 2-naphthol (**3a**), naphtharylamides (**3b**) and 3-hydroxy-2-naphthoic acid (βONA) (**3c**).

As with diazotisation, careful control of experimental conditions during azo coupling is essential to ensure that the azo pigments are obtained in high yield and purity and to minimise the formation of side products. Temperature control, which is so critical in diazotisation, is generally less important in the case of azo coupling. The reactions are normally carried out at, or just below, ambient temperatures. There is usually little advantage in raising the temperature, other than in a few special cases, since this tends to increase the rate of diazonium salt decomposition more than the rate of azo coupling. The experimental factor that requires most careful control in azo coupling is pH. There is usually an optimum pH range for a specific azo coupling

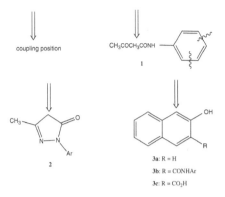

**Figure 3.3:** Coupling components commonly used in azo pigment synthesis.

reaction, which is principally dependent on the coupling component used. Phenols, such as 2-naphthol (**3a**) and its derivatives, are usually coupled under alkaline conditions, in which case the phenol (ArOH) is converted predominantly to the phenolate anion (Ar-O⁻). There are two reasons why this facilitates the reaction. This first is a practical reason in that the anionic species is more water-soluble than the phenol itself. A second reason is that the -O⁻ group is more powerfully electron-releasing than the OH group and hence more strongly activates the system towards electrophilic substitution. Highly alkaline conditions are usually avoided as they promote diazonium salt decomposition. Also, this can cause conversion of the diazonium cation ($ArN_2^+$) to the diazotate anion (Ar-N = N-O⁻), a species that is less reactive than the diazonium cation in azo coupling. Reactive methylene-based coupling components undergo azo coupling via the enolate anion, the concentration of which increases with increasing pH. These components are frequently coupled at weakly acidic to neutral pH values, where a sufficiently high concentration of enolate anion exists for reaction to proceed and side-reactions due to diazonium salt decomposition are minimised. It is generally desirable to perform the reaction at the lowest pH at which coupling takes place at a reasonable rate. The rate of addition of the diazonium salt solution to the coupling component is usually controlled carefully to ensure that an excess of diazonium salt is not allowed to build up in the coupling medium, in order to minimise side-reactions due to diazonium salt decomposition, especially when higher pH conditions are required.

A range of disazo pigments encompasses products that are highly important industrially. Most of these products, invariably symmetrical in structure, are synthesised by tetrazotisation (bis-diazotisation) of the appropriate diamine, most commonly 3,3'-dichlorobenzidine (DCB) and subsequent reaction with two equivalents of the coupling component. Since the two azo coupling reactions effectively proceed simultaneously, the process is essentially the same as in the synthesis of monoazo pigments. A few disazo pigments require a rather different strategy involving coupling of two moles of diazotised amine with a bisacetoacetanilide coupling component. Metal salt azo pigments are prepared by traditional diazotisation/coupling processes to form the sodium salt of an azo dyestuff which may show some solubility in water. This is treated with a solution of an appropriate salt of the divalent metal, which displaces the sodium to form the insoluble pigment.

There are two groups of high-performance azo pigments. Benzimidazolone-based azo pigments are prepared by traditional diazotisation of a primary aromatic amine, followed by azo coupling with a coupling component, either of the acetoacetanilide or naphtharylamide type, containing the benzimidazolone group that is primarily responsible for their superior performance [11]. Disazo condensation pigments are prepared by connecting two monoazo compounds using a condensation reaction [11]. Details are given in the relevant individual papers.

## 3.5 Applications

Azo pigments find extensive use across a wide range of application areas, including printing inks, paints, and plastics. They generally provide strong bright colours, mostly yellows, oranges, and reds, and are economical in use. There is an extensive range of structural types of azo pigment available although they vary significantly in their technical properties. In applications where moderate levels of fastness properties are acceptable, classical azo pigments are likely to be the pigments of choice. More demanding applications may require the use of high-performance pigments, such as the benzimidazolone azos and disazo condensation pigments, or pigments of other chemical classes, invariably at higher cost. The applications of the commercial products belonging to specific azo pigment chemical types are discussed at length in the individual papers.

**Acknowledgement:** We thank the Royal Society of Chemistry for permission to reprint certain passages from the publication given in reference [1].

## References

1.  Christie RM. Colour chemistry. 2nd ed. London: RSC Ch 3, 2015.
2.  Zollinger H. Color chemistry. 3rd ed. Weinheim: Wiley-VCH Verlag GmbH Ch 7, 2003.
3.  Christie RM. The organic and inorganic chemistry of pigments, surface coatings reviews. London: OCCA, 2002.
4.  Hunger K, Schmidt MU. Industrial organic pigments. 4th ed. Weinheim: Wiley-VCH Verlag GmbH Ch 2, 2019.
5.  Griess P. Vorlaufige Notiz uber die Einwirkung von saltpetriger Saure auf Amidodinitro und Aminonitrophenyl saure. Liebigs Ann. Chem. 1858:106;123–125.
6.  VM. The 125th anniversary of the griess reagent. J Anal Chem. 2004;59:1002–5.

7.  Whitaker A. Crystal structures of Azo pigments based on acetoacetanilides. J Soc Dyers Colour. 1988;104:294.

8.  Barrow MJ, Christie RM, Lough AJ, Monteith JE, Standring PS. The crystal structure of CI pigment yellow 12. Dyes Pigm. 1989;1:109.

9.  Zollinger H. Diazo chemistry I: aromatic and heteroaromatic compounds. Weinheim: VCH, 1994.

10. Rys P, Zollinger H. Fundamentals of the chemistry and applications of dyes. London: Wiley-Interscience, 1972.

11. Faulkner EB, Schwartz RJ. High performance pigments. 2nd ed. Weinheim: Wiley-VCH Verlag GmbH, 2009.

Gerhard Pfaff

# 4 Bismuth vanadate pigments

**Abstract:** Bismuth vanadate pigments belong to the most important inorganic yellow pigments since their market introduction in the 1970s. They have substituted the greenish yellow lead chromate and cadmium sulfide pigments in a considerable number of applications. Bismuth vanadate pigments are based on bismuth vanadate with monoclinic or tetragonal structure. Their composition ranges from pure $BiVO_4$ up to the mixed pigment $4BiVO_4 \bullet 3Bi_2MoO_6$, where molybdenum is incorporated in the structure. Bismuth vanadate pigments are characterized by excellent optical and application characteristics in particular regarding brightness of shade, hiding power, tinting strength, weather fastness, and chemical resistance.

**Keywords:** bismuth vanadate pigments, bismuth molybdate

## 4.1 Fundamentals and properties

Pigments based on bismuth orthovanadate, $BiVO_4$, represent a class of colorants with interesting greenish yellow characteristics [1–4]. They belong to the yellow inorganic pigments, such as iron oxide yellow, chrome yellow, cadmium yellow, nickel rutile yellow, and chromium rutile yellow. Their coloristic properties allow the substitution of the greenish yellow lead chromate and cadmium sulfide pigments. Bismuth vanadate pigments are registered in the Color Index as C.I. Pigment Yellow 184.

The minerals pucherite (orthorhombic), clinobisvanite (monoclinic) and dreyerite (tetragonal) are natural forms of bismuth vanadate. All these deposits are not suitable for the production of bismuth vanadate pigments. $BiVO_4$ based pigments were developed and produced for the first time in the 1970s [5, 6]. Their development was driven by the demand to find suitable non-toxic alternatives to existing yellow pigments, such as lead chromate or cadmium sulfide. Besides $BiVO_4$, pigments containing other compositions, e. g. $Bi_2XO_6$ (X = Mo or W), have been investigated and developed [7].

The commercially available bismuth vanadate pigments are based on bismuth vanadate with monoclinic or tetragonal structure. The pigments are characterized by brilliance in the greenish yellow range of the color space. Their composition ranges from pure $BiVO_4$ up to the mixed pigment $4BiVO \bullet 3Bi_2MoO_6$, where molybdenum is incorporated in the structure. The bismuth vanadate component crystallizes in such

This article has previously been published in the journal Physical Sciences Reviews. Please cite as: G. Pfaff, Bismuth Vanadate Pigments *Physical Sciences Reviews* [Online] 2021, 6. DOI: 10.1515/psr-2020-0150

https://doi.org/10.1515/9783110588071-004

phases in the tetragonal scheelite structure, whereas bismuth molybdate occurs in the orthorhombic perovskite crystal lattice [1].

## 4.2 Production of bismuth vanadate pigments

Bismuth vanadate pigments are industrially synthesized either by solid state processing using appropriate starting materials, e. g. $Bi_2O_3$ and $V_2O_5$, or by co-precipitation from aqueous solutions [4–11].

The solid state reaction takes place at temperatures of about 500 °C according to the equation

$$Bi_2O_3 + V_2O_5 \rightarrow 2\ BiVO_4 \tag{4.1}$$

More suitable for the synthesis of bismuth vanadate pigments is the precipitation process starting from solutions of bismuth nitrate and sodium vanadate respectively ammonium vanadate. The $Bi(NO_3)_2$ solution is produced by reaction of bismuth metal with nitric acid. An amorphous precipitate consisting of oxide and hydroxides is formed by addition of sodium hydroxide to the strongly acidic nitrate and vanadate containing solution. The precipitate is washed salt-free after the end of the reaction. The pigment formation can be summarized by the equation:

$$Bi(NO_3)_3 + NaVO_3 + 2\ NaOH \rightarrow BiVO_4 + 3\ NaNO_3 + H_2O \tag{4.2}$$

In the next step, the formed suspension is heated to reflux at a controlled pH-value. A transformation occurs under these conditions producing a fine crystalline product. Controlling the co-precipitation conditions allows the selective formation of particular $BiVO_4$ modifications, but only two of the four bismuth vanadate polymorphs are of interest for pigments, the monoclinic and the tetragonal modifications, which exhibit brilliant yellow colors. The process has therefore to be steered in such a way, that these $BiVO_4$ crystal structures are formed. The precipitation conditions, such as concentration, temperature and pH-value, have a strong influence on color shade and brilliance of the pigments formed.

The formation of the pigment composition $4BiVO_4 \bullet 3Bi_2MoO_6$ needs ammonium molybdate in addition to the starting components described for pure $BiVO_4$.

$$\begin{aligned} 10\ Bi(NO_3)_3 + 4\ NaVO_3 + 3\ (NH_4)_2MoO_4 + 20\ NaOH \\ \rightarrow 4BiVO_4 \cdot 3Bi_2MoO_6 + 24\ NaNO_3 + 6\ NH_4NO_3 + 10\ H_2O \end{aligned} \tag{4.3}$$

The pigment properties for all types of bismuth vanadate pigments can be improved by an additional annealing step for the washed and dried precipitate at temperatures of about 500 °C.

Bismuth vanadate pigments can be stabilized with additional layers to improve their characteristics, e. g. weathering and acid resistance. Surface treatments are often used for this purpose. They contain in particular calcium, aluminum or zinc phosphate, but also oxides like aluminum oxide. For the use in plastics, the pigments are specifically coated with dense layers of silica or other silicon containing components to increase the stability in polymers up to 300 °C [10]. The pigments may also be offered as fine granulates to avoid dust during handling. The production of such free flowing granulates proceeds in spray tower dryers combined with an automatic packaging unit. The last steps in the manufacture of bismuth vanadate pigments in form of powders are conventional drying using a belt dryer or another drying equipment and grinding.

## 4.3 Pigment properties and uses

The outstanding properties of bismuth vanadate pigments are excellent brightness of shade, very good hiding power, high tinting strength, very good weather fastness, and high chemical resistance. Moreover, they are easy to disperse and environmentally friendly. Bismuth vanadate pigments are regarding the color properties close to cadmium sulfide and lead chromate. Their exhibit a greenish yellow color shade. Figure 4.1 illustrates the coloristic situation for different inorganic yellow pigments by means of the remission curves of bismuth vanadate, cadmium sulfide, nickel rutile yellow, lead sulfochromate, and iron oxide yellow. The color saturation of bismuth

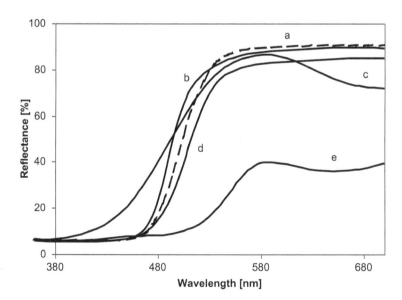

**Figure 4.1:** Reflectance spectra of inorganic yellow pigments. (a) $BiVO_4$, (b) CdS, (c) $(Ti,Ni,Sb)O_2$, (d) $Pb(Cr,S)O_4$, (e) $\alpha$-FeOOH [2].

vanadate pigments is significantly higher than for iron oxide yellow and nickel rutile yellow and close to cadmium sulfide and lead sulfochromate.

Unstabilized bismuth vanadate pigments show photochromism. They exhibit reversible color changes when irradiated intensely with light. This behavior can be reduced or nearly prevented by using a suitable surface treatment of the pigment for stabilization. Bismuth vanadate pigments have a density of $5.6\,g/cm^3$, a refractive index of 2.45 and a specific surface area (BET) of up to $10\,m^2/g$. In combination with titanium dioxide, $BiVO_4$ pigments have a very good weather resistance in full shade.

The application of bismuth vanadate pigments includes the manufacture of cadmium-free and lead-free, weather resistant, brilliant yellow colors for automotive OEM and re-finishes, industrial and decorative paints, partly for powder coatings and coil-coating systems. The combination with additional colorants allows the generation of a broad series of yellow, orange, red and green color tones.

$BiVO_4$ pigments can be produced with a heat stability of up to 300 °C. Their very good fastness to light and weathering allows the application in plastics also under outdoor conditions. The thermostable types can be readily incorporated into polyolefins and ABS at 260–280 °C, and even in polyamide at 280–320 °C [3].

Bismuth vanadate pigments do not exhibit acute toxicity ($LD_{50}$ value rat oral: > 5000 mg/kg). They are not irritating to skin or mucous membranes. Animal studies gave some indication of inhalation toxicity, which may have been due to the vanadium content in the pigments [1, 3, 4]. Toxic effects are observable for rats only when the concentration in the lungs reaches levels that do not occur under the usual conditions of industrial hygiene. For risk reduction, some producers supply bismuth vanadate pigments in a free flowing, low-dusting form. Such preparations make the pigments inaccessible to the lungs. Therefore, the dust-free pigments can be handled under usual hygienic working conditions.

# References

1. Endriss H. Aktuelle anorganische Buntpigmente, Zorll U, editor. Hannover: Vincentz Verlag, 1997:139.
2. Etzrodt G. In Kittel - Lehrbuch der Lacke und Beschichtungen, Spille J, editor. vol. 5. 2nd ed. Stuttgart/Leipzig: S. Hirzel Verlag, 2003:93.
3. Seeger O, Wienand H. In industrial inorganic pigments, Buxbaum G, Pfaff G, editors. 3rd ed. Weinheim:Wiley-VCH Verlag, 2005:123.
4. Pfaff G. Inorganic pigments. Berlin/Boston: WalterdeGruyterGmbH, 2017:133.
5. Patent US 4,026,722 (DuPont) 1976; Patent US 4,063,956 (DuPont) 1976; Patent US 4,115, 141 (DuPont) 1977. Patent US 4,115,142(DuPont). 1977.
6. Patent US 4,251,283 (Montedison) 1978; Patent US 4,230,500 (Montedison) 1978; Patent US 4,272,296 (Montedison) 1979. Patent US 4,316,746 (Montedison). 1980.
7. Patent EP 0 074 049 (BASF) 1981. Patent EP 0 271 813 (BASF). 1986.

8. Patent DE 33 15 850 (Bayer) 1983; Patent DE 33 15 851 (Bayer) 1983; Patent EP 0 492 224 (Bayer) 1991. Patent EP 0 723 998 (Bayer). 1995.
9. Patent EP 0 239 526 (Ciba) 1986; Patent EP 0 304 399(Ciba) 1987. Patent EP 0 430 888, (Ciba) 1989; Patent US 5,123,965 (Ciba). 1989.
10. Patent EP 0 551 637 (BASF) 1992. Patent EP 0 640 566 (BASF). 1993.
11. Patent US 5,399,197 (Colour Research Compnay) 1990. Patent EP 0 650 509 (Colour Research Company). 1992.

Gerhard Pfaff

# 5 Black pigments

**Abstract:** Black pigments are inorganic or organic pigments, whose optical action is predominantly based on non-selective light absorption in the range of visible light. Most of the black pigments are inorganic pigments. The main representatives of black pigments are carbon black pigments followed by iron oxide black and spinel blacks.

**Keywords:** black pigments, carbon black pigments, iron oxide black, mixed metal oxide pigments, aniline black

Black pigments are pigments, whose optical action is mainly based on non-selective light absorption. The large majority of the black pigments are of inorganic nature. The most important inorganic black pigments are carbon black, iron oxide black and spinel blacks, such as manganese black or spinel black. Carbon black pigments are of greatest importance by far among these inorganic blacks. Aniline black is the oldest synthetic organic pigment. Compared with carbon black, aniline black has only little importance.

Main application fields of black pigments are coatings, plastics, and printing inks. They are also used in building materials, cosmetics, and artist's colors [1, 2].

Table 5.1 contains a summary of inorganic black pigments. Carbon black as the most important black pigment is described in an own chapter. Iron oxide black (magnetite) is described in the chapter "Iron Oxide Pigments". Black spinel type pigments are discussed in the chapter "Mixed Metal Oxide Pigments".

**Table 5.1:** Summary of inorganic black pigments [1].

| Name/Color Index | Chemical composition | Structure |
|---|---|---|
| Carbon Black (C.I. Pigment Black 6 and 7) | C | Amorphous |
| Magnetite (C.I. Pigment Black 11) | $Fe_3O_4$ | Spinel |
| Pigment Black 30 | Ni(II), Fe(II,III), Cr(III) oxide | Spinel |
| Pigment Black 26 | Mn(II), Fe(II,III) oxide | Spinel |
| Pigment Black 22/ | Cu(II), Cr(III) oxide | Spinel |
| Pigment Black 28 | Mn(II), Cu(II), Cr(III) oxide | Spinel |
| Pigment Black 27 | Co(II), Cr(III), Fe(II) oxide | Spinel |

https://doi.org/10.1515/9783110588071-005

# References

1. Pfaff G. Inorganic pigments. Berlin/Boston: WalterdeGruyterGmbH, 2017.
2. Buxbaum G, Pfaff G, editors. Industrial inorganic pigments. Weinheim: Wiley-VCH Verlag, 2005.

Gerhard Pfaff

# 6 Cadmium sulfide / selenide pigments

**Abstract:** Cadmium sulfide and selenide pigments (cadmium pigments) belong to the inorganic yellow, orange and red pigments. Cadmium sulfide pigments are based on the wurtzite lattice, where cadmium can be partially substituted by zinc or mercury and sulfide by selenide. Cadmium pigments are characterized by excellent optical and application characteristics in particular regarding brightness of shade, hiding power, tinting strength, and weather fastness. The declining use of cadmium-containing materials in the last decades is a result of the environmental discussion and the development of less problematic substitute products, especially of bismuth vanadate and high-value organic, temperature-stable yellow and red pigments.

**Keywords:** cadmium sulfide pigments, cadmium selenide pigments

## 6.1 Fundamentals and properties

Cadmium sulfide and selenide pigments belong to the yellow, orange and red inorganic pigments. They are also referred to as cadmium pigments and consist of cadmium sulfides and sulfoselenides as well as zinc containing sulfides of cadmium. Cadmium pigments are characterized by brilliant colors. The yellow pigments vary from pale primrose to deep golden yellow, the red pigments from light orange via deep orange, light red, crimson to maroon [1–5].

Cadmium sulfide, which crystallizes in the hexagonal wurtzite structure, occurs in nature in form of cadmium blende or greenockite. Natural sources of cadmium sulfide have, however, no importance for the production of cadmium pigments. Parameters used for the color adjustment of the pigments are the composition and the size of the primary particles.

The wurtzite lattice represents the basic structure of all cadmium pigments. Cadmium ions fill half of the tetrahedral vacancies in a hexagonal close packed arrangement of sulfide ions. The cations as well as the anions of this lattice can be replaced within certain limits by chemically related elements. Zinc and mercury as cations and selenide as anion have gained importance as exchangeable ions. Development work for the substitution of selenide by telluride was not successful because the so-obtained pigments have inferior coloristic properties [1].

This article has previously been published in the journal Physical Sciences Reviews. Please cite as: G. Pfaff, Cadmium Sulfide /Selenide Pigments *Physical Sciences Reviews* [Online] 2021, 6. DOI: 10.1515/psr-2020-0151

https://doi.org/10.1515/9783110588071-006

Zinc incorporation leads to greenish-yellow pigments, mercury and selenide change the color shade to orange and red.

Four cadmium pigments have gained technical importance:

(Cd,Zn)S (cadmium zinc sulfide): pigment cadmium yellow C.I. Pigment Yellow 35, greenish-yellow.

CdS (cadmium sulfide): pigment cadmium yellow C.I. Pigment Yellow 37, reddish-yellow.

Cd(S,Se) (cadmium sulfoselenide): pigment cadmium orange C.I. Pigment Orange 20, orange.

Cd(S,Se) (cadmium sulfoselenide): pigment cadmium red C.I. Pigment Red 108, red.

Cadmium zinc sulfide pigments contain 59–77 wt % Cd, 13–0.2 wt % Zn and in addition 1–2 wt % $Al_2O_3$ for lattice stabilization. Cadmium orange and cadmium red have 76–66 wt % Cd and 1–14 wt % Se in the composition.

Cadmium sulfides and selenides belong to the semiconductors. Their color is explained using the band model by the distance between the valence and conduction bands in the crystal lattice. Distinct wavelengths in the range of the visible light are sufficient to lift electrons from the valence to the conduction band. Pigments based on this color formation principle often exhibit an extremely high color purity.

The first production of cadmium sulfide and cadmium sulfoselenide pigments in industrial scale took place around 1900. Early applications of the pigments were in paints, artists' colors and ceramics. The use in other application fields was limited because of relatively high costs for raw materials and production, which led to high pigment prices. The stronger demand for colored plastics since the 1920s was the reason for an increased use of cadmium pigments. Cadmium pigments exhibit excellent properties in polymers and offer a broad range of bright, intermixable, dispersible and light fast shades. They have almost no problems with the processing temperatures necessary for the engineering of polymers.

The public discussion concerning cadmium-containing materials and the development of less problematic substitute products, especially of bismuth vanadate and high-value organic, temperature-stable yellow and red pigments, have led to a significant reduction of the cadmium pigment quantities.

## 6.2 Production of cadmium sulfide / selenide pigments

Cadmium pigments need for their production pure grades of chemicals, because the final pigment powders must be free of transition metal compounds, which form deeply colored sulfides (e. g. copper, iron, nickel, cobalt, lead) [3, 6]. First step of the manufacturing process is the dissolution of high-purity cadmium metal (99.99%) in sulfuric, hydrochloric or nitric acid. It is also possible to use mixtures of these acids.

Alternatively, the cadmium metal is first oxidized to cadmium oxide, which is diluted in a next step in mineral acids. Cadmium oxide is formed by melting the cadmium metal, vaporization of the molten cadmium at about 800 °C and oxidation of the vapor with air [6]. Zinc or zinc oxide may be added depending on the desired color tone of the final pigments. After dissolution of the cadmium metal respectively the cadmium oxide, cadmium sulfide is precipitated by adding sodium sulfide solution. Very fine-crystalline cadmium or cadmium zinc sulfide precipitates are obtained, which have a cubic crystal form. They have not yet pigment properties. Main parameters having an influence on particle size and shape of the precipitate are concentration, temperature, pH value and mixing conditions. eq. (6.1) demonstrates the CdS precipitation using a $CdSO_4$ solution as cadmium source:

$$CdSO_4 + Na_2S \rightarrow CdS + Na_2SO_4 \qquad (6.1)$$

A filter press or a belt filter is used to separate the precipitated particles from the suspension. Sodium sulfate and unreacted cadmium salts are washed out in a next step of the procedure. Remaining salts would have an adverse effect on the pigment properties. Soluble cadmium must be removed also to fulfill the regulatory requirements. This removal is mostly achieved by filtration and washing or by acidification with diluted mineral acids followed by decantation and washing.

The obtained filter cake is dried in an oven, crushed into small lumps and then calcined in a rotary, tunnel or static kiln. The calcination step is done at about 600 °C. The cubic crystal lattice of the precipitated and dried sulfide is transformed into the hexagonal structure during calcination. The particles grow to a size, which is the basis for suitable color properties of the final pigment. Depending on the chosen composition, sulfur dioxide and/or selenium and selenium dioxide are emitted. Careful fume removal through a scrubber or similar equipment is necessary.

A second manufacturing process for cadmium pigments is the so-called powder process. Finely divided cadmium carbonate or cadmium oxide is subjected in this process to intensive mechanical mixing with sulfur and mineralizers and then calcined in the absence of oxygen. Zinc or selenide addition produces the same color effects as in the case of the precipitation process. The products obtained after calcination are washed, dried and calcined at about 600 °C.

The production of selenium-containing cadmium red and cadmium orange pigments is carried out with sodium sulfide solutions containing the amount of selenium required for the final pigment. Selenium is added in powder form to the sodium sulfide solution, where it is dissolved. The sulfide ions react with the cadmium ions first to form cadmium sulfide [6]. Cadmium sulfoselenide is only formed during the calcination step. An example for the formation of a sulfoselenide pigment is shown in the eq. (6.2):

$$CdSO_4 + 0.8\,Na_2S + 0.2\,Na_2Se \rightarrow Cd(S_{0.8}Se_{0.2}) + Na_2SO_4 \qquad (6.2)$$

The now following processing stages are similar to that described for the pure cadmium sulfide pigments.

## 6.3 Pigment properties and uses

Cadmium sulfide and sulfoselenide pigments belong to the most intensely colored inorganic pigments. They are lightfast, stable at high temperatures, migration-resistant. and practically insoluble in water, alkaline solutions and organic solvents. Disadvantages are the limited weather fastness of cadmium pigments and the instability against acids. It is to be noted here that cadmium sulfide is oxidized slowly by atmospheric oxygen and under the influence of light, especially sunlight, to form cadmium sulfate.

The density of cadmium pigments is between 4.2 and 5.6 g/cm$^3$. The average particle size of the powders lies typically in the range from 0.2 to 0.5 µm. Their hiding power in most application systems is very good. The high color purity of the pigments is due to the steep reflectance spectra, which is typical for a semiconductor-related pigment (Figure 6.1).

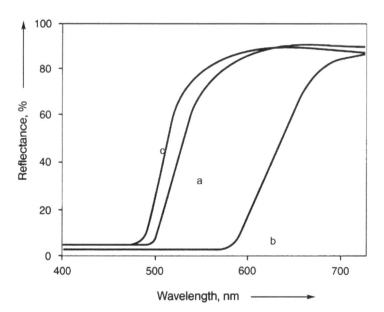

**Figure 6.1:** Reflectance spectra of the pigments CdS (a), Cd(S$_{0.53}$Se$_{0.47}$) (b), and (Cd$_{0.82}$Zn$_{0.18}$)S (c) [5].

Cadmium pigments are used in form of pure powders but also as pigment preparations. The color strength in such preparations is adjusted by addition of barium sulfate. The pigments are placed on the market also in a dust-free and dispersed form as highly concentrated plastic granulates (masterbatch pellets), as pasty concentrates and as liquid colorants.

The most important application media for cadmium pigments are plastics with 90% and ceramics with 5%. The use in coating is of minor importance. The application of the pigments in polymers is broad and almost universal. Polyethylene, polystyrene, polypropylene and ABS are typical polymers for the application of cadmium sulfides and sulfoselenides.

Typical applications for the pigments in ceramics are glazes and enamels. The high temperatures used in both cases for the burning process do not cause larger problems. Organic pigments and several inorganic colored pigments cannot be used here for stability reasons.

Users of cadmium pigments have to take into account the sensitivity to friction. Especially red cadmium sulfoselenides have problems with excessive shear forces, which can cause a color change of the pigments.

Cadmium sulfide and sulfoselenide pigments do not exhibit acute toxicity ($LD_{50}$ value rat oral: $> 5000$ mg/kg). They are not irritating to skin or mucous membranes. The pigments are compounds with low solubility, but small amounts of cadmium are however solved in diluted hydrochloric acid comparable with gastric acid of similar concentration. Animal feeding studies over three months with dogs and rats show a cadmium intake and accumulation in the internal organs, especially in the kidneys [7]. There was no indication concerning carcinogenic effects after a long-term animal feeding study [1]. The use of cadmium pigments is not seen any longer as optimal due to the principal possibility to solve at least small amounts of the pigments uncontrolled and the potential danger that cadmium compounds can get unintended into the cycle of nature. Recycling of cadmium pigment containing materials demands special care and advanced processing technologies.

The legal requirements for the use of cadmium sulfide and sulfoselenide pigments must be adhered to in all applications. Waste materials containing cadmium pigments should never be incinerated in by the fireside, in the oven or on the open fire. They have to be treated as special refuse, which is disposed by a certified waste management company. Sophisticated incineration plants for domestic waste are equipped with effective filter systems. These are able to withhold cadmium particles originating from the waste material completely.

# References

1. Endriss H. Aktuelle anorganische Bunt-Pigmente, Zorll U, editor. Hannover: Vincentz Verlag, 1997:121.
2. Etzrodt G. In Kittel - Lehrbuch der Lacke und Beschichtungen, Spille J, editor. vol. 5. 2nd ed. Stuttgart/Leipzig: S. Hirzel Verlag, 2003:116.
3. Pfaff G. In Winnacker Küchler - Chemische Technik. Band 7. Weinheim: Wiley-VCH Verlag, 2004:336.
4. Etzrodt G. In industrial inorganic pigments, Buxbaum G, Pfaff G, editors. 3rd ed. Weinheim: Wiley-VCH Verlag, 2005:121.
5. Pfaff G. Inorganic pigments. Berlin/Boston: WalterdeGruyterGmbH, 2017:149.
6. Dunning P. In high performance pigments, Faulkner EB, Schwartz RJ, editors. Weinheim: Wiley-VCH Verlag, 2009:13.
7. Klimisch, H. J. Lung deposition, lung clearance and renal accumulation of inhaled cadmium chloride and cadmium sulfide in rats. Toxicology 1993;84:103.

Gerhard Pfaff

# 7 Carbon black pigments

**Abstract:** Carbon black pigments are manufactured today mainly by modern chemical processes in industrial scale production. They are the most important representatives of black pigments. Carbon black pigments have a number of advantages compared with other inorganic black pigments and with black organic colorants. Hiding power, color stability, solvent resistance, acid and alkali resistance as well as thermal stability are excellent good properties that are not achieved from other blacks. Carbon black pigments are applied in most of the pigment relevant systems, such as printing inks, paints and coatings, plastics, and cosmetics. They are produced by several industrial processes. Furnace blacks, channel blacks and gas blacks have the highest importance among the various carbon blacks. Particle size, particle size distribution, surface quality and structure determine the coloristic and application technical properties of the individual pigments. Oxidative aftertreatment is used in many cases to modify the surface of the pigments concerning the stability and the compatibility with the application system. Particle management, aftertreatment and the provision of pigment preparations are suitable ways for the improvement of the pigments and the optimization of the dosage form.

**Keywords:** carbon black pigments, channel black process, furnace black process, gas black process, oxidative aftertreatment

## 7.1 Fundamentals and properties

Inorganic black pigments include carbon black, iron oxide black and spinel blacks. Carbon black pigments are by far most important among the blacks. The term "carbon black" stands for a number of well-defined industrially manufactured products, which are manufactured under exactly controlled conditions.

Carbon black consists of highly dispersed carbon particles with almost spherical shape. These particles are produced by incomplete combustion or thermal decomposition of gaseous or liquid hydrocarbons [1,2,3,4,5]. Carbon blacks do not consist of pure carbon. They still contain considerable amounts of chemically bound hydrogen, oxygen, nitrogen, and sulfur depending on the manufacturing conditions and the quality of the raw materials. Carbon black is not only applied as a pigment, but also as active filler material in rubber, particularly in car tires. Nearly 80 carbon

This article has previously been published in the journal Physical Sciences Reviews. Please cite as: G. Pfaff, Carbon Black Pigments *Physical Sciences Reviews* [Online] 2021, 6. DOI: 10.1515/psr-2020-0152

https://doi.org/10.1515/9783110588071-007

black types are used today as pigments in a broad variety of applications. More than 35 types are applied as fillers in rubber. The majority of the carbon black produced worldwide (90%) is used in the rubber industry [4].

Carbon black pigments consist of extremely small primary particles with the tendency to form aggregates among each other, which in turn consist of chains and clusters. The degree of aggregation, the so-called "structure" of carbon black, is attributed to the high attraction of the aggregates among each other. The diameters of the primary particles range from 5 to 500 nm. These particles are mainly amorphous but contain also microcrystalline subregions. Diffraction patterns show that the spherical primary particles consist of relatively disordered nuclei surrounded by concentrically deposited carbon layers [6]. The degree of order increases from the center of the particles to their peripheral areas. The particle size distribution as well as the degree of aggregation can be adjusted within relatively wide ranges by the parameters chosen for the production process of carbon black.

Structural units of carbon black particles are very similar to those of graphite, which is crystallizing in a layer structure of hexagonally bound carbon atoms. The layers of the graphite structure are connected among each other only by loose van der Waals forces. The arrangement of the carbon atoms within the layers of carbon black is similar to that of graphite. The orientation of the carbon layers to each other is nearly parallel. On the other hand, the relative position of these layers in carbon black is arbitrary in contrast to graphite. A consequence of this is that there is no order in the structural c direction for carbon black. Crystalline regions in carbon black are therefore only small. They are typically in the range of 1.5–2 nm in length and 1.2–1.5 nm in height, corresponding to 4–5 carbon layers [6]. Figure 7.1 shows the simplified structures of carbon black and graphite [5]. Graphite and carbon black are electrically conductive, in contrast to diamond. This physical behavior can be understood by the explained structural relationships.

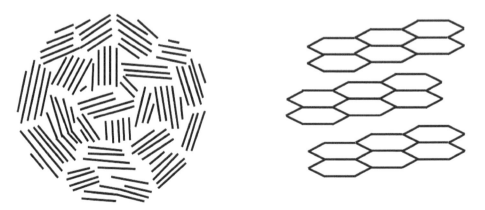

**Figure 7.1:** Simplified layer structure of carbon black (left) and graphite (right) [5].

Carbon black is provided in form of powders, pellets, or dispersions. Six industrial processes are used today for the production of carbon black. The products obtained are called furnace black, channel black, gas black, lampblack, thermal black, acetylene black, corresponding to the eponymous processes. The chemical and physical properties of carbon black powders can differ considerably depending on the manufacturing method.

## 7.2 Production of carbon black pigments

The most important processes for the manufacture of carbon black are listed in Table 7.1. The processes are basically classified in those with partial combustion of the raw materials (thermal-oxidative decomposition) and those, which are based on pure pyrolysis (thermal decomposition). The main difference of the two processes is the procedure for the thermal decomposition step. The thermal-oxidative decomposition uses air for the combustion of a part of the starting materials. This reaction provides also the high temperatures necessary for the pyrolysis. The pure thermal decomposition is using external power supply to achieve the energy for the carbon black synthesis.

**Table 7.1:** Production methods for carbon black pigments [5].

| Chemical process | Manufacturing methods | Main raw materials |
|---|---|---|
| Thermal-oxidative decomposition | Furnace black process | Aromatic oils on coal tar basis or mineral oil, natural gas |
| | Gas black and channel black process | Coal tar distillates |
| | Lamp black process | Aromatic oils on coal tar basis or mineral oil |
| Thermal decomposition | Thermal black process | Natural gas or mineral oils |
| | Acetylene black process | Acetylene |

Heavy oils are used as raw materials for most of the carbon black production processes. These oils consist mainly of aromatic hydrocarbons, which deliver the highest carbon to hydrogen ratio and maximize the available carbon for the carbon black synthesis. The use of such oils is the most efficient way to achieve high yields of carbon black.

Oily distillates from coal tar and residual oils that are created by catalytic cracking of mineral oil fractions are other sources of raw materials for carbon black. Hydrocarbons produced by thermal cracking of naphtha or petrochemical oil are further sources of feedstock. The chemical and physical properties of the raw materials (density, distillate residue, viscosity, carbon/hydrogen ratio, asphaltene content, specified impurities, etc.) play an important role for the process management and for the final product quality.

### 7.2.1 Furnace black process

The furnace black process is used for the manufacture of carbon black rubber grades as well as for carbon black pigments. This process can be used for nearly all types of carbon black required today in various industries. Carbon blacks produced by the furnace black process meet also the high demands concerning economy and ecology. Most of the semi-reinforcing rubber blacks with specific surface areas of 20–60 m$^2$/g and of the active reinforcing blacks with specific surface areas of 65–150 m$^2$/g are manufactured by the furnace black process. A large amount of carbon black pigments with much greater specific surface areas and smaller particle sizes is also produced using this process [4]. The most important product properties such as specific surface area, particle size distribution, structure, and absorption can be controlled by adjusting the reaction parameters of the process. Application-relevant properties of carbon black such as abrasion resistance, tear strength, jetness and tinting strength can also be affected by adjusting the operating parameters.

The furnace black process is carried out continuously in closed reactors. The central unit of the production plant is the special furnace in which the carbon black is generated. The starting components are injected as an atomized spray into the high-temperature zone of the furnace. The necessary high temperatures are achieved by burning a fuel (natural gas or oil) in the presence of air. An excess of oxygen is used in regard to the fuel. The feedstock is pyrolyzed to form carbon black at temperatures of 1200–1900 °C. The resulting products are quenched with water and cooled further in heat exchangers. Suitable filter systems are used to collect the formed carbon black. Rotary pumps transport the heavy oil (petrochemical or carbochemical oils are preferably used) to the reactor via heated pipes and a heat exchanger. The oils are heated to 150–250 °C. They get a viscosity, which is appropriate for atomization. Specific spraying devices are used to introduce the oils and the fuel into the reaction zone.

Alkali metal compounds, such as potassium hydroxide or potassium chloride, are often added to the oil in the oil injector with the aim to affect the carbon black structure [7]. In another configuration, the additives are sprayed separately into the combustion chamber. Alkaline-earth metal compounds may also be added in this way to increase the specific surface area of the formed carbon black powders.

The necessary high pyrolysis temperatures are obtained in most cases with natural gas as the fuel source. Gases like coke oven gases or vaporized liquid gas and various oils are also possible for the use as fuel. Special burners adapted to the gas or the oil type are used to achieve a complete combustion (Figure 7.2) [4]. The air required for the combustion is compressed in the beginning by rotating piston compressors or turbo blowers. Moreover, the air is preheated using hot gases in heat exchangers to 500–700 °C. This pretreatment of the air is advantageous with respect to the energy utilization and the carbon black yield.

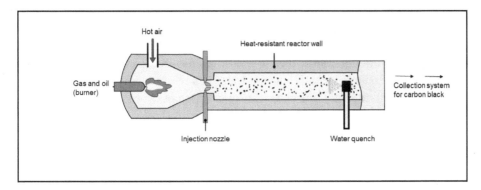

**Figure 7.2:** Scheme of furnace black production process (source: MagentaGreen, Wikimedia Commons).

Carbon black plants vary significantly in flow characteristics, internal geometry and the way in which feedstock and fuel are introduced. The basic steps, on the other hand, are similar for all process variations, in particular the provision of hot combustion gases in a combustion chamber, the injection of the feedstock and its rapid mixing with the combustion gases, vaporizing of the oil, pyrolyzing it in the reaction zone, and rapid cooling of the reaction mixture in the quenching zone to temperatures of 500–800 °C.

The pyrolysis reaction can be summarized in a simplified manner with the following equation:

$$C_n H_m + \text{energy} \rightarrow n\,C + m/2\,H_2 \tag{7.1}$$

The morphology of carbon black primary particles formed during the pyrolysis reaction allows the conclusion that the first carbon nuclei condense from the gas phase. Further carbon layers are adsorbed now onto the surface of the growing particles. Precursor components can also be adsorbed in the growing layers, but typically only to a small extent. The layers formed in the following are orientated parallel to the surface of the nuclei. Aggregates are generated by further carbon deposits on initially formed loose agglomerates. A conversion of the carbon layers to a graphitic arrangement takes place at temperatures above 1200 °C.

The reactors used for the furnace black process are typically horizontally arranged and have a length of up to 18 m and an outer diameter of up to 2 m. For the manufacture of certain semi-reinforcing blacks, also vertical reactors are in use [8].

The ratio of feedstock, fuel and air is very important for the characteristics of the resulting carbon black. Increasing amounts of excess air in relation to the amount needed for the complete combustion of the fuel, for example, lead often to smaller particle sizes of the formed carbon black. The nucleation velocity and the number of particles formed can be higher in this case [4].

Other technical parameters with a significant influence on the quality of the carbon black formed are the way in which the oil is injected, atomized, and mixed with the combustion gases, the type and the quantity of additives, the preheating temperature of the air, and the quench position. Various reactions with the hot surrounding gases are possible at the surface of the freshly formed carbon black particles. The Boudouard reaction and the water gas reaction play a relevant role in this context. Consequently, the chemical composition of the carbon black surface is modified with increasing residence time. These reactions are abruptly stopped by quenching the reaction products to temperatures below 900 °C. A certain state of surface activity for the carbon black is fixed at this point.

The mixture of carbon black and gas leaving the furnace reactor is cooled down in heat exchangers to temperatures of 250–350 °C and conducted into a collection system (typically one high-performance bag filter with several chambers). This system is periodically purged by counter-flowing filtered gas or by pulsejets. The carbon black is pneumatically transferred at this point from the filter into a first storage tank.

Freshly synthesized carbon black has a bulk density of only 20–60 g/l. Compaction of the powdered material is beneficial to facilitate handling and further processing. A possibility for the compaction is the weak densification of the powder using a process by which the carbon black is conducted over porous, evacuated drums [9]. Most of the carbon black types used in paints, inks and plastics are compacted in this manner, which ensures a good dispersibility of the pigment in the application system.

The carbon black contains ca. 50 wt-% water at this point of the process. The material is dried in the following step in dryer drums, indirectly heated by burning tail gas. The dried carbon black powder is transported via conveyor belts and elevators to storage tanks or packing stations.

### 7.2.2 Channel black and gas black process

The channel black process is the oldest industrially used method for producing small-particle-size carbon blacks. Low profitability and environmental regulations were the reasons of the closure of the last production plant in the United States in 1976. Natural gas was used as the feedstock for the process [4].

The gas black process shows similarities to the channel black process, but also differences. One of these is the use of coal tar oils instead of natural gas. Much higher yields and production rates are achievable by using of oil-based raw materials. Various carbon black qualities for a broad variety of applications are produced today using the gas black process.

A characteristic of the gas black process is that the raw materials are partially vaporized in the first area of the reactor (Figure 7.3). The residual coal tar oil is continuously removed. The oil vapor is transported to the production vessel by combustible carrier gases, such as hydrogen, coke oven gas, or methane. The addition of air to the oil-gas mixture leads to the formation of very small carbon black particles. Although the gas black process is technically not as flexible as the furnace black process, it is possible to produce a large number of different gas black types by the variation of the relative amounts of carrier gas, oil, and air.

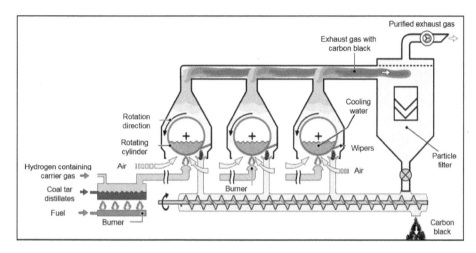

**Figure 7.3:** Scheme of gas black production process (source: MagentaGreen, Wikimedia Commons).

The second part of the manufacturing facility for the gas black process is the burner pipe. It is approximately 5 m long and is equipped with 30 to 50 diffusion burners. The flames are in contact with a water-cooled rotating cylinder. About half of the carbon black formed during the process is deposited at this cylinder. The deposited carbon black is scraped off from the cylinder and transported by a screw to a pneumatic conveying system. The exhaust gas is sucked by fans at the top of the apparatus into particle filters, which collect the carbon black suspended in the gas.

The air amount used in the gas black process is regulated by valves in the exhaust pipes. A combination of several gas-black units is possible to install a larger production plant. It is possible in this case that one oil vaporizer feeds the whole plant. The yield for a typical carbon black production plant is 60%, the production rate is 7–9 kg/h [4].

The use of air in the process leads to the contact of oxygen with the freshly formed carbon black particles at high temperatures. Acidic oxides are formed on the surface of the particles under these conditions. As a result, gas blacks react acidic when suspended in water, unlike furnace blacks.

### 7.2.3 Lamp black process

The lamp black process is the oldest commercially used production method for carbon black. The main part of the equipment for this process is a cast-iron pan on which the liquid feedstock is applied. The pan is surrounded by a fire-proof flue hood equipped with refractory bricks. The pyrolysis conditions are regulated mainly by the stream of the incoming air, the gap between the pan and the hood, and a vacuum present in the system. The control of these conditions allows the fine tuning of the properties of the resulting carbon blacks. The vaporization of the raw materials is achieved using radiant heat from the hood. This heat contributes also partially to the pyrolysis and thereby to the formation of the carbon black. The carbon black containing exhaust gases are cooled down and reach finally a system of filters. The carbon black is collected by this means. It is processed further in a similar way as described for the furnace black process.

Carbon black powders manufactured using the lamp black process are mostly used in special applications. They show typically a broad primary particle size distribution ranging from 60 to over 200 nm [4].

### 7.2.4 Thermal black process

The thermal black process is in contrast to the other carbon black production methods a noncontinuous or cyclic process. This process works without the inflow of air and does therefore not belong to the thermal-oxidative processes. The most common feedstock for the thermal black process is natural gas. Higher hydrocarbon oils are used to a lesser extent. Tandem plants are typical for the production of thermal black in order to improve the efficiency. These plants consist of two reactors operating alternately in cycles of 5 to 8 minutes. One of the reactors is heated up with natural gas or an oil-air mixture whereas the other one is fed at the same time with pure feedstock, which undergoes thermal decomposition. The carbon black formation occurs in the absence of oxygen and at decreasing temperatures. The properties of the carbon black powders formed by the thermal black process differ considerably from those produced using thermal-oxidative processes. The formation of thermal blacks happens usually relatively slow and coarse particles in the diameter range from 300 to 500 nm are obtained. A possibility to produce thermal blacks with smaller particle sizes is the use of natural gas as feedstock diluted with inert gases.

Thermal blacks are mainly used in mechanical rubber goods with high filler content. It should be noted, however, that the importance of thermal blacks for this application is decreasing for a long time. Clays, milled coals, and cokes are used more and more as cheaper substitutes for thermal blacks. This development has led to a declining significance of thermal blacks and to lower production volumes in general.

### 7.2.5 Acetylene black process

The acetylene black process is based on the pyrolysis of mixtures of acetylene with light hydrocarbons. This pyrolysis is an exothermic process, in contrast to the thermal decomposition of other hydrocarbons. This means that energy for the carbon black formation process is provided by the pyrolysis reaction.

Acetylene black is formed in a continuous production process. Acetylene or acetylene-containing gases are fed into a preheated, cylindrical reactor with a ceramic inner liner. The pyrolysis of the acetylene and the light hydrocarbons starts after an ignition and is maintained by the decomposition heat, which is released by the exothermic reaction. The formed carbon black is collected in settling chambers and cyclones. Yields of more 95% can be achieved with the acetylene black process.

The primary particles of acetylene black are different compared with other carbon blacks in relation to the shape. An increased order in the c-direction of the crystalline regions can be detected. Main structural units of acetylene black are folded sheets of carbon layers. Carbon blacks produced with the acetylene black process have found only limited applications, e. g., in dry cells [4].

### 7.2.6 Other production processes

Vaporized hydrocarbons can be converted almost quantitatively into high-purity carbon and hydrogen in a plasma arc at temperatures of about 1600 °C by means of electricity [4, 10]. This process could in principle be used to produce carbon blacks with small particle sizes, but it is not yet economically efficient.

The carbon black industry is highly dependent on raw materials from the petrochemical industry. In order to reduce this dependency, several approaches have been made to replace existing raw materials with other ones. Methods were investigated in this context to obtain carbon black directly from coal or to isolate it from old tires [4, 11]. These and other approaches have not been successful with regard to commercial importance up to now. On the other hand, the use of clay, milled coal and coke has found limited interest as substitute for very coarse carbon blacks [4].

### 7.2.7 Oxidative aftertreatment

Functional groups on the surface have a strong influence on the application properties of carbon blacks. Oxygen-containing functional groups are of special interest here. High amounts of surface oxides are responsible for the decrease of the vulcanization rate, the improvement of the flow characteristics of inks, and the gloss increase of coatings. The color tone of carbon black changes under the influence of such functional groups from brownish to bluish.

The color properties of carbon black are typically adjusted by oxidative aftertreatment. Different types and amounts of oxygen-containing groups are formed on the surface of the particles depending on the oxidizing agent and the reaction conditions (e. g., carbonyl, ether, ketone, peroxide, phenol, lactone, hemiacetal, and anhydride groups).

The oxidation of carbon black surfaces is possible with air in a temperature range from 350 to 700 °C. Higher amounts of surface oxides can be better achieved by using nitric acid, mixtures of nitrogen dioxide and air, ozone, or sodium hypochlorite solutions as oxidizing agents [12–14]. Elevated temperatures are typically used for the oxidation reactions.

The oxygen content of oxidative treated carbon black powders reaches values of up to 15 wt%. The surfaces are strongly hydrophobic. This is the reason why some of the powders form spontaneously colloidal solutions in water. The surface oxidation contributes to a better wettability and dispersion behavior of carbon black pigments in coatings and polar printing inks.

The oxidative aftertreatment of carbon black powders with nitrogen dioxide and air can industrially be carried out in a fluidized-bed reactor [15]. The carbon black is fluidized at the beginning in a preheating vessel. It is further heated in the reaction vessel to carry out the surface oxidation. The adsorbed nitric oxide is finally removed in a desorption vessel. The oxidative reaction occurs at temperatures of 200 to 300 °C. Oxygen from the air acts as the oxidizing agent, the nitrogen dioxide works primarily as a catalyst [4].

The surface oxidation is technically also possible with carbon black pellets. Nitric acid is used as pelletizing agent in this case. The oxidation takes place in this case while the wet pellets are dried at elevated temperatures [16].

## 7.3 Pigment properties and uses

Industrially manufactured carbon blacks have specific surface areas in a broad range from 8 m$^2$/g for coarse thermal blacks up to 1000 m$^2$/g for the finest pigment grades. Carbon black types with specific surface areas of more than 150 m$^2$/g are normally porous with pore diameters of less than 1 nm. Carbon black powders with a very large surface area have often an inner pore surfaces area of the particles, which exceeds the outer (geometrical) one. The large specific surface area is the reason for the high adsorption capacity of carbon blacks for water, organic solvents and binders. Chemical and physical adsorption of these substances at the particle surfaces is possible. The nature of the adsorption is important for the wettability and the dispersibility of the carbon black particles in the different application systems. It has a great significance for the decision on the use of a carbon black type as a pigment or as a filler in rubber. An interesting fact is that carbon black qualities with a large specific surface area adsorb up to 20 wt% of water when stored in an atmosphere of humid air [4].

Corresponding with the high specific surface areas, the particle sizes of carbon black powders are often very small. There are carbon black grades that consist mainly of nanoparticles with sizes smaller than 100 nm. Typical particle size distributions for channel and gas blacks are in the range from 10 to 30 nm, for furnace blacks from 10 to 110 nm, for lamp blacks from 60 to 200 nm, for thermal blacks from 100 to 500 nm, and for acetylene blacks from 30 to 50 nm.

Visible light is absorbed by carbon black up to a rate of 99.8%. The broad use of carbon blacks as black pigments is based on this property. Many carbon black pigments are pure black, others show a bluish or brownish color tone. The specific color variation depends on the structure of the carbon black, the light conditions, and the application system in which the carbon black particles are incorporated. Carbon black absorbs not only visible light, but also light from the infrared and ultraviolet range of the electromagnetic spectrum. Some special carbon black grades can be used therefore as UV stabilizers in polymers.

Carbon black is characterized by a high electroconductivity. However, the conductivity of carbon black is inferior compared with graphite and depends on the structure, the exact composition and the specific surface area. The electrical conductivity in a paint, a printed surface or a plastic material depends strongly on the distance between conductive neighboring particles in the system. Only if the distance of a large number of particles is very close and the concentration of particles in the application medium is high enough to establish a conduction path, a high electrical conductivity can be achieved. Special carbon black types with suitable conductivity are produced to equip application systems with antistatic or electrically conductive characteristics. A better conductivity of carbon black pigments is the result of a higher structure. High-structured pigment blacks are used for the production of electrically conductive printing inks.

The chemical analysis of carbon blacks varies significantly: 95.0–99.5 wt% carbon, 0.2–1.3 wt% hydrogen, 0.2–3.5 wt% oxygen, 0–0.7 wt% nitrogen, 0.1–1.0 wt% sulfur, <1.0 wt% residual ash. Production process, raw materials, and aftertreatment are important factors for the specific composition of a carbon black type. The residual ash is due to some salts and to the raw materials. Salts, particularly potassium salts, are added during the manufacturing process for the control of the carbon black structure. A part of the salts in the residual ash comes from the process water, from where it is not completely washed out.

The oxygen content is of special importance for the application properties of carbon black powders. Bonding of oxygen at the surface can occur in form of acidic or basic functional groups. There is a significant influence of the manufacturing process and of the aftertreatment on the amount of surface oxides and their chemical functionality. Furnace and thermal blacks contain only 0.2–2 wt% oxygen, which is bound here preferably in form of almost pure, basic surface oxides. Gas and channel blacks have mainly surface oxides and only small amounts of basic oxides. They contain up to 3.5 wt% oxygen.

Hydrogen is bound on the surface of the carbon black particles in two different ways. A part of the hydrogen forms together with the oxygen surface bound functional groups, another part is directly fused to the carbon. Identified oxygen- and hydrogen-containing groups at the surface of carbon black particles are carbonyl, carboxyl, pyrone, phenol, quinone, lactone and ether groups [4].

Nitrogen in carbon black is usually incorporated in the graphite-like lattice. Sulfur appears in a variety of forms: as a bound molecule, in an oxidized state and in its elementary form.

Carbon black is comparatively stable and does not spontaneously ignite at temperatures below 600 °C. It glows slowly when ignited in air. As a consequence, ignition sources must be excluded during processing and storage of carbon black powders and pellets.

Carbon blacks are divided with regard to their structure in types with a "high structure" and with a "low structure." Basis for this classification is the three-dimensional arrangement of primary particles in aggregates. Extensive interlinking or branching is characteristic for a "high structure." Less pronounced interlinking or branching, on the other hand, is typical for a "low structure." It is nearly impossible to determine the structure of carbon black directly. A method for the structural characterization, which is widely used and accepted among experts is the absorption of dibutyl phthalate (DBP). This measurement gives the desired information on the degree of aggregation of individual carbon blacks. The relationship between the absorption and the structure can be explained as follows: the greater the DBP absorption, the higher the carbon black structure [4].

High-resolution electron microscopy provides a true picture of the primary particles and of the aggregates of a specific carbon black quality. The inclusion of X-ray analytical techniques leads to the recognition that the primary particles consist of concentrically arranged, graphite-like crystallites. The results show that the graphitic layers are often twisted into each other with the result of a nonordered state. Up to 1500 of the graphite-like crystallites form one single primary particle. Electron microscopic studies using a scanning tunnel microscope have led to the additional information that the primary particles consist of superimposed, scale-like layers of graphite [4].

Table 7.2 gives an overview of the most important carbon black applications. Main consumer of carbon black is with about 90% the rubber industry. Carbon black is mainly used in this field for reinforcing fillers in tires, tubes, conveyor belts, cables, rubber profiles, and other mechanical rubber goods. Carbon blacks are classified in active, semi-active and inactive blacks relating to their reinforcing capacity. Active blacks are very fine powders with particle sizes in the range of 15 to 30 nm, which are characterized by a high reinforcing capability. They are mainly used in tire treads. Semi-active blacks with a lower reinforcing capability consist of particles in the size range from 40 to 60 nm. They find application in the tire carcass and in technical rubber components. Inactive blacks have only a limited reinforcing capability. They are characterized by particle sizes of more than 60 nm and need high filling levels for the use as reinforcing materials.

**Table 7.2:** Applications for carbon black [4].

| Area | Application |
| --- | --- |
| Rubber | Reinforcing filler in tires and mechanical rubber components |
| Printing inks | Pigmentation, rheology |
| Coatings | Full black and tinting application |
| Plastics | Black and gray pigmentation, tinting, UV protection, conductivity, conductor coatings |
| Fibers | Pigmentation |
| Paper | Black and gray pigmentation, conductivity, decorative and photo-protective papers |
| Construction | Cement and concrete pigmentation, conductivity |
| Power | Carbon brushes, electrodes, battery cells |
| Metal reduction, compounds | Metal smelting, friction compound |
| Metal carbide | Reduction compound, carbon source |
| Fire protection | Reduction of mineral porosity |
| Insulation | Graphite furnace, polystyrene and PU foam |

Pigment blacks are high sophisticated powdered materials. They are used in many applications, particularly in printing inks, paints, coatings, plastics, fibers, and paper. Depending on the application, they are referred to as printing blacks, coating blacks, or plastic blacks.

The main advantages of carbon black pigments compared with other black pigments (inorganic and organic) and black dyes are the strong hiding power, the high color stability, the excellent solvent, acid and alkali resistance as well as the sufficient thermal stability.

A widely accepted international classification system for carbon black pigments was established, in which four groups of products are differentiated: high color (HC), medium color (MC), regular color (RC), and low color (LC). The manufacturing process of a specific carbon black type is expressed by a third character: (F) for furnace black and (C) for channel or gas black. Oxidative aftertreatment is indicated by the suffix (o) for "oxidized" (Table 7.3) [4].

Jetness and tinting strength are of highest importance for the use of carbon black pigments. Jetness is the achievable intensity of blackness. Tinting strength stands for the coloring ability of a carbon black pigment and is measured against a white pigment, typically titanium dioxide or zinc oxide.

**Table 7.3:** Pigment black classification [4].

| Designation | | Particle size range (nm) |
|---|---|---|
| *Gas blacks* | Furnace blacks | |
| HCC | HCF | 10–15 |
| MCC | MCF | 16–24 |
| RCC | RCF | 25–35 |
| | LCF | >36 |
| Gas blacks oxidized | | |
| HCC (o) | | 10–17 |
| MCC (o) | | 18–24 |
| RCC (o) | | >25 |

Carbon black pigments are used in printing inks, paints and other applications in three respects: achievement of pure black color shades (use of carbon black alone) and darkening of any color (mixtures of colored pigments with carbon black) or generation of grey effects (mixtures of white pigments with carbon black).

Printing blacks have to fulfill special requirements in the printing ink itself and in the printed product. Good wettability and dispersibility of the pigments are very important for the use in printing inks. Suitable compatibility with the pigment, optimum viscosity, good flow characteristics, and storage stability are requirements for the pigmented printing ink. A blue hue of the black, achieved by the use of carbon black with finer particle sizes, is desired for most of the printed products. Carbon black pigments in printing inks act as color giving (inking) components and/or as rheological additives. Inking includes properties such as jetness, undertone and gloss, rheology represents viscosity, flow properties and tack. The inking characteristics depend strongly on the particle size of the carbon black pigment used. The rheological properties, on the other hand, dependent on the particle size, the surface area, the structure and the surface chemistry of the carbon black particles. Very fine pigment blacks with a high surface area generate strong thickening effects. Increasing particle sizes of the pigments exhibit reduced thickening effects and lower viscosity.

Carbon black pigments find only limited application in UV curing printing because the black particles strongly absorb not only visible light but also the incident light from the ultraviolet spectral range. The consequence is that blacker printing inks exhibit a greater delay in curing. Special carbon black qualities are however able to fulfill the requirements of UV curing printing.

Most of the carbon black containing inks are specifically produced for one of the industrially used printing processes. They are referred to as offset inks, letterpress inks, newspaper printing inks, gravure printing inks for magazines and decoration, and packaging printing inks. Each of these inks has to fulfill special requirements to work in the printing process in an optimal and cost-efficient way.

Carbon black pigments are by far the mostly used black colorants in the paint and coatings industry. They are applied in nearly all industrial and automotive coatings systems. The pigments appear to have the best overall performance with respect to requirements of black properties (normal black, deep black, black with blue or brown shade), color characteristics (tinting strength, hiding power, jetness), stability (alkali and acid resistance, usability in water-based systems), and weathering behavior. Black characteristics and color properties are the most important factors for the decision to use a specific pigment in a paint or coating system. Both factors depend strongly on the average particle size of the carbon black pigment. The jetness of a coating is related to the average size of the primary particles. A deeper jetness is achieved with a carbon black pigment of smaller particle size. The extremely small particle sizes of many carbon black powders, together with their large surface areas are the reason, why these pigments are more difficult to disperse in paint and coating formulations than most of the other pigments. Basically, carbon blacks are dispersed with all common grinding units for powders, such as sand mills, pearl mills, ball mills, and triple roll mills.

Carbon blacks are also in the plastics industry the most important black pigments. Their application in plastics allows the generation of various coloristic effects and modifies in addition the electrical and mechanical properties of the polymers (filler function). The pigments provide also heat and UV resistance to the plastic material. The requirements to achieve a black or gray tinting for polymers (polyethylene, polypropylene, polyvinyl chloride, polystyrene, ABS polymers, polyurethane) are best fulfilled with carbon black pigments. The pigments are typically incorporated in plastics in two steps to achieve a good distribution of the particles. The first step consists in the preparation of a carbon black plastic concentrate (master batch) in a kneader or another suitable device. Master batches with carbon blacks have a pigment content of 20 to 50 wt%. Dispersion of the pigment in the plastic material happens already during the preparation of the concentrate. In the second step, the concentrate is diluted with the appropriate polymer amount to obtain the required final carbon black content. The final master batches are normally offered in form of chips or pellets.

Many polymers have severe problems with degradation by UV radiation. The use of carbon black pigments can provide long-term stability to such polymers. The pigments absorb a part of the incident UV radiation and act as free-radical acceptors. They contribute to the deactivation of active intermediate species formed during the degradation process. The use of pigments with larger particle sizes and higher carbon black concentrations in the polymer enhance the stabilizing effect.

Carbon black is also used as conductive filler in plastics in order to achieve antistatic and conductive properties. Antistatic (resistivity $10^6$–$10^9$ $\Omega$) or conductive ($< 10^6$ $\Omega$) characteristics of the pigmented polymer material can be generated by the use of sufficient amounts of carbon black. Carbon black fillers with small particle sizes and high degree of aggregation together with a sufficient high carbon black concentration lead to high conductivities of plastic materials. The pigment particles should be uniformly distributed in the polymer and many of them must have close contact to each other in order to achieve good electrical conductivity. Bridges of conductivity between the particles are formed under these conditions and the flow of electrons is promoted. Important for the conductivity is also the composition and the quality of the polymer. High-structure furnace blacks with relatively fine particles and low content of volatile components are mostly used to generate electrical conductivity and antistatic behavior of plastics. Conductive polymers contain 10 to 40 wt % carbon black. The concentration of carbon black in antistatic plastics used for floor covering and cable sheathing is in the range from 4 to 15 wt % [4, 5].

Decades of experience in the industrial production and processing of carbon black allow the conclusion, that there are under conditions of normal use no significant harmful effects on humans have to be expected. This assessment has also been confirmed by numerous epidemiological studies [2].

Carbon black pigments do not exhibit acute toxicity (LD50 value rat oral: >10,000 mg/kg). They are not irritating to skin or mucous membranes. Long-term inhalation studies and investigations on animals, particularly on rats, have shown that under so-called "lung overload" conditions chronic inflammation, pulmonary fibrosis and formation of tumors are possible. The influence of the animal species as well as the mechanism of the tumor formation is still not sufficiently understood. There is also uncertainty concerning the role of dust in this context. The International Agency for Research and Cancer (IARC) stated with respect to the carcinogenicity that carbon black is possibly carcinogenic to humans [17].

Carbon blacks are not explosive under the common conditions used during handling of the pigments. An explosion of carbon black/air mixtures can only happen when these are close to a sufficient ignition source. Normal precaution is necessary in closed silos or poorly ventilated locations because carbon monoxide from the carbon black manufacturing process can still be present in small quantities. It is recommended that ignition sources are strictly kept away and suitable respiratory protective devices are installed. It is recommended to store carbon black in clean, dry, uncontaminated areas away from exposure to high temperatures, open flame sources and strong oxidizers.

# References

1.  Ferch H. Pigmentruße. Hannover: Vincentz Verlag, 1995.
2.  Mathias J. In Kittel - Lehrbuch der Lacke und Beschichtungen. Spille J, editor. vol. 5. 2nd ed. Stuttgart/Leipzig: S. Hirzel Verlag, 2003:214.
3.  Pfaff G. In Winnacker – Küchler, Chemische Technik, Prozesse und Produkte. Dittmeyer R, Keim W, Kreysa G, Oberholz A, editors. vol. 7. 5th ed. Weinheim: Wiley-VCH Verlag, 2004:358.
4.  Stroh P. In industrial inorganic pigments. Buxbaum G, Pfaff G, editors. 3rd ed. Weinheim: Wiley-VCH Verlag, 2005:163.
5.  Pfaff G. Inorganic pigments. Berlin/Boston: WalterdeGruyterGmbH, 2017:167.

6.   Boehm HP. Struktur und Oberflächeneigenschaften von Rußen. Farbe + Lack. 1973;79:419.

7.   Patent US. 3,010,794 (Cabot Corp.) 1958; Patent US 3,010,795 (Cabot Corp.). 1958.

8.   Patent DE. 15 92 853 (Cities Service Co.). 1967.

9.   Patent DE. 895 286 (Degussa) 1951; Patent DE 11 29 459 (Degussa). 1960.

10.  Patent US. 3,649,207 (Ashland Oil & Refining Co.) 1969; Patent US 3,986,836 (Phillips Petroleum Co.). 1974.

11.  Kühner G, Dittrich G. Untersuchungen zum Furnaceruß-Prozeß an einem Modellreaktor. Chem Ing Tech. 1972;44:11.

12.  Patent DE. 742 664 (Degussa) 1940; Patent US 2,420,810 (Cabot Corp.). 1941.

13.  Patent GB. 895 990 (Degussa). 1958.

14.  Patent US. 2,439,442 (Cabot Corp.). 1943.

15.  Patent US. 3,383,232 (Cabot Corp.) 1968; Patent US 3,870,785 (Phillips Petroleum Co.). 1975.

16.  Bode R, Ferch H, Koth D, Schumacher W. Schwarzgradskala für Pigmentruße. Farbe + Lack. 1979;85:7.

17.  Kuempel ED, Sorahan T. In views and expert opinions of an IARC/NORA expert group meeting, Lyon, France, 30 June – 2 July 2009. IARC technical publication No. 42. vol. 42. Lyon, France: International Agency for Research on Cancer, 2010:61.

Robert Christie and Adrian Abel

# 8 Carbonyl pigments: general principles

**Abstract:** This chapter describes the general features of the chemical class of pigments designated as carbonyl pigments. These pigments are characterized by the presence of carbonyl groups linked to one another via an extended conjugated system, often forming polycyclic aromatic structures. Carbonyl pigments have experienced distinct phases in their histories. Certain carbonyl colorants, notably anthraquinonoids, were discovered in the early twentieth century and subsequently used as vat dyes for textiles, but their potential as high-performance pigments was not realized until the mid-twentieth century when demand began to emerge for pigments of the quality that they could provide. After conversion to a suitable physical form, several vat dyes were then introduced as vat pigments. Several other carbonyl pigment types did not originate as vat dyes but were developed specifically for pigment use. Carbonyl pigments provide a wider diversity of structural arrangements. The broad carbonyl chemical class may be categorized into several sub-types, each with its own characteristic structural features. These categories, which are discussed in separate chapters, include anthraquinonoids, quinacridones, diketopyrrolopyrroles, perylenes, perinones, indigoids, isoindolines, isoindolinones, and quinophthalones. These products generally owe their high levels of technical performance to their large molecular size and high molecular planarity, which lead to highly compact crystal structures and, in many cases, to the ability of the carbonyl group to participate in strong intra- and intermolecular hydrogen bonding.

**Keywords:** Carbonyl pigments, polycyclic, vat dyes, vat pigments, conjugated system, high-performance pigments, anthraquinonoids, quinacridones, diketopyrrolopyrroles, perylenes, perinones, indigoids, isoindolines, isoindolinones, quinophthalones, automotive paints, plastics

## 8.1 Fundamentals

Carbonyl pigments constitute a chemical class of pigments that are characterized by the presence of the carbonyl (C = O) group. Most carbonyl pigments contain two or more carbonyl groups which, as illustrated in Figure 8.1, are linked to one another via an extended conjugated system, usually constructed with multiple aromatic ring

This article has previously been published in the journal Physical Sciences Reviews. Please cite as: R. Christie, A. Abel, Carbonyl Pigments: General Principles *Physical Sciences Reviews* [Online] 2021, 6. DOI: 10.1515/psr-2020-0153

https://doi.org/10.1515/9783110588071-008

**Figure 8.1:** The general structural arrangement in most carbonyl pigments.

systems [1–4]. The carbonyl groups are contained within the essential chromophoric unit. In many application classes of dyes for textiles, carbonyl dyes, especially anthraquinones, are second in industrial importance to azo dyes. There is also a wide range of industrially important carbonyl pigments, most of which can be classified as high-performance products. An alternative way that is often used in the broad classification of organic pigments according to chemical structure is to use the term "polycyclic" because the structures mostly contain multiple aromatic ring systems, either carbocyclic or heterocyclic [5]. However, the products encompassed by this term can include not only carbonyl pigments but also phthalocyanine and dioxazine pigments, both of which are addressed in separate chapters.

A particular textile dye application class where carbonyl dyes play an essential role, is the vat dye class, dominated by anthraquinonoid structures. This group of dyes are highly insoluble products, in contrast to most other textile dye classes which have at least some solubility in water, the medium from which they are generally applied to textile substrates. In the vat dyeing of cellulosic fibers, notably cotton, the insoluble dyes are subjected to reduction, usually by an alkaline solution of sodium dithionite, to a provide a water-soluble (leuco) form, which can be taken up by the fibers. Subsequently, it is re-oxidized either by air or an oxidizing agent such as hydrogen peroxide. This process regenerates the original insoluble structure, the particles formed becoming trapped mechanically within the textile material. A final stage in the sequence involves heating to the boil in the presence of a surface-active agent, in a process described as "soaping", which develops the crystalline particle structure of the colorant thus maximizing the coloristic and technical performance. In certain ways, this stage serves a similar function to the conditioning after-treatment processes used in the manufacture of pigments. The chemical processes involved in vat dyeing are illustrated in outline in Figure 8.2, demonstrating that the carbonyl groups linked through the conjugated system are an essential feature for the process. The insolubility of vat dyes and their generally very good fastness properties stimulated considerable effort into the selection of suitable vat dyes for use as pigments, following conversion to a physical form that is appropriate for paint, printing ink and plastics applications. Of the vast numbers of known vat dyes, about 25 have been fully converted to pigment use, although fewer are of real commercial significance. These products are commonly referred to as vat pigments. In addition to the anthraquinonoids, certain perylene and perinone pigments originated as vat dyes. There are, however, several other types of carbonyl pigment that have been developed specifically for pigment use, mostly as high-performance pigments. Individual chapters in this series are devoted to anthraquinonoid, quinacridone and diketopyrrolopyrrole pigments. Perylenes and perinones are described in a single chapter because of their structural commonality, and a further chapter encompasses indigoids, isoindolines, isoindolinones, and quinophthalones, to complete the coverage.

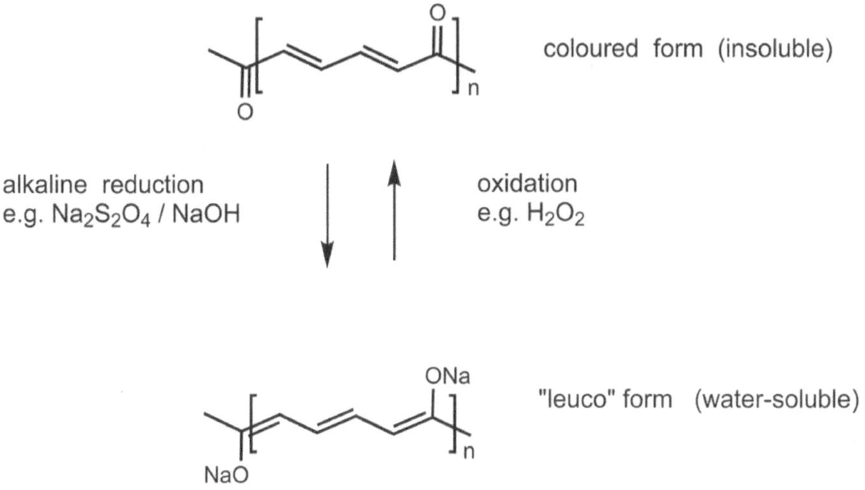

coloured form (insoluble)

alkaline reduction
e.g. $Na_2S_2O_4$ / NaOH

oxidation
e.g. $H_2O_2$

"leuco" form (water-soluble)

**Figure 8.2:** The chemistry of vat dyeing.

## 8.2 History

This section provides an overview of the historical developments over more than a century that have led to the introduction of the diverse types of industrial carbonyl pigments that are currently in commercial use. Further details are discussed in separate chapters on the individual pigment types. In contrast to the azo dye class, carbonyl dyes are found in nature. Madder, the principal component of which is alizarin (1,2-dihydroxyanthraquinone) [6], and indigo [7], were amongst the most important natural dyes that were used for centuries to dye textiles. In the period after Mauveine was discovered by Perkin in 1856 and introduced as the first synthetic dye manufactured on an industrial scale, and at a time when organic chemistry was coming of age as a science, a range of synthetic dyes were developed quite quickly to replace the natural products, with synthetic pigments emerging soon after. The first commercial process for the synthesis of indigo was developed in 1897, followed by thioindigo in 1905. Vat dyes, as described in the previous section, were the application class of most interest as pigments because of their insolubility. Indigo, which is still extensively used as a vat dye, has never achieved real significance as a pigment, and thioindigoid pigments, while important for a time, are now of little significance commercially. Several types of carbonyl pigment have experienced two distinct phases in their histories, whereby they were discovered in the early twentieth century, but their potential as high-performance pigments was not realized until the mid-twentieth century. Before then, there was little demand for pigments of the quality that they could provide. Several anthraquinonoid vat dyes were eventually developed for use as pigments after conversion to a physical form appropriate for pigment applications, including indanthrone which was discovered in 1901. Many of these products, including indanthrone, became important high-performance pigments. In more recent years, however, many original anthraquinonoid pigments have declined in importance, and relatively few remain as current commercial products. The linear *trans* quinacridones were first described in 1915 [8], their first documented synthesis reported in 1935 [9], and they were introduced as commercial pigments in the late 1950s. The histories of perylene and perinone pigments followed a similar pattern. Isoindoline pigments were introduced in the 1950s as a group of high-performance yellow pigments. In recent years, some early isoindoline pigments have declined significantly in importance, although CI Pigment Yellow 139 has grown in stature. The structurally related isoindolinones were introduced to the market in the mid-1960s. The first two pigments of this type, CI Pigments Yellow 109 and 110, set new standards for levels of technical performance and remain important today. The chemistry of the quinophthalone system has seen intense research activity since its original discovery in the late nineteenth century. However, CI Pigment Yellow 138 is currently the only commercial pigment in this group. The most recently introduced group of carbonyl pigments are the diketopyrrolopyrroles. The formation of a diketopyrrolopyrrole (DPP) derivative was first reported in 1974. In the 1980s, researchers from Ciba-Geigy (now BASF) reinvestigated the product and recognized its potential as a pigment and, following the development of an efficient synthetic route, products were launched on to the market, eventually to become one of the most important groups of high-performance organic pigments, especially in the red shade area.

## 8.3 Structures and properties

Carbonyl pigments adopt a much wider diversity of structural arrangements than is the case with azo and phthalocyanine pigments. The broad carbonyl chemical class, with the essential feature of carbonyl groups connected through a conjugated system, may be separated into several sub-types each with its own characteristic structural features. Figure 8.3 illustrates the molecular structures (**1–8**) of important industrial pigments from each of the subclasses, with reference to their identification in Table 8.1. As demonstrated by the Colour Index designations in Table 8.1, carbonyl pigments can provide colors throughout the entire spectrum. A major reason underlying the industrial importance of carbonyl textile dyes is their dominance of the vat dye application class, for which azo dyes are not appropriate. Also, in other dye application classes they can provide bright violet, blue and green colors, while azo dyes available in these colors tend to be duller in shade. While carbonyl pigments can provide these colors,

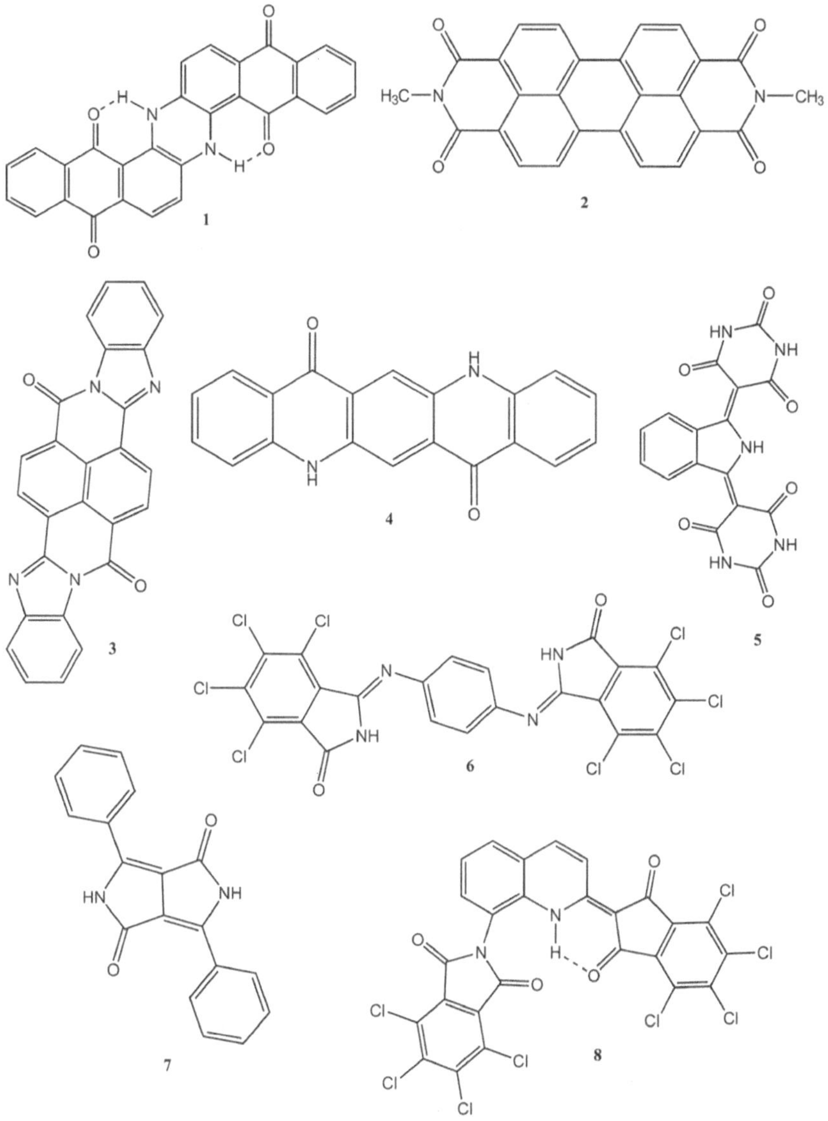

**Figure 8.3:** The structures of important representative examples of carbonyl pigment sub-classes.

the most important violet organic pigments are of the dioxazine class, and blue and green pigments are provided by the phthalocyanines. As a general class, carbonyl pigments provide technical performance that is superior to classical azo pigments, most being described as high-performance pigments and often offering properties similar to copper phthalocyanine. These products owe their high fastness to light, heat, and solvents to the large molecular size of the polycyclic systems and high molecular planarity, which provides solid state crystalline structures with highly compact, stacked

**Table 8.1:** Carbonyl pigments, as illustrated in Figure 8.3.

| Compound | Carbonyl sub-class | CI Pigment |
|---|---|---|
| 1 | anthraquinononoid | Blue 60 (indanthrone) |
| 2 | perylene | Red 179 |
| 3 | perinone | Orange 43 |
| 4 | quinacridone | Violet 19 |
| 5 | isoindoline | Yellow 139 |
| 6 | isoindolinone | Yellow 110 |
| 7 | diketopyrrolopyrrole | Red 255 |
| 8 | quinophthalone | Yellow 138 |

arrangements and, in many cases, to the ability of the carbonyl group to participate in intra- and intermolecular hydrogen bonding with other functionality in the molecules. Further details can be found in the chapters devoted to individual sub-classes.

## 8.4 Synthesis and manufacture

The chemical sequences involved in the manufacture of the range of carbonyl pigments are generally more elaborate and much less versatile than is the case with azo pigments. Often the synthetic sequence involves multiple stages and the use of specialist intermediates. Consequently, the number of commercial products is more restricted, and they tend to be significantly more expensive than classical azo and phthalocyanine pigments. The structures of the pigments are so diverse, as illustrated by the structures selected for inclusion in Figure 8.3, that it is difficult to detect commonality, except that in many cases the conditioning processes used to convert the crude products obtained from the synthesis into physical forms appropriate for pigment applications are similar. In view of these features, in the chapters on the individual sub-classes, the characteristic structural features of the most important carbonyl pigments are discussed in relation to performance, and the more important synthetic routes are presented.

## 8.5 Applications

Carbonyl pigments frequently offer technical and coloristic advantages over other organic pigment types, notably their high levels of fastness performance and bright colors. These features are essential to justify the high price required for commercial viability. Thus, they are often aimed at applications where the high cost may be justified, such as in automotive paint finishes and the coloration of plastics. Detailed description of the applications for which the different sub-classes are suited, and of the individual pigments within each sub-class, are provided in the relevant chapters.

## References

1.  Hallas G. In colorants and auxiliaries: organic chemistry and application properties. In: Shore J,editor. Bradford: society of dyers and colourists. vol. 1, ch.6 1990.
2.  Zollinger H. Color chemistry: syntheses, properties and applications of organic dyes and pigments, 3rd ed. Weinheim: Wiley-VCH, Ch. 8 2003.
3.  Gordon PF, Gregory P. Organic chemistry in colour. New York: Springer-Verlag, Ch. 4 1983.
4.  Christie RM. Colour chemistry, 2nd ed. London: RSC, Ch 4 2015.
5.  Hunger K, Schmidt MU. Industrial organic pigments. 4th ed ed. Weinheim: Wiley-VCH Verlag GmbH, 2019.
6.  Chenciner R. Madder red: a history of luxury and trade, plant dyes and pigments. Richmond: Curzon, 2000.
7.  Balfour-Paul J. Indigo. London: British Museum Press, 1998.
8.  Sharvin VV. J Rus Phys Chem Soc. 1915;47:1260.
9.  Liebermann H. Über die Bildung von Chinakridonen aus *p*-Di-arylamino-terephtalsäuren. 6. Mitteilung über Umwandlungsprodukte des Succinylobernsteinsäureesters. Annalen Der Chemie. 1935;518:245.

Robert Christie and Adrian Abel

# 9 Carbonyl pigments: miscellaneous types

**Abstract:** Carbonyl pigments are characterized by the presence of one or more carbonyl ($C = O$) groups in their structures, generally as a component of the chromophoric grouping and as part of an extended conjugated $\pi$-electron system. Structurally, they constitute a diverse group of pigments that offer a wide range of colors throughout the spectrum, and most of them provide high levels of technical performance. This paper provides a description of the historical development of thioindigoid, isoindoline, isoindolinone, and quinophthalone pigment types, and discusses their molecular and crystal structures in relation to their properties, the synthetic procedures used in their manufacture and their principal applications. They provide some of the most important high-performance yellow organic pigments for demanding applications in paints, inks, and plastics. Separate individual chapters in this series are devoted the anthraquinonoid, quinacridone, diketopyrrolopyrrole, perylene, and perinone carbonyl pigment subclasses.

**Keywords:** carbonyl pigments, indigoid pigments, indigo, thioindigoid pigments, thioindigo, quinophthalone pigments, quinophthalones, isoindolinone pigments, isoindolinones, isoindoline pigments, isoindolines, tetrachloroisoindolinones, high-performance pigments, intermolecular hydrogen bonding, CI Pigment Red 88, CI Pigment Red 181, CI Pigment Yellow 138, CI Pigment Yellow 139, CI Pigment Yellow 109, CI Pigment Yellow 110, CI Pigment Yellow 173, CI Pigment Yellow 185, CI Pigment Orange 6

## 9.1 Fundamentals

The principal chemical subclasses of industrial carbonyl pigments, namely the anthraquinonoids, quinacridones, diketopyrrolopyrroles, perylenes and perinone pigments, are dealt with in separate individual chapters in this series. However, there are other types of carbonyl pigments not included in these chapters. These miscellaneous types are collected here for convenience and to complete the broad coverage of the range of carbonyl pigments, but also because they exhibit some similarities in their chemical structures, performance, and applications. The types of pigments included here are indigoid (mainly thioindigo derivatives), isoindoline, isoindolinone, and quinophthalone pigments. Indigo itself is of interest historically

This article has previously been published in the journal Physical Sciences Reviews. Please cite as: R. Christie, A. Abel, Carbonyl Pigments: General Principles *Physical Sciences Reviews* [Online] 2021, 6. DOI: 10.1515/psr-2020-0154

https://doi.org/10.1515/9783110588071-009

and remains today an important vat dye. However, while highly insoluble, it has been of little interest as a commercial pigment. A few thioindigo derivatives played an important role in the range of pigments during the twentieth century but are now no longer used to any significant extent in the major organic pigment applications. The isoindolines, isoindolinones, and quinophthalones provide high-performance yellow pigments, with a few oranges. While the products currently manufactured are relatively few in number, these pigment types contain some of the most important yellow pigments for highly demanding paint, ink, and plastics applications.

## 9.2 History

Indigo has a long history as a natural dye [1]. Its use in Peru is reported to date back 6 millennia but it also extended much further worldwide. The name of the colorant is based on a plant species from which it is derived (*Indigofera tinctoria*), well-known in India. It was traded in Egypt and, especially, in the Roman empire where it was rather unromantically described by Pliny the elder as "being a slime that adheres to the scum upon reeds. When it is sifted out it is black, but in dilution it yields a marvelous mixture of purple and blue" [2]. Nobel laureate, Adolf von Baeyer, began his work on the synthesis of indigo in 1865. His original synthesis (from isatin) was described in 1878 but proved impractical, and so he developed a second synthetic route in 1880 from 2-nitrobenzaldehyde. Based on his studies, he ultimately reported the chemical structure in 1883 [3]. The first commercial process for the synthesis of indigo was developed by BASF in 1897, followed shortly after by MLB (Hoechst). In 1905, Friedländer, an assistant of von Baeyer, discovered thioindigo, the sulfur analogue of indigo. Although the parent compound has not experienced significant industrial exploitation, substituted derivatives were developed and introduced commercially as pigments in the 1950s to mid-1960s [4]. Isoindoline pigments were developed in the 1950s by BASF and Ciba (now part of BASF), with Bayer and Dainippon also playing a part in their development, and several products were introduced commercially. In recent years, a number of isoindoline pigments have declined significantly in importance. In contrast, one isoindoline, CI Pigment Yellow 139, has grown in stature and is now offered as a high-performance pigment by several manufacturers. The structurally related isoindolinones were first patented by ICI in 1946, although the original patent does not mention their use as pigments [5]. However, the market for high-performance organic yellow pigments barely existed when the discovery was made, as lead chromate yellow satisfied most industrial needs and at a fraction of the price at which the isoindolinone pigments could be manufactured. The first commercial pigments were patented in 1956 by Pugin and associates at Geigy (prior to its merger with Ciba) and were introduced to the market under the Irgazin designation in the mid-1960s. When the two earliest pigments CI Pigments Yellow 109 (Irgazin Yellow 2GLT) and 110 (Irgazin Yellow 3RLT), were introduced, they set new standards for lightfastness matching or even surpassing other more expensive carbonyl-based vat pigments. The quinophthalones are a group of colorants the origin of which may be traced back to 1882 when Jacobsen obtained the parent compound by reacting quinaldine with phthalic anhydride. Initially, quinophthalones were used as textile dyes. Quinoline yellow is a water-soluble acid dye, a food dye in Europe (E104) and a cosmetic colorant in the US (D&C Yellow 10). Non-ionic disperse dyes and solvent dyes based on the quinophthalone structure were also developed. The first pigments were described in the 1960s by BASF, Sandoz and Ciba. However, the first commercial product, CI Pigment Yellow 138, was launched by BASF in 1973. This pigment is similar in coloristics and fastness properties to CI Pigment Yellow 109. Although quinophthalone chemistry has been an active area of research over the years, with several products proposed, CI Pigment Yellow 138 is currently the only commercial pigment in this group.

## 9.3 Structures and properties

### (a) Indigoid pigments

Indigo (**1**) is one of the oldest known natural dyes and was the first vat dye [1]. Natural indigo dyeing starts with extracts from the leaves of certain plants of the *Indigofera* and *Isatis* species, subjected to a fermentation process, and the process is completed by air oxidation. The structure of indigo is given in Figure 9.1. Although natural dyeing of textiles with indigo is still practiced to an extent in certain communities and settings, for more than a century it has been manufactured synthetically on a large scale. Synthetic indigo remains a highly important blue vat dye, by far its most important use being to dye denim. Indigo is an intriguing compound in that it absorbs at such long wavelengths for such a small molecule. Although quite insoluble due to strong association by intermolecular hydrogen bonding and π-π stacking in its crystal structure, demonstrated by X-ray crystal structure analysis, indigo has never assumed real commercial significance as a pigment [6, 7]. Thioindigo (**2a**), the sulfur analogue illustrated in Figure 9.1 with reference to Table 9.1, is itself not regarded as important but a series of substituted derivatives were introduced industrially, initially as vat dyes. At least seven of these products were later to be commercialized as pigments. The chlorinated derivatives (**2b**) and (**2c**), CI Pigments Red 88 and 181, as illustrated in Figure 9.1 with the substituent pattern identified in Table 9.1, exhibit fastness to light and solvents that meet the requirements of high-performance applications and have been of some importance for application in paints and plastics although the former is now discontinued, and the latter finds only specialty applications.

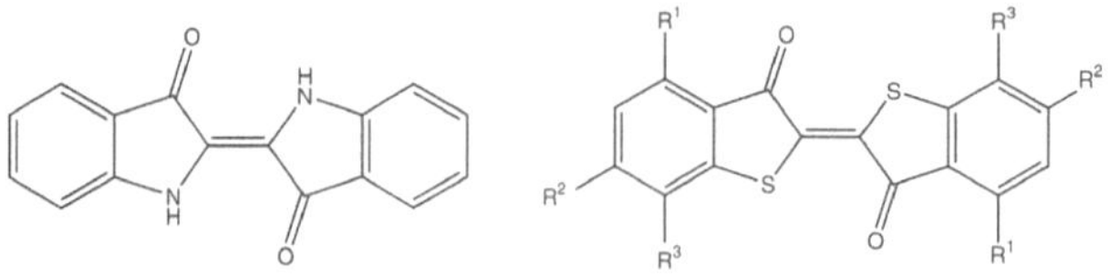

Figure 9.1: The molecular structure of indigo (1), and thioindigo derivatives (2).

Table 9.1: The substituent pattern in thioindigoid colorants.

| Compound | CI Pigment Red | $R^1$ | $R^2$ | $R^3$ |
|---|---|---|---|---|
| 2a | – | H | H | H |
| 2b | 88 | Cl | H | Cl |
| 2c | 181 | $CH_3$ | Cl | H |

## (b) Isoindoline pigments

The molecular structures of isoindolines are derived from a central isoindole unit connected by two methine (-C=) links to other units [8,9,10]. Isoindoline pigments of a variety of related structural types (3) – (7) (Figure 9.2), have been introduced commercially. These pigments provide high levels of technical performance in yellow through orange to yellowish-red shades. The most important product in the series is the reddish yellow CI Pigment Yellow 139 (3). This pigment is particularly noteworthy in that it can produce high opacity, almost matching that of inorganic pigments, when prepared in an optimum particle size form. Other isoindoline pigments that have been introduced commercially include CI Pigment Yellow 185 (4), CI Pigments Orange 66 (5a) and 69 (6), and CI Pigment Red 260 (7). The low solubility of these pigments is attributed to their highly ordered crystalline assemblies, the main driving force being strong intermolecular hydrogen bonding. As an example of the intermolecular association in isoindoline structures, X-ray crystallography has demonstrated that compound (5b) adopts an intermolecular hydrogen bonding arrangement between cyano and amide substituents, as illustrated in Figure 9.3 [11].

Isoindolinone pigments are related structurally to the isoindolines, with a carbon atom attached to the isoindole unit replaced by nitrogen [9, 12]. Structures of products that have been commercialized are illustrated in Figure 9.4. They constitute a range of yellow and orange pigments that show outstanding properties, providing excellent lightfastness and resistance to solvents and chemicals, suited to the most demanding paint, ink, and plastics applications. There are certain similarities in the structures of the isoindolinones compared with disazo condensation pigments, since they are mostly symmetrical structures derived from two isoindole units linked using a condensation reaction with an aromatic diamine. The most important industrial examples of this

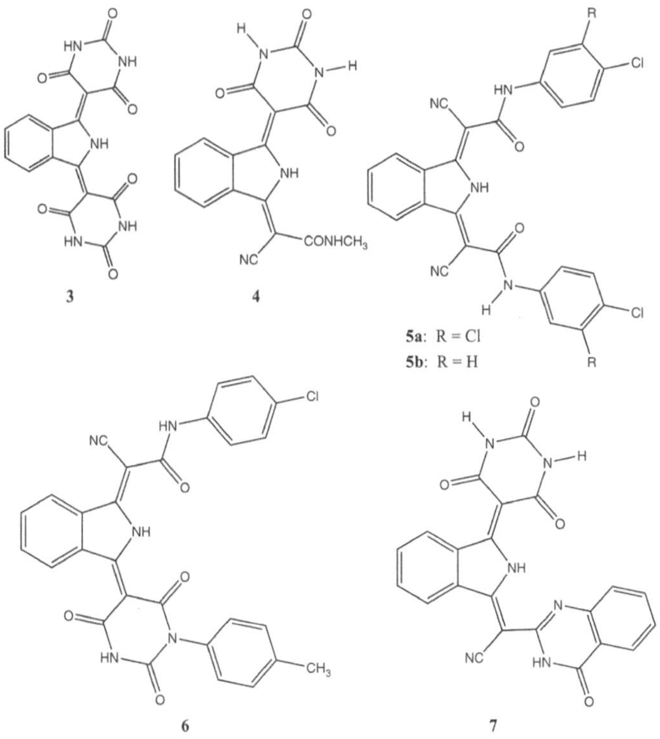

**Figure 9.2:** The structures of isoindoline pigments.

group of pigments are the greenish-yellow CI Pigment Yellow 109 (**8**) and the reddish-yellow CI Pigment Yellow 110 (**9**). CI Pigment Orange 61 (**10**), which contains an azo group in its molecular structure, is produced commercially for plastics applications. A red product, CI Pigment Red 180, is no longer of commercial importance. These pigments have traditionally been referred to as tetrachloroisoindolinones based on the four chlorines in each isoindole benzene ring (so that the molecules contain eight chlorines in total). Sandoz (now Clariant) introduced the greenish-yellow CI Pigment Yellow 173 (**11**), which does not contain chlorine in the isoindole benzene rings but has two chloro substituents in the central bridging ring. The pigment is a dull greenish yellow with slightly lower levels of fastness properties compared with the much more highly chlorinated products. It has recently been withdrawn by Clariant. Evidently, the complete chlorination of the isoindole benzene rings plays an essential part in determining the shade and high levels of fastness properties of this group of pigments and, in contrast to many other chlorine-rich products, they exhibit reasonably high tinctorial strength.

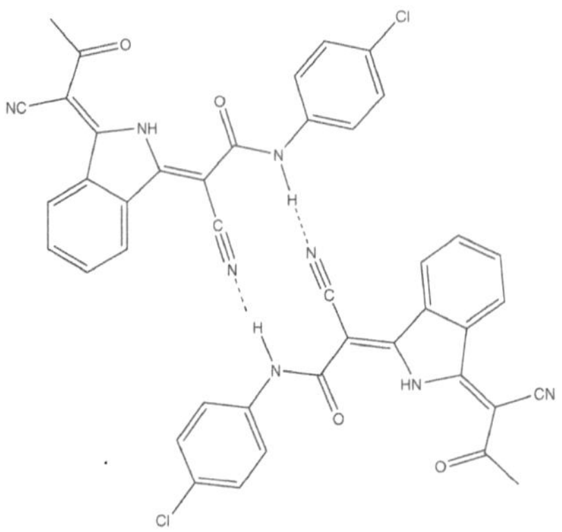

**Figure 9.3:** Intermolecular hydrogen bonding in isoindoline (**5b**).

### (c) Quinophthalones

Quinophthalone pigments have been extensively investigated over the years but appear to have been less extensively commercialized. The only significant commercial pigment is CI Pigment Yellow 138 (**12**), a high-performance greenish-yellow pigment, as illustrated in Figure 9.5 [13]. The crystal structure of CI Pigment Yellow 138 has been determined from X-ray powder diffraction data with Rietveld refinement. The tautomeric form was established by solid-state multinuclear NMR studies as the NH-tautomer, the NH hydrogen atom attached to the quinoline unit forming an intramolecular resonance-assisted hydrogen bond with the neighboring indandione unit. The indandione unit is almost coplanar with the quinoline unit, whereas the phthalimide group forms an angle of 57° with this unit, caused by intramolecular steric hindrance [14].

## 9.4 Synthesis and manufacture

The most important synthesis route to thioindigo pigments (Figure 9.6) involves the treatment of the appropriately substituted thiophenol (**13**) with chloroacetic acid to give the arylthioacetic acid (**14**). Cyclization of this intermediate using chlorosulfonic acid gives thioindoxyl (**15**), which is then oxidatively dimerized either with air or a variety of inorganic oxidizing agents to the thioindigo (**2**). Conditioning of the crude product can be achieved by milling with salt or reprecipitation from sulfuric acid, followed by solvent treatment.

  Isoindolines may be prepared from condensation reactions of 1-amino-3-iminoisoindoline (**17**) (1 mol), an intermediate proposed in the formation of copper phthalocyanine, with compounds containing reactive methylene groups (2 mol), barbituric acid (**18**) in the case of CI Pigment Yellow 139 (**3**) and a cyanoacetanilide (**19**) in the case of

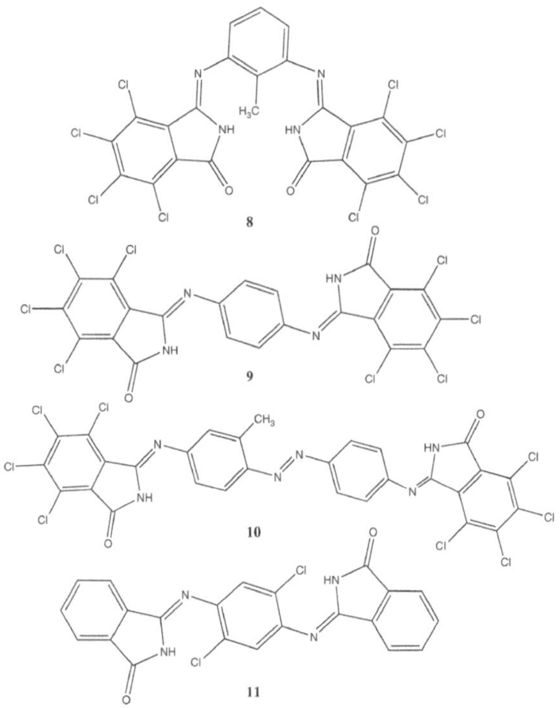

**Figure 9.4:** The structures of isoindolinone pigments.

compounds (**5a**) and (**5b**), as illustrated in Figure 9.7. As an example, in a synthetic sequence leading to pigment (**3**), compound (**16**), formed by treatment of phthalonitrile (**15**) with ammonia gas in a water miscible organic solvent, is added without isolation to an aqueous solution of barbituric acid (**18**), formic acid and a surfactant. After a prolonged period at the boil, the pigment is filtered, washed, and dried. It is important that the two condensation reactions may be carried out in separate stages to allow the efficient preparation of unsymmetrical isoindolines, such as compounds (**4**), (**6**) and (**7**).

The tetrachloroisoindolinones, CI Pigments Yellows 109 (**8**) and 110 (**9**), and CI Pigment Orange 61 (**10**) may be synthesized by the condensation of a suitable chlorinated precursor, for example either of compounds (**20a**), (**20b**) or (**21**) (2 mol), with the appropriate aromatic diamine (1 mol) in a high boiling organic solvent at around 160–180°C, as illustrated in Figure 9.8. When reaction is complete, the pigment is filtered hot, washed, dried, and milled. CI Pigment Yellow 173 may be prepared from a similar precursor that is unchlorinated in the benzene ring.

The quinophthalone pigment, CI Pigment Yellow 138 (**12**), may be prepared by the acid catalysed condensation of 8-amino-2-methylquinoline (**22**) (1 mol) with tetrachlorophthalic anhydride (**23**) (2 mol), as illustrated in Figure 9.9. The process is conducted for prolonged times at around 180°C leading after filtration to the crude product. Conditioning can involve organic solvent or alkali treatments or milling

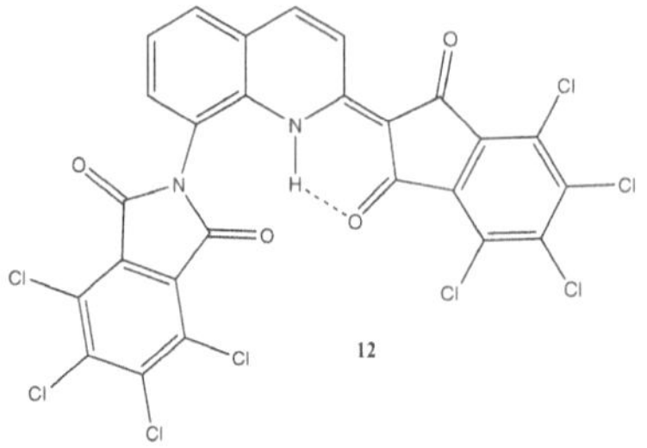

**Figure 9.5:** The structure of quinophthalone pigment (**12**).

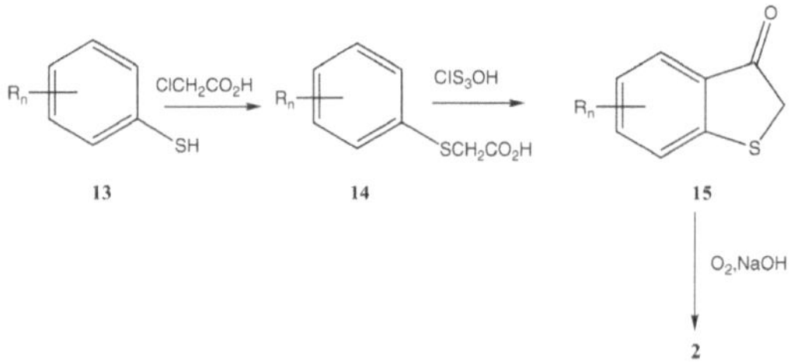

**Figure 9.6:** Synthesis of thioindigo pigments.

with surfactants. To produce a finer particle size grade which is greener and more transparent, the crude product is conditioned at a lower temperature. A coarser version is obtained by conditioning at higher temperatures leading to a slightly redder shade but with much higher opacity.

## 9.5 Applications

### Thioindigoid pigments

### CI Pigment Red 88 (2b)

For many years this was an important product, mainly for the paint industry, where it was used in high-performance paints, including automotive original finishes. It was often preferred to the quinacridone reddish violet, CI Pigment Violet 19 (β-form), for its deeper more intense bordeaux shade. It provided very good to excellent fastness to light and weathering. It was stable to 200°C, and fastness to oversphaying was excellent in paints stoved at 120°C, but a little less fast at 160°C. Its resistance to solvents is less than perfect, especially ketones, in this respect being significantly inferior to the quinacridone pigment. The lack of brightness of the product when used in white reduction limited its use in such applications, in which quinacridone pigments are preferred. The pigment used to have a serious defect in that it could generate hydrogen sulfide in acidic conditions, resulting in the paint having a "rotten egg"

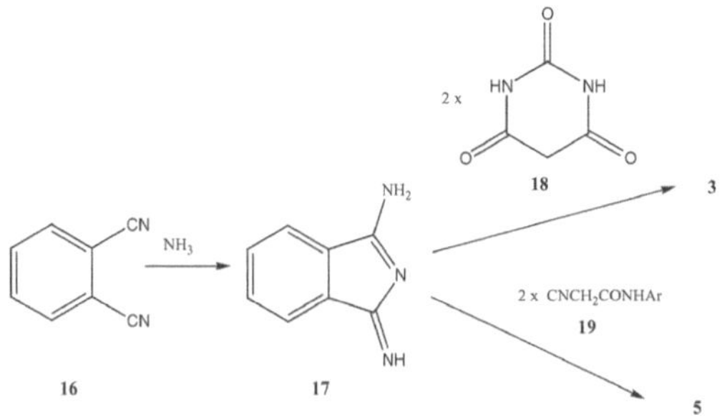

**Figure 9.7:** The synthesis of isoindoline pigments.

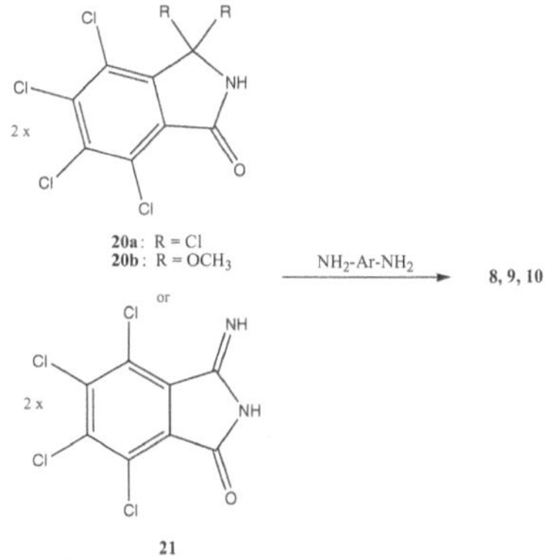

**Figure 9.8:** Synthesis of isoindolinone pigments.

smell. Just before the pigment was generally withdrawn, a new grade was introduced that did not show this defect. In plastics, it could be used in PVC, but lacked excellent fastness to migration in plasticized PVC, although as purer grades were introduced, migration fastness improved. In polyolefins heat stability could vary. While the best grades were stable to 300°C, others were stable only to 260°C. The pigment produced quite severe dimensional instability problems, leading to warpage in injection moldings. It was frequently used for spin dyeing of polypropylene. However, all of this is history. Environmental concerns involved in the manufacturing process have led to the major manufacturers ceasing the marketing of this pigment and there are no longer any products registered in the Colour Index.

## CI Pigment Red 181 (2c)

This pigment is still marketed in modest quantities despite its high price. It has two main specialty applications. When incorporated into polystyrene and other styrenic co-polymers it dissolves to produce a very bright pink shade, which becomes trapped in the polymer matrix on cooling, thus providing excellent fastness properties. The other

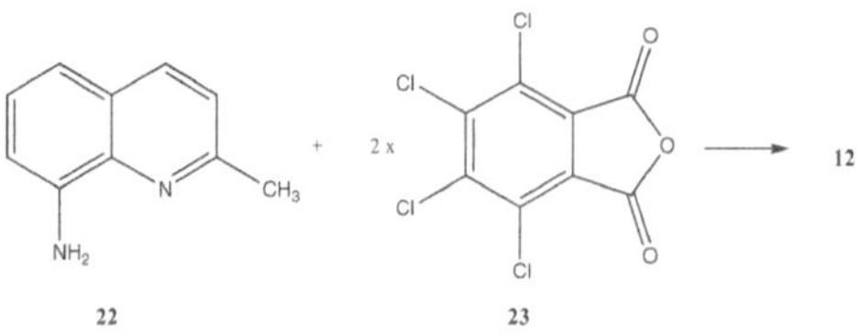

**Figure 9.9:** Synthesis of quinophthalone pigment, CI Pigment Yellow 138 (**12**).

application is as a cosmetic colorant (D&C Red 30 in the USA, CICN 73360 in the EU). It provides the red stripe frequently seen in toothpaste.

## Isoindoline Pigments

### CI Pigment Yellow 139 (3)

This pigment is available as a high color strength, transparent product and also as a highly opaque reddish yellow product. It has found markets mainly in paint and plastics on account of its reddish-yellow shade and generally excellent fastness properties, especially the highly opaque, redder grade which is ideally suited to replace lead chromate yellows, either alone or in combination with alternative inorganic pigments. It has excellent fastness to light and weather especially in deep shades, although it darkens when used alone in full and near full shades. It provides very good flow behavior in applications which allows it to be used at high concentrations, up to 1:4 parts pigment to solid binder). Its weakness is inferior resistance to alkali, which may limit its use in some water-based systems such as emulsion (latex) paints. The opaque form is considered good enough for original equipment automotive paints. The more transparent version fades at nearly twice the rate. It is highly insoluble and does not dissolve in any of the solvents normally used in the paint industry. It is stable to 200°C. It is widely used in the plastics industry for the coloration of PVC and polyolefins but is not recommended for styrenic co-polymers or engineering polymers. In PVC it is resistant to bleeding. In full shade it has excellent lightfastness, but this decreases rapidly in reductions. In polyolefins it does not migrate and has generally good lightfastness, although it darkens quite severely in full shade. It is stable up to 250°C, but above that temperature it starts to decompose becoming much darker. It has a low warpage effect in injection moldings and can be used in polypropylene fibers. In inks, it can replace disazoacetoacetanilide (diarylide) yellows and chrome yellows, when high-quality prints are demanded such as in metal decorating and laminate printing. It is also highly recommended for security printing. The pigment is highly recommended for use in colored toners for photocopying and laser printing and can be used in solvent-based inkjet inks. Recent rationalizations have seen the range of isoindoline pigments severely pruned but, in contrast, CI Pigment Yellow 139 has grown in stature and is now offered in both high color strength and opaque reddish yellow versions by several leading manufacturers.

### CI Pigment Yellow 185 (4)

This pigment is a green shade yellow isoindoline pigment used mainly in the ink industry, especially for packaging and digital printing inks. Its lightfastness is inferior to CI Pigment Yellow 139 (3) but is considerably superior to most disazoacetoacetanilide (diarylide) yellow pigments, which it can replace coloristically, if good alkali stability is not required.

Other isoindoline pigments that are no longer actively marketed include CI Pigment Orange 66 (5a), a yellowish orange product that was mainly used in paints, CI Pigment Orange 69 (6), also a yellowish orange that was supplied as a high opacity pigment with excellent fastness properties, and CI Pigment Red 260 (7).

### Isoindolinone pigments

### CI Pigment Yellow 109 (8)

This is a clean, greenish-yellow shade pigment, which is suitable for use across most pigment applications, but its use is limited to areas where cost is less important than a high level of fastness properties. It is very insoluble in most organic solvents, and highly heat stable, providing the pigment does not dissolve in the polymer in which case the color becomes unstable. Its main applications are in the paint and plastics industry. In paints, it can be used for most automotive and industrial finishes, in full and strong shades, often in combination with copper phthalocyanines and inorganic pigments. Its lightfastness decreases quite rapidly in paler shades and it darkens features that limit its application. The classical grade is rated as opaque, so that it is unsuitable for metallic finishes. Its dominance in the greenish yellow shade area for paint applications has been reduced by competition from benzimidazolone azo pigments CI Pigments Yellow 151 and 154. The latter is about 20% stronger, while the former lacks the excellent alkali stability of the isoindolinones, and thus is less suitable for aqueous paints. Neither of these two benzimidazolone azo pigments matches the outstanding heat stability of the isoindolinones. In plastics, CI Pigment Yellow 109 is mainly used in polyolefins, where its heat stability is its main advantage, stable up to 300°C in full shade, although decreasing somewhat as the proportion of titanium dioxide increases. Its tinctorial strength is quite good. It has an adverse effect on dimensional stability leading to warpage in many injection moldings. In PVC, it is fast to migration at temperatures rising from a limit of 160°C for pale reductions up to 200°C in full shade. Its lightfastness is good to very good, but that is not considered quite sufficient for most coil coating applications. While CI Pigment Yellow 109 competes with the benzimidazolone azo pigments in paints, in plastics it competes more with the greener shade disazo condensation yellow pigments, where its lightfastness is usually slightly inferior. In inks, it has good heat stability, making it suitable for demanding applications such as metal decorative finishes and laminate paper printing.

### CI Pigment Yellow 110 (9)

Because CI Pigment Yellow 110 has a much redder shade, it rarely competes with CI Pigment Yellow 109 (**8**), although it is generally found to have superior fastness properties. In paints, its very high lightfastness and weatherfastness meet the requirements for automotive original equipment, usually considered as one of the most demanding paint applications. It has excellent fastness to most solvents and meets most requirements for heat stability. There are numerous grades available, differing mainly in particle size distribution, the finest of which (Irgazin Yellow L2040) is transparent and can be used for metallic and pearlescent finishes, whereas the coarsest grade (Irgazin Yellow L2060) is opaque and is suitable for opaque, full shades with high color saturation. In plastics, it is widely used in polymers, especially polyolefins, PVC, and styrene-based polymers. In polyolefins it is stable up to 300°C, decreasing only slightly in reductions. It has excellent lightfastness in all but the palest of reductions, but it causes dimensional instability in crystalline polymers leading to warpage in injected molded articles, especially those produced at lower temperatures. In PVC it is fast to migration and has excellent lightfastness, even for demanding applications like window profiles and coil coatings. It can be used in polystyrene copolymers, including ABS. The pigment can be used for high value inks, where its excellent properties, especially heat stability, make it an ideal choice. These applications include laminate printing inks and metal decorative inks.

### CI Pigment Orange 61 (10)

This pigment provides a yellowish shade of orange and was used widely in paints, plastics, and inks. However, it has only moderate solvent fastness, a feature that limits its application in high-performance paints and it does not offer lightfastness comparable with the yellow isoindolinone pigments. It is now only offered in the BASF plastics range under the Irgazin designation. In plastics, it is suitable for polyolefins, where it has very good heat stability up to 300°C but provides low tinctorial strength. Its lightfastness is rated as excellent and it has slightly better weatherfastness than the DPP pigment (CI Pigment Orange 71), which has a similar color and tinctorial strength. It has a high tendency to cause warping of injection moldings. The pigment is widely used in PVC, where it provides excellent fastness to migration and lightfastness.

**CI Pigment Yellow 173 (11)**

The dull, greenish-yellow shade of this pigment has limited its applications and it is no longer offered by Clariant, the only company to register it in the Colour Index. Its main use was in coatings, being essentially insoluble in hydrocarbons, but having some solubility in polar solvents. Its high transparency made it useful for metallic effects in combination with other transparent pigments. It was tinctorially stronger than other isoindolinone pigments and its heat stability was high enough for most applications, 160°C or even higher in strong shades. Its lightfastness was rated as very good to excellent, even in pale reductions. Although it was not recommended for plastics, it had good heat stability up to 300°C, but tended to dissolve as it approached this maximum, leading to solvent dye behavior. When it dissolved it became stronger and cleaner in shade but returned to its pigmentary form on cooling, while retaining its coloristic performance.

**Quinophthalone Pigments**

While quinophthalones have been extensively investigated, with several potential products proposed, CI Pigment Yellow 138 remains the only commercial pigment with a quinophthalone structure.

**CI Pigment Yellow 138 (12)**

The pigment is greenish yellow in color and has found an extended range of applications. It is available in high color strength and high opacity grades depending on the processes used in conditioning the crude pigment. The coarser grade has a slightly redder shade but with much higher opacity. This grade has excellent fastness properties and is used mainly in paints and plastics but with some limited applications in inks. The stronger, less opaque version is greener, and has good to very good lightfastness. In paints it has heat stability up to 220°C. The opaque grade has very good to excellent lightfastness and weatherfastness in full shade, but these properties decrease even at quite strongly colored reductions, while the higher strength versions exhibit lower levels of fastness performance. It is slightly superior in performance to the much redder CI Pigment Yellow 139 (**3**) but does not reach the weatherfastness levels of CI Pigment Yellow 110 (**9**). Its resistance to solvents is excellent. It has been widely used to replace lemon chrome pigments for demanding applications, including automotive finishes. It can be used in decorative paint tinting systems, but its stability to alkali in decorative paints can be problematic. The pigment finds wide application in plastics, including styrenic polymers and while not recommended for engineering polymers, it is sometimes used. In PVC it is fast to migration and gives excellent lightfastness in stronger shades. Its heat stability in polyolefins is up to 270°C with very good to excellent fastness to light and weather. It does have quite a strong tendency to affect the dimensional stability of injection moldings, especially when molded at lower temperatures. Economics limit its use in inks, but it is recommended for metal decorative inks and can be used for inkjet inks and toners.

# References

1.  Balfour-Paul J. Indigo. London: British Museum Press, 1998.
2.  https://www.loebclassics.com/view/pliny_eldernatural_history/1938/pb_LCL394.295.xml?readMode=recto. Accessed: 31 Jan 2021.
3.  Baeyer A. Ueber die Verbindungen der Indigogruppe. Berichte der Deutschen chemischen Gesellschaft zu Berlin. 1883:16;2188–204.
4.  Fisher W. Thioindigo pigments. In: Hoboken PT, editor. Pigment handbook, vol. l. Wiley Interscience, 1973:Vol 1;676.
5.  Jones W. GB 2,537,352. (7th September 1946). (ICI).
6.  Kettner F, Hüter L, Schäfer J, Röde K, Purgahn U, Krautscheid H. Selective crystallization of indigo B by a modified sublimation method and its redetermined structure. Acta Cryst. 2011: E67;2867.
7.  Mizuguchi J, Endo A, Matsumoto S. Electronic structure of intermolecularly hydrogen-bonded indigo: comparison with quinacridone and diketopyrrolopyrrole pigments. J Imaging Soc Jpn. 2000:39;94–102.
8.  Christie RM. Colour chemistry, 2nd ed. London: RSC, Ch 9, 2015.
9.  Hunger K, Schmidt MU. Industrial organic pigments, 4th ed. Weinheim: Wiley-VCH Verlag GmbH, Ch 2, 2019.
10. Radtke V, Erk P, Sens B. In: Faulkner EB, Schwartz RJ. High performance pigments, 2nd ed. Weinheim: Wiley-VCH Verlag GmbH, ch 14, 2009.
11. Von Der Crone. New isoindolines for high quality applications. J Coat Technol. 1985:57;67–72.
12. Iqbal A, Herren F, Wallquist O. In: Faulkner EB, Schwartz RJ. High performance pigments, 2nd ed. Weinheim: Wiley-VCH Verlag GmbH, ch 15, 2009.
13. Radtke V. In: Faulkner EB, Schwartz RJ. High performance pigments, 2nd ed. Weinheim: Wiley-VCH Verlag GmbH, ch 19, 2009.
14. Gumbert SD, Körbitzer M, Alig E, Schmidt MU, Chierotti MR, Gobetto R, et al. Crystal structure and tautomerism of Pigment Yellow 138 determined by X-ray powder diffraction and solid-state NMR. Dyes Pigm. 2016:31;364–72.

Robert Christie and Adrian Abel

# 10 Cationic (Basic) dye complex pigments

**Abstract:** Cationic (or basic) dye complex pigments are classical organic pigments obtained from water-soluble cationic dyes for textiles, mainly of triarylmethine (arylcarbonium ion) types, which are precipitated using large inorganic counterions, especially those derived from heteropolyacids such as phosphotungstomolybdic acid or, to a certain extent, using the counteranion derived from copper ferrocyanide. This range of pigments includes red, violet, blue and green products, offering brilliant shades, high color strength and good transparency. They are well suited to printing ink applications, although they provide only moderate levels of fastness properties. The pigments are synthesized by treating aqueous solutions of the dyes under highly controlled conditions with solutions of the heteropolyacids, prepared in situ. The copper ferrocyanide salts are obtained by treatment of potassium ferrocyanide with sodium sulfite in water, and subsequently with solutions of the cationic dye and copper (II) sulfate. The pigments are primarily used in inks for packaging and advertising materials. However, they have little use outside printing inks. Reflex or alkali blue pigments are structurally related cationic dye derivatives which are inner salts of the dye structures and are also used in printing inks.

**Keywords:** cationic dye complex, basic dye complex, triarylmethine, arylcarbonium ion, heteropolyacids, phosphotungstomolybdate, phosphomolybdate, phosphotungstate, reflex blue, copper ferrocyanide, alkali blue

## 10.1 Fundamentals

This long-established group of pigments is variously described as cationic dye complex pigments, basic dye complex pigments, arylcarbonium ion pigments, or triarylmethine pigments [1, 2]. The traditional terminology, triarylmethane, is also still often used to an extent in the chemical classification, although this term is strictly not correct. The pigments are obtained from water-soluble cationic (basic) dyes, which are well-known as dyes for textile applications, by converting them to insoluble salts using large counterions anions derived from certain inorganic acids, especially those known as heteropolyacids [3]. Alternatively, counteranions derived from copper ferrocyanide may be used. The pigments offer a range of bright and very strong colors, well suited to printing ink applications, although with only moderate levels of fastness

This article has previously been published in the journal Physical Sciences Reviews. Please cite as: R. Christie, A. Abel, Cationic (Basic) Dye Complex Pigments *Physical Sciences Reviews* [Online] 2021, 6. DOI: 10.1515/psr-2020-0155

https://doi.org/10.1515/9783110588071-010

properties. Reflex or alkali blue pigments are structurally related cationic dye derivatives that are inner salts of the dye structures and are also used in certain printing ink applications.

## 10.2 History

Historically, cationic (basic) dyes based on triarylmethine (arylcarbonium ion) chemical structures were the earliest synthetic textile dyes, introduced in the nineteenth century for the dyeing of cotton, wool, and, especially, silk. Mauveine, a purple dye discovered by Perkin in 1856 and the first synthetic dye to be manufactured on an industrial scale, belonged to this group [4]. Nowadays, cationic dyes are used principally to dye acrylic fibers, in which they provide intense, bright colors covering virtually the complete shade range. Among the earliest synthetic organic pigments were *lakes* derived from the known water-soluble cationic dyes precipitated using counteranions derived from certain organic acids in the presence of colorless insoluble inorganic substrates such as alumina hydrate. In these products, the dyes not only formed salts, but also were absorbed into the substrates. One of the best-known colors of this type is nigrosine, a black dye precipitated with oleic acid. The applications of these products were limited, although they were used to an extent in early flexographic inks. The early products retained the brilliance and intensity of color that is characteristic of the cationic dyes but were fugitive [5]. It was subsequently found that products with better lightfastness could be obtained by precipitation of the cationic dyes with large polymeric counteranions. The earliest pigments were developed by BASF in 1913 and commercialised under the designation Fanal. In 1917, Immerheiser and Beyer patented the use of inorganic acids containing tungsten or molybdenum to insolubilise the cationic dyes. The products were not immediately viable on cost grounds, as molybdenum and tungsten were in short supply due to the demand for these metals especially for munitions manufacture in World War I. Indeed, through subsequent history, the cost of these metals has fluctuated due to supply and demand issues, for example during the Great Depression of the 1930s. In 1921, Lendle patented the use of phosphomolybdate and phosphotungstate ions derived from the so-called heteropoly acids with various cationic dyes, and in 1927 Hartman introduced mixed products. These pigments proved to have superior lightfastness compared with earlier products and several have stood the test of time and remain as commercially viable products. It was also found that the copper ferrocyanide anion may be used as the counteranion.

The history of the structurally related reflex (or alkali) blues dates back to the earliest days of synthetic dye chemistry and cuts across international boundaries. Two years after Perkin had discovered Mauveine, Verguin, a French chemist patented Fuchsine in 1859, a brilliant scarlet dye, named after the plant genus "fuchsia" which in turn was named after the German botanist Fuchs [2].

In 1861, Girard and deLaire found that when Fuchsine was heated with aniline, a blue compound, known as Lyon Blue, was formed. In the UK in 1862, Nicholson investigated the sulfonation of Lyon Blue aiming to make the compound water soluble by incorporating sulfonic acid groups. The monosulfonated compound, now known as reflex blue, unexpectedly turned out to be insoluble, as it had formed an insoluble inner salt. It was found that it could be converted into a soluble dye by forming the sodium salt, hence alkali blue.

## 10.3 Structures and Properties

The most important cationic dyes used to prepare the pigments are represented by structures (**1**), (**2**) and (**3**), as illustrated in Figure 10.1, together with the counteranions and substituent patterns identified in Tables 10.1–10.3. In terms of chemical class, the dyes are referred to as triarylmethine or arylcarbonium ion dyes. The essential structural feature of the dyes is a central carbon atom attached to three aromatic rings [6, 7]. With dyes (**3**), there is bridging across the *ortho-ortho'* positions of two of the aromatic rings thus forming a heterocyclic ring. These dyes are of the xanthene chemical class, and the most important compounds are referred to as rhodamines. Formally, dyes (**1**) – (**3**) contain a carbonium ion (carbocation) center, the central carbon atom, hence the reference to arylcarbonium ion colorants, although the molecules are resonance-stabilized by delocalization of the positive charge on to *p*-amino nitrogen atoms. Cationic dyes used for the coloration of textiles have unparalleled color strength and can offer very bright, indeed sometimes fluorescent, colors. The range of commercial pigments prepared from these dyes provides reds, violets, blues, and greens that show high brilliance and intensity of color, together with high transparency. Thus, they are well-suited to printing ink applications, especially when good fastness to light and heat, and stability to alkali are not essential.

The counteranions primarily responsible for the insolubility of the pigments are large polymeric oxoanions of acids of a metal, either molybdenum or tungsten and a non-metallic element, either phosphorus or silicon [3]. These metals and non-metals are bonded to multiple oxygen atoms, some of those attached to acidic hydrogens. The central non-metal (P or Si) forms a tetrahedral arrangement with oxygens and the metal (Mo or W) forms octahedral arrangements that connect together by sharing corners or faces. The exact nature of the species formed depends on the elements used, their proportions and the conditions, notably pH and temperature, used in their preparation. The salts with three elements in the composition (PTM, STM) are referred to as triple salts, and from two (PM, PT, SM) as double salts. There are no standard proportions of the components, so they can be varied to meet the required coloristic properties, the economic demands, or as a compromise. Copper ferrocyanide complex salts, formed from copper (I) hexacyanoiron (II) acid, $HCu_3[Fe(CN)_6]$ (CF) may be used as alternatives. As a general principle, PTM salts produce brighter shades and better

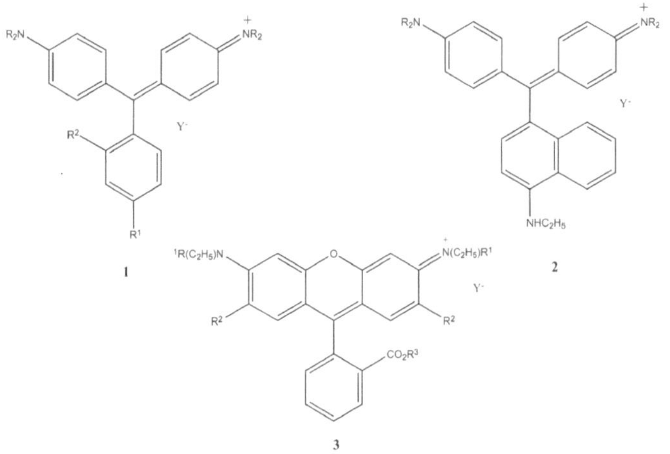

**Figure 10.1:** Structures of the cationic (basic) dyes used to prepare the complex pigments.

**Table 10.1:** Substituent pattern in cationic dye complex pigments (1).

| Dye | CI pigment | R | $R^1$ | $R^2$ | *Counteranion (Y⁻) |
|---|---|---|---|---|---|
| **1a**, Methyl Violet | Violet 3 | $CH_3$ | $NHCH_3$ | H | PTM or PM |
| **1a**, Methyl Violet | Violet 27 | $CH_3$ | $NHCH_3$ | H | CF |
| **1b**, Crystal Violet | Violet 39 | $CH_3$ | $N(CH_3)_2$ | H | PTM or PM |
| **1c**, CI Basic Blue 1 | Blue 9 | $CH_3$ | H | Cl | PTM or PM |
| **1d**, Diamond Green G | Green 1 | $CH_2CH_3$ | H | H | PTM or PM |

**Table 10.2:** Substituent pattern in cationic dye complex pigments (2).

| Dye | CI pigment | R | *Counteranion (Y⁻) |
|---|---|---|---|
| **2a**, Victoria Pure Blue B | Blue 1 | $CH_3$ | PTM or PM |
| **2b**, Victoria Pure Blue R | Blue 62 | $CH_2CH_3$ | CF |

lightfastness, an effect that has been attributed to the counteranion acting as a UV filter, thus inhibiting photo-oxidation [8]. The PM salts are less lightfast. PTM salts are more expensive due to the higher cost of tungsten. However, they tend to have a softer texture and are thereby usually easier to disperse. Silicate-based derivatives were introduced as lower cost alternatives. Silicomolybdate complexes (SM) give rise to particularly good texture. However, their manufacturing process is more complex. This group of pigments can be considered structurally as the converse of *lakes* (metal salt pigments) derived from colored anionic species, mainly based on monoazonaphthols, rendered insoluble as salts of alkaline earth metals (notably Ca and Ba) or transition metals, especially manganese [1]. More recently, pigments have

**Table 10.3:** Substituent pattern in cationic dye complex pigments (**3**).

| Dye | CI pigment | R$^1$ | R$^2$ | R$^3$ | *Counteranion |
|---|---|---|---|---|---|
| **3a**, Rhodamine 6G | Red 81 | H | CH$_3$ | CH$_2$CH$_3$ | PTM, PM, or SM |
| **3a**, Rhodamine 6G | Red 169 | H | CH$_3$ | CH$_2$CH$_3$ | CF |
| **3b**, Rhodamine B | Violet 1 | CH$_2$CH$_3$ | H | H | PTM, PM, or SM |
| **3c**, Rhodamine 3B | Violet 2 | CH$_2$CH$_3$ | H | CH$_2$CH$_3$ | PTM |

*The counteranions in Tables 10.1–10.3 are referred to by appropriate initials, phosphomolybdate (PM), phosphotungstomolybdate (PTM), silicomolybdate (SM), silicotungstomolybdate (STM), and copper ferrocyanide (CF).

been reported in which cationic dye complex pigments have been precipitated on to the inorganic pigment ultramarine [9]. CI Pigment Blue 88, which appears to be based on this concept, has been registered in the Colour Index [10].

A different type of cationic dye derivative gives rise to pigment products referred to as reflex or alkali blue. The pigments may be represented by the general structure (**4**), which are inner salts of the sulfonated dye structures, i.e., with both cationic and ionic centers within the structure, as illustrated in Figure 10.2. The structure is based in the parent compound, known as parafuchsin, the unsulfonated derivative with R$^1$ = R$^2$ = R$^3$ = H. The pigments are intense reddish blue and are used in certain printing ink applications. The two pigments of this type registered in the Colour Index are CI Pigment Blue 56, a greener shade product, and CI Pigment Blue 61, a redder shade product. Although the Colour Index designates the constitution of the pigments specifically, they are in fact complex mixtures of chemically similar species with varying substituents and degrees of sulfonation. The color can be modified towards greener shades by increasing the number of phenyl ring substituents. Compounds with three phenyl rings (R$^1$ = R$^2$ = R$^3$ = C$_6$H$_4$CH$_3$/C$_6$H$_5$) and two phenyl rings (R$^1$ = R$^2$ = C$_6$H$_4$CH$_3$/ C$_6$H$_5$, R$^3$ = H) are currently important products. The CH$_3$ groups are predominantly positioned *meta* to the secondary amino groups [2, 11]. The term "reflex" arises from the red sheen (reflex) that it provides in prints, used to advantage to give a black ink a more jet-black appearance. This contrasts with use of the inorganic pigment Prussian Blue which gives a green sheen, referred to as bronzing.

## 10.4 Synthesis and manufacture

The dyes used to manufacture the cationic dye complex pigments are long established. Since the synthetic procedures used for the industrial scale manufacture of the dyes for textile applications are also long established and well documented in the literature on textile dye chemistry [2, 6, 12], they are not covered further here. The focus in this paper is thus on the methods used to insolubilize them to form pigments. The pigments are precipitated when aqueous solutions of the water-soluble dyes are treated under

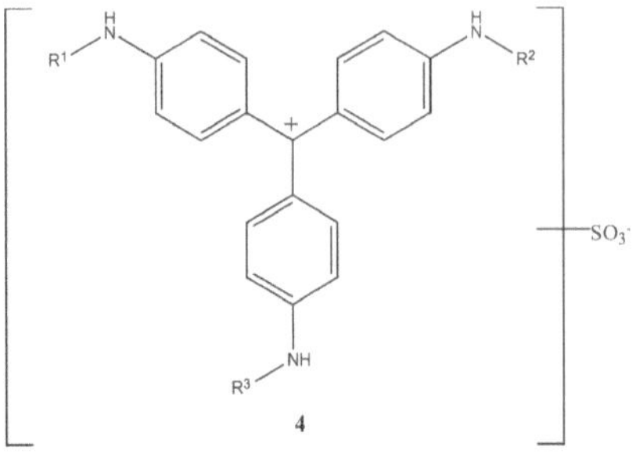

**Figure 10.2:** Structures of reflex (alkali) blue pigments.

appropriate conditions with solutions of the heteropolyacids, prepared *in situ*. In the manufacture of the pigments based on phosphomolybdic acid (PM), aqueous solutions of sodium molybdate (from molybdenum trioxide and aqueous sodium hydroxide) are added to aqueous solutions of disodium hydrogen phosphate. This solution is then acidified with either hydrochloric or sulfuric acid to the required pH, and then added to an aqueous solution of the cationic dye at around 65°C. In the case of the pigments based on phosphotungstomolybdic acid (PTM), a proportion of tungsten trioxide is incorporated into the process. The acid on which the copper ferrocyanide salts are based is obtained by reaction of an aqueous solution of potassium ferrocyanide, $K_4[Fe(CN)_6]$ with sodium sulfite as a reducing agent, and the resulting solution is added to a solution of the cationic dye. Finally, a copper (II) sulfate solution is added at 70°C. In all cases, the conditioning process in the pigment manufacture involves heating the aqueous reaction mixture to the boil, optionally in the presence of a surfactant. The pigment is collected by filtration and washed. The pigment may be finished by drying and milling to a powder. However, the pigment particles, because of their high polarity and fine particle size, tend to form hard agglomerates and aggregates and can be notoriously difficult to disperse. While dispersible powder grades are now available, commonly the pigments are supplied in the form of flush pastes, obtained by mixing the aqueous pigment press-cake with an ink binder in the presence of surfactant, whereby the pigment transfers from the aqueous phase to the oil phase to give a flush paste containing up to 40% pigment. The medium used to manufacture the flush paste is selected to conform to the requirements of the ink maker, for example products based on linseed oil, mineral oil, or heat-setting resins.

The most important synthetic route to reflex blue pigments is outlined in Figure 10.3. *P*-chlorobenzotrichloride (1 mol) and chlorobenzene (2 mol) undergo Friedel Crafts arylation reactions to form intermediate (**5**), which is then condensed with *m*- or *p*-toluidine and aniline to form the free base (**6**) after treatment with alkali. It is possible to carry out this phase of the reaction sequence stepwise so that two or three differently substituted arylamino groups can be incorporated. The crude pigment (**4**) is then obtained by sulfonation with 20–40% sulfuric acid in a solvent such as chlorobenzene

[2]. The pigmentary form is obtained by dissolving in alkali and reprecipitating with a mineral acid in the presence of a surfactant, and it is then filtered, dried, and milled, or converted to a flush paste.

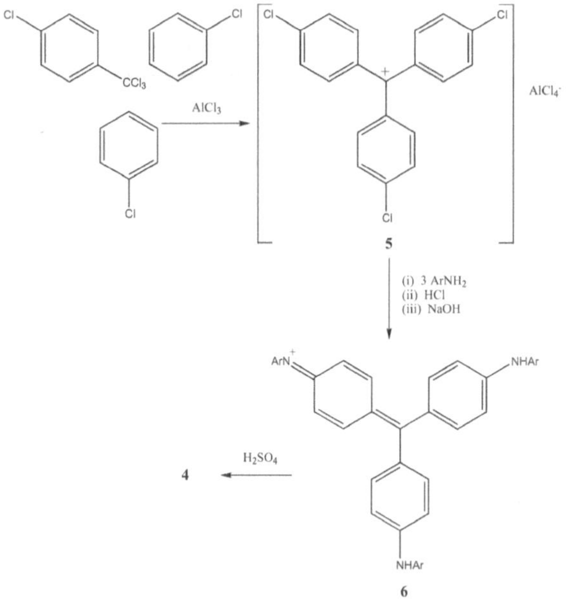

**Figure 10.3:** Synthesis of reflex (alkali) blue pigments.

## 10.5 Applications

### 10.5.1 Cationic dye complex pigments

The current range of cationic dye complex pigments based on heteropolyacids (PM, PTM) offer brilliance of shade, high color strength and good transparency, but generally low fastness to alkalis, polar solvents, and even to water. They are especially valued by the packaging industry and printers of advertising materials, although they have little use outside printing inks. Their lightfastness is at best moderate, so that when used in the three or four color processes they are invariably the first colors to fade. However, they command a fairly high price due mainly to the high costs of the metals used in their manufacture. The copper ferrocyanide (CF) pigments have lower brightness and poorer lightfastness, fading at double to four times the rate of the heteropolyacid based products (PM, PTM). They can also lead to premature drying when used in oil-based inks, forming a hard skin. However, they have the advantage of being significantly more stable in water-based inks, especially at higher pH values.

### 10.5.1.1 CI Pigment Red 81 (3a)

The most common grades registered in the Colour Index are the PTM salts (CI Pigment Reds 81 and Red 81:4) or the more recent SM salts (CI Pigment Red 81:5). These pigments all provide a bright pink shade that cannot be matched by any other commercial pigment. Therefore, they are used as a standard for certain types of printing inks, known as Process Red in the US. The pigment lacks fastness to polar organic solvents but have quite good stability to aliphatic and aromatic hydrocarbons. They have only moderate fastness to light as they darken and fade. Several marketed pigments are registered in the Colour Index as simply CI Pigment Red 81 (i.e., without the colon number suffix) but are not the PTM salt. Indeed, some manufacturers even indicate in their trade name that their commercial pigment registered as CI pigment Red 81 has a different anion, e.g., SM or CF.

### 10.5.1.2 CI Pigment Red 169 (3a)

This pigment is the copper ferrocyanide (CF) version of CI Pigment Red 81. It is generally cheaper than the other versions, but with much lower lightfastness. However, its shade is similar and is recommended and widely used for water-based inks.

### 10.5.1.3 CI Pigment Violet 1 (3b)

This pigment offers higher color strength with a bluer and purer shade than CI Pigment Violet 2 (**3c**). Its lightfastness is slightly lower, and prints have lower resistance to solvents. Only the PTM type is currently registered in the Colour Index.

### 10.5.1.4 CI Pigment Violet 2 (3c)

Although of relatively little importance, this group of pigments offer a redder shade than CI Pigment Violet 1 (**3b**). It is probably more accurately described as bluish red than as violet. It has slightly better lightfastness than CI Pigment Violet 1. Only the PTM type is registered in the Colour Index.

### 10.5.1.5 CI Pigment Violet 3 (1a)

This pigment has a very different shade, almost blue. It has the best lightfastness of the group, and also provides better resistance to reagents such as soap, fats and detergents. However, its solvent fastness is poor. All currently registered products are of the PTM type.

### 10.5.1.6 CI Pigment Violet 27 (1a)

This pigment is based on the same dye as CI Pigment Violet 3 but uses copper ferro-cyanide. It is an important pigment, especially in water-based inks. It provides tinc-torially strong, bright shades on the blue side of violet. Its lightfastness properties are inferior to versions based on the complex anions, especially in the full shades in which it is generally used. Prints also have less resistance to reagents such as alkalis, including soap and fats.

### 10.5.1.7 CI Pigment Blue 1 (2a)

This pigment is a bright reddish blue. The PTM version predominates as it offers the purest shade, due to the tungsten content, but this also reduces the tinctorial strength and increases the price. The properties are similar to other cationic dye complexes, with moderate lightfastness, as well as darkening with exposure. It has poor solvent fastness, especially to polar solvents. It is used mainly in oil-based bind-ers but is usually difficult to disperse. It is sometimes used to improve the jetness of black inks.

### 10.5.1.8 CI Pigment Blue 62 (2b)

This pigment is based on copper ferrocyanide producing a reddish blue shade. Its tinctorial strength is higher but its lightfastness is much lower than the CI Pigment Blue 1 types. Its main use is in water-based inks, in which it finds extensive use.

### 10.5.1.9 CI Pigment Blue 9 (1c)

This is a bright greenish blue pigment, mainly used in the US, providing a close match to standard cyan process colors. It is no longer of significant interest as a simi-lar shade can be achieved with copper phthalocyanine blue, which is advantageous both economically and in terms of fastness properties.

### 10.5.1.10 CI Pigment Green 1 (1d)

Green cationic dye complex pigments have never been commercially as important as other colors, mainly due to their lack of brilliance in shade, although they do have high tinctorial strength. However, more brilliant shades can be obtained using copper phthalocyanine blue or green pigments combined with organic yellows.

It is always necessary for application chemists and technologists to consult ma-terial safety data for any of the products under consideration. Differences may exist between chemically identical pigments from different manufacturers, especially as the proportions of the components can vary. Many of the cationic dye complexes are based on dyes that have health restrictions, so that manufacturers do not generally recom-mend them for sensitive applications such as food packaging or toys. Their concerns are not only due to direct health considerations, but also on what may occur when the printed paper is recycled. One international food manufacturer has prohibited their use in any of its packaging [12].

### 10.5.2 Reflex (alkali) blue pigments

Reflex blue provides various shades of reddish blue, much redder than α- copper phthalocyanine. For optimum jetness of black inks, it is used at 20–35% of total pigment, depending on the grade of carbon black pigment employed. When used in this way, high tinctorial strength, approximately double that of copper phthalocyanine blue, is the critical property. Their lightfastness is at best moderate, although not usually so important as black inks are not normally required to retain their jetness for a long time, and the UV-absorbing properties of carbon black provide some protection. However, when the pigment is used on its own or in white reduction, it fades fairly quickly. The pigments have good resistance to water and to acids, but they are not stable to alkalis. In prints they show good thermal resistance, when assessed for 30 minutes at 140°C, adequate for most applications. They have poor resistance to polar solvents, and only moderate resistance to aromatic hydrocarbon solvents, a feature that tends to prohibit their use in publication gravure inks containing toluene. Reflex blues were widely used in office requisites such as carbon paper and typewriter ribbons, a demand that as virtually disappeared with the development of desk top printers. Dispersible powder grades are available, but most ink manufacturers prefer to use flush pastes [13]. Pigments of this type are also proposed for use as colorants and charge control agents in toners used for laser printers and photocopiers and for electrostatically applied powder coatings [14].

## References

1. Christie RM. Colour chemistry, 2nd ed, London: RSC, Ch 9 2015.
2. Herbst W, Hunger K. Industrial organic pigments. Weinheim: Wiley-VCH Verlag GmbH, Ch 3 1993.
3. Cotton FA, Wilkinson G. Advanced inorganic chemistry. 4th ed. New York: John Wiley & Sons, 1980:852.
4. Garfield S. Mauve: how one man invented a colour that changed the world. New York: WW Norton, 2001.
5. Sanders JD. Pigments for inkmakers. London: SITA Technology, 1989:120.
6. Zollinger H. Color chemistry, 3rd ed, Weinheim: Wiley-VCH, Ch 4 2003.
7. Hunger K. Industrial dyes: chemistry, properties, applications. Weinheim: Wiley-VCH Verlag GmbH, Ch 2 2007.
8. Smith WE. Metals and metal ions in pigmentary systems. J Oil Col Chem Assoc. 1985;68:170.
9. Subramaniam Balasubaramaniam K, Thamby D. EP 167,826,0B1. 12 Dec 2007.
10. http://Www.colour-index.com.

11.  Buckwalter GR. In: In pigment handbook, Patten TC, editor, vol. 1, New York: Wiley-Interscience, 1973:617.
12.  https://www.nestle.com.pe/sites/g/files/pydnoa276/files/nosotros/informacion-provee dores-nestle/documents/actualizacion%202019/guidance%20note%20on%20packaging% 20inks%20-%20version%202018.pdf. Accessed: 24 Mar 2021.
13.  Iyengar DR. US 4,373,962A. 15 Feb 1983.
14.  Eduard M, Baur R, Macholdt HT, Kohl N. US 621967. 24 Nov 2009.

Gerhard Pfaff

# 11 Ceramic colors

**Abstract:** Ceramic colors or stains consist mainly of pigments, glaze or body, and opacifiers. They are used for the decoration of porcelain, earthenware bone china and other ceramics. Glazes and enamels are the main application systems for ceramic colors. Pigments are the color giving components in the composition of a ceramic color. High temperature and chemical stability as well as high tinting strength are characteristics of stains. Technically important ceramic colors are cadmium sulfide and sulfoselenides (occluded in zircon), metals such as gold, silver, platinum, and copper (as colloidal particles), metal oxides ($\alpha$-$Fe_2O_3$, $Cr_2O_3$, $CuO$, $Co_3O_4/CoO$, $MnO_2/Mn_2O_3$, and $NiO/Ni_2O_3$), mixed metal oxides and silicates, zirconia-based and zircon-based compositions. Ceramic colors are often produced using solid state reactions.

**Keywords:** ceramic colors, composite pigments, stains, zircon-based colorants, zirconia-based colorants, zircon, zirconia

## 11.1 Fundamentals and properties

Ceramic colors are used for the decoration of porcelain, earthenware bone china and other ceramics. The application systems in which they are used are glazes and enamels (underglaze, inglaze, onglaze colors). Ceramic colors are also called stains and consist mainly of pigments, glaze or body, and opacifiers. The pigments are the color giving components in the composition of a ceramic color. Pigments used in ceramic colors are characterized by extremely high temperature and chemical stability, but also by high tinting strength when dispersed and fired with glazes or ceramic matrices [1, 2].

Colored surfaces of porcelain, earthenware bone china and other ceramics are achieved by dissolution of colored metal oxides directly in the ceramic material or by the use of ceramic colors in glazes or enamels. The colorants are introduced in the ceramic materials before sintering or with the glaze or enamel using an additional step. The use of ceramic colors is a very effective method to achieve defined color tones in a reproducible manner for a glaze or enamel composition. The chemical stability of the ceramic color in the glaze or enamel including the color giving pigment has to be high enough that the color is not essentially affected by the melt

This article has previously been published in the journal *Physical Sciences Reviews*. Please cite as: G. Pfaff, Ceramic Colors Physical Sciences Reviews [Online] 2021, 6. DOI: 10.1515/psr-2020-0156

https://doi.org/10.1515/9783110588071-011

generated during the firing procedure. As a general rule, ceramic colors look pretty much the same before and after firing [1–7].

The classification of ceramic pigments is mostly based on their chemical composition (Table 11.1).

**Table 11.1:** Ceramic pigments for glazes [1].

| Chemical composition | Examples | Remarks |
|---|---|---|
| Nonoxides | $Cd(S_xSe_{1-x})$ | |
| Metals | Au, Ag, Pt, Cu | |
| Metal oxides | $Cu_2O$, CuO, NiO, $Fe_2O_3$, $Cr_2O_3$, $CO_3O_4$, MnO, $MnO_2$, $SnO_2$, $TiO_2$, $SnO_2$ | |
| Complex compositions | Spinel Pyrochlore Olivine Garnet Phenacite Periclase | Intrinsically colored |
| | Zircon Baddeleyite Corundum Rutile Cassiterite Sphene | Colored by addition of colored substances |

## 11.2 Types and applications of ceramic colors

### 11.2.1 Cadmium sulfide and sulfoselenides

Sulfides and sulfoselenides of cadmium are of great interest for yellow, orange and red colored glazes. Basically, these chalcogenides are not suitable for firing to higher temperatures, but they can be stabilized by encapsulation in a vitreous or crystalline matrix. The colored cadmium sulfide and sulfoselenide pigments are thus occluded in a colorless matrix using a sintering process. Finally, this kind of ceramic colors consists of two phases (composite pigments).

Zirconium silicate ($ZrSiO_4$, zircon) is the most relevant compound used as the matrix phase [8–10]. It is initially formed in the process at about 900 °C starting from $SiO_2$ and $ZrO_2$. Mineralizers, respectively, modifiers support the formation of zircon. Cadmium sulfide is added in its pure form to the mixture of $SiO_2$ and $ZrO_2$. Cadmium sulfoselenide, on the other hand, is formed during the process simultaneously with $ZrSiO_4$. Hexagonal crystals of the sulfoselenide are generated by the reaction of CdS and selenium or of $CdCO_3$, sulfur and selenium. A liquid vitreous phase of low melting compounds is formed as an intermediate under the influence of the mineralizers. Zircon grows around the sulfoselenide crystals under these conditions. Composites formed in this way are sold as ceramic pigments.

Relevant information on toxicology and occupational health for cadmium sulfide and sulfoselenides are available in the chapter "Cadmium sulfide /Selenide pigments."

### 11.2.2 Metals

Metals are used as ceramic colors in form of metallic colloids. Gold, silver, platinum, and copper play a relevant role for coloring of glazes and enamels. Selenium is also used, but only for glasses. Pink is the most interesting color in this application segment. It is achieved by the use of colloidal gold. Colloidal metallic gold is produced by the addition of tin(II) chloride to an acidic solution of gold chloride. The colloidal particles formed can have different colors ranging from pink to violet. A decisive factor for the achieved color is the ratio of tin and gold (Cassio's purple). One of the main objectives of the colloid formation is to produce colors, which are stable at high temperatures. In order to fulfill this requirement, the settling of the colloid particles is performed in a slip of kaolin or clay to avoid coagulation. The advantage hereby is that the metallic gold particles are separated by the clay particles. The color can be shifted towards reddish by adding of silver chloride. Additions of cobalt oxide, on the other hand, lead to a color change towards violet.

Gold, silver, platinum, and copper used for ceramic colors in form of their colloids do not exhibit acute or chronic toxicity. They are regarded as harmless when the usual precautionary measures are followed. The requirements of the regulations for chemicals, safety, and environment have to be respected during transport and storage [1].

### 11.2.3 Metal oxides

Metal oxides used as ceramic colors are typically dissolved in the vitreous matrix. Their application leads to a colored transparent appearance of glazes, which is based on the formation of metal ions in the system. The most suitable metal oxides used as ceramic colors are $\alpha\text{-}Fe_2O_3$, $Cr_2O_3$, $CuO$, $Co_3O_4/CoO$, $MnO_2/Mn_2O_3$, and $NiO/Ni_2O_3$. The following colors can be achieved by the addition of these oxides to the matrix composition [1].

$\alpha$-iron(III) oxide $Fe_2O_3$
- yellow – pink (iron coordination VI), stable color at low temperatures
- red – brown (iron coordination IV)

Chromium(III) oxide $Cr_2O_3$
- green, synthesis in the absence of zinc oxide to avoid the formation of brown $ZnCr_2O_4$ spinel, stable at low temperatures, limited importance due to easy reaction with silicate matrix

Copper(II) oxide CuO
- blue and green possible (copper coordination in most cases VI)

Cobalt(II,III) oxide / cobalt(II)oxide $Co_3O_4$/CoO
- blue (cobalt coordination IV), $Co_3O_4$ decomposes to CoO and $O_2$ at about 900 °C, the coordination changes to VI at higher temperatures in boric and phosphate glasses and a purple color appears

Manganese(IV) oxide / manganese(III) oxide $MnO_2$ / $Mn_2O_3$
- brown

Nickel(II) oxide / nickel(III) oxide NiO / $Ni_2O_3$
- yellow-purple, color changes due to the effect of retro-polarization of alkaline oxides

The use of these metal oxides as colorants in ceramics and enamels is challenging in some cases. This applies, for example, to $MnO_2$ and $Co_3O_4$. These oxides decompose with liberation of oxygen during the firing process. Defects in the glaze can be the result if the process is not controlled in a suitable way. The necessary high temperatures lead to changes of most of the oxides during the firing procedure. Different oxidation states are formed under these conditions. In general, temperature and oxidation/reduction conditions affect the redox balance strongly and are therefore decisive for the color tone generated.

Information on toxicology and occupational health for α-iron(III) oxide and chromium(III) oxide can be found in the chapters Iron Oxide Pigments and Chromium Oxide Pigments. When using copper(II) oxide, cobalt(II/III) oxide, cobalt(II) oxide, manganese(IV) oxide, manganese(III) oxide, nickel(II) oxide, and nickel(III) oxide for ceramic applications, the necessary safety measures are to be kept. All these oxides are more or less critical in regard to toxicity and occupational health, especially when they are inhaled or pass into solution. Several health-damaging effects are associated with these metal oxides. They are regarded as water-polluting. Their release into the environment must absolutely be avoided [1].

## 11.2.4 Mixed metal oxides and silicates

Colored mixed metal oxides belong to the complex compositions used for coloring of ceramics and enamels. Solid state reactions at high temperatures starting from the single oxides or salts are mostly used for their production (see chapter "Mixed metal oxide pigments"). The manufacture of mixed metal oxides takes place commonly in the presence of mineralizers such as alkali chlorides, fluorides, borates or carbonates [11–13]. Formation temperatures for mixed metal compositions are typically in the range from 800 °C to 1400 °C. The pigments are stable in the glaze during thermal treatment.

One of the most important mixed metal oxides used for the decoration of porcelain, earthenware bone china and other ceramics is cobalt alumina blue (cobalt aluminate, $CoAl_2O_4$), which crystallizes in the spinel structure. The composition consists basically of cobalt and aluminum, but may also include lithium, magnesium, titanium or zinc. Cobalt alumina blue is used as an underglaze colorant and for coloring of glass, pottery and vitreous enamels. The variation of the composition by incorporation of chromium in the spinel structure, possibly along with magnesium, silicon, titanium, zinc or strontium leads to blue-green pigments that are also suitable for colored glazes.

Several intrinsically colored silicates belong also to the complex compositions used for ceramics and enamels. The most important representatives of these silicates crystallize in structures belonging to the pyrochlore, olivine, garnet, phenacite, or periclase type. Victoria green is an example for an industrially used colored garnet silicate. It is a complex calcium chromium silicate based on chromium oxide, fluorspar, quartz and other minor substances and is mainly used for coloring ceramic glazes. A colored silicate based on the olivine structure is cobalt silica blue. The incorporation of zinc in the composition generates a lighter blue shade, while phosphorous in the structure changes the shade from deep blue to a violet-blue color tone. Another example for the olivine structure is nickel silicate green, which is mainly used for coloring clay bodies.

## 11.2.5 Zirconia-based compositions

Ceramic colorants based on zirconia ($ZrO_2$) and on zircon ($ZrSiO_4$) are the most important representatives of ceramic colors. The monoclinic crystal phase of $ZrO_2$ is stable at room temperature. Above 1200 °C, a transformation into the tetragonal phase takes place accompanied by a lattice contraction. Heating to temperatures above 2370 °C leads to the transformation into the cubic phase. Simultaneously, a lattice expansion occurs. Zirconia melts at about 2600 °C, and during cooling down the reverse crystallographic changes are observed [9, 10].

The incorporation of vanadium pentoxide into the $ZrO_2$ lattice leads to yellow ceramic colors. The manufacture starts with intimate mixing of zirconia (80 to 99% by weight) and ammonium metavanadate (1 to 20% by weight). The powder mixtures are calcined in a second step at temperatures of about 1400 °C under formation of yellow products. More intense yellow colors are achieved by the addition of small amounts of indium oxide. Other colorless metal oxides can be added together with vanadium pentoxide in order to modify the zirconia. They contribute to color shades ranging from greenish-yellow to orange-yellow. Vanadium containing zirconia-based colorants are stable in most types of glazes and ceramic bodies for temperatures up to 1350 °C.

### 11.2.6 Zircon-based compositions

The calcination of intense mixtures of zirconia and silica powders in the presence of vanadium at suitable temperatures leads to zircon, which has a blue color. The use of praseodymium instead of vanadium allows the formation of bright yellow zirconium silicate. A coral colored zircon is obtained when iron is introduced in the composition. The encapsulation of cadmium sulfoselenide leads to orange and red products, as described in Section 11.2.1.

The zircon-based ceramic colors vanadium zircon blue, praseodymium zircon yellow, and iron zircon coral fulfill the requirements for high-temperature applications in a very effective manner. They are stable up to temperatures of 1350 °C in all types of glazes. Cross-mixing of the three colorants allows a wide range of high-temperature stable colors.

Vanadium zircon blue is one of the most frequently used ceramic colorants. The chemical composition of vanadium zircon blue pigments can be expressed by the formula V-$ZrSiO_4$, respectively (Zr,V)$SiO_4$. The pigments are produced by calcining a mixture of zirconia, silica, a vanadium compound and an alkali metal halide. The high-temperature reaction occurs in the range of 550 to 1200 °C. The presence of alkali metal halides in the reaction mixture promotes the zircon formation and the color generation. Sodium fluoride is added for this purpose in most cases. The color of the final products depends on the proportions of vanadium pentoxide and sodium fluoride, the purity of the raw materials used, the particle sizes of the components, and the atmosphere used for the high-temperature reaction. The color strength of vanadium zircon blue depends especially on the amount of vanadium in the zircon lattice. The quality of the incorporation of the vanadium in the lattice is dependent mainly on the type and the amount of the mineralizer component used in the reaction mixture. Vanadium zircon blue pigments are exceptionally suitable for coloring of ceramic glazes and clay bodies.

Praseodymium zircon yellow is another colored composition based on the zircon lattice, which is suitable for high-temperature applications. The chemical composition of praseodymium zircon yellow pigments can be described by the formula $Pr$-$ZrSiO_4$, respectively $(Zr,Pr)SiO_4$. The pigment contains praseodymium as the coloring component. Sodium chloride and sodium molybdate act as mineralizers in the manufacturing process. Praseodymium zircon yellow is produced by high-temperature calcination of mixtures consisting of zirconia, silica, and praseodymium (III, IV) oxide in an appropriate ratio. Orange shades can be achieved by the addition of cerium oxide. Lead compounds as mineralizers instead of alkali metal halides generate more intense colors. The addition of ammonium nitrate to the alkali metal halide mineralizer is also beneficial for the achievement of stronger colors. Praseodymium zircon yellow pigments are brighter and cleaner colorants compared with other high-temperature stable yellow pigments. They are especially suitable for coloring of ceramic glazes and clay bodies.

Iron zircon coral is suitable for high-temperature applications in the pink and red color segment. The chemical composition of iron zircon coral pigments is specified by the formula $Fe$-$ZrSiO_4$, respectively $(Zr, Fe)SiO_4$. The pigments are produced by calcining zirconia and silica together with iron(III) oxide and suitable mineralizers at temperatures in the range of 1000 to 1100 °C. The composition of the resulting pigments may include small amounts of alkali or alkaline earth halides, which are used as mineralizers. The pigments are exceptionally suitable for coloring of ceramic glazes

The synthesis of zircon-based colorants requires chemicals of special purity to achieve products with highest chroma. An older, established process for zircon of high purity is based on the separation of $ZrO_2$ from mineral zircon sand. Meanwhile, a new process has been developed to manufacture zircon-based colorants directly from zircon sand. Equimolar proportions of $ZrO_2$ and $SiO_2$ are used for the high-temperature synthesis of $ZrSiO_4$. In a specific execution of the process, an alkali compound is added to the two oxides to form an alkaline zirconate silicate. The treatment of this product with acids leads to decomposition and formation of an intense zirconia-silica mixture. After calcination of this mixture together with one of the coloring components, a zircon-based ceramic colorant is formed. This process is suitable for the manufacture of all zircon-based colorants of technical importance.

Zirconia and zircon-based compositions used for ceramic colors are basically noncritical in regard to toxicity, occupational health, and environment. The situation is different when the composition contains one or more heavy metals, which may cause toxic or carcinogenic effects. However, the effects of heavy metals in complex inorganic matrices show special features. The heavy metals are in this case closely incorporated in the $ZrO_2$ or $ZrSiO_4$ lattice where they are insoluble and not bioavailable [1].

# References

1.  Pfaff G. Inorganic pigments. Berlin/Boston: Walter de Gruyter GmbH, 2017:190.
2.  Monros G. Encyclopedia of color science and technology. New York: Springer Science +Business Media, 2013. DOI:10.1007/978-3-642-27851-8_181-3 #.
3.  Mettke P. Über das Färben von Glasuren: Wechselwirkung von farbgebenden Komponenten mit Glasurbestandteilen. Keramische Zeitschrift. 1984;36:538.
4.  Eppler RA. In Ullmann's encyclopedia of industrial chemistry. vol. A5. 5th ed. Weinheim: VCH Verlagsgesellschaft, 1986:545.
5.  Eppler RA. Selecting ceramic pigments. Ceram Bull. 1987;66:1600.
6.  Bell BT. Ceramic colorants. J. Soc. Dyers Colourists. 1993;109:101.
7.  Etzrodt G. In Lehrbuch der Lacke und Beschichtungen. vol. 5. Kittel H, editor. Stuttgart/ Leipzig: S. Hirzel Verlag, 2003:101.
8.  Batchelor RW. Modern inorganic pigments. Trans Br Ceram Soc. 1974;73:297.
9.  Bayer G, Zirkon - vom Edelstein zum mineralischen Rohstoff. Wiedemann H-G. Chem unserer Zeit. 1981;15:88.
10. Kleinschmitt P. Zirkonsilikat-Farbkörper. Chem unserer Zeit. 1986;20:182.
11. Brussaard H. In industrial inorganic pigments. Buxbaum G, Pfaff G, editors. 3rd ed. Weinheim: Wiley-VCH Verlag, 2005:116.
12. White J. In high performance pigments. Faulkner EB, Schwartz RJ, editors. 2nd ed. Weinheim: Wiley-VCH Verlag, 2009:41.
13. Maloney J. In high performance pigments. Faulkner EB, Schwartz RJ, editors. 2nd ed. Weinheim: Wiley-VCH Verlag, 2009:53.

Gerhard Pfaff

# 12 Cerium sulfide pigments

**Abstract:** Cerium sulfide pigments belong to the inorganic orange and red pigments. They were developed to substitute cadmium and lead containing red and orange colored pigments, in particular in plastics. Cerium(III) sulfide exists in three allotropic modifications, but only $\gamma$-Ce$_2$S$_3$, which is colored dark red, is suitable for the use as a pigment. Cerium sulfide pigments are characterized by high color strength, strong hiding power and sufficient thermal stability. Their limited stability in water containing systems and in humid atmospheres is not without problems. The pigments tend to degradation in the presence of water or humidity to form hydrogen sulfide, which limits the application possibilities significantly.

**Keywords:** cerium sulfide pigments, solid–gas reaction, solid-state reaction

## 12.1 Fundamentals and properties

Cerium sulfide pigments consist of more than 90% of Ce$_2$S$_3$. They have been introduced into the market in the 1990s as orange and red pigments with the objective to substitute existing inorganic red and orange pigments. The specific aim of pigment developments on the basis of rare earth metal sulfides was the replacement of cadmium and lead containing red and orange colored pigments, which are in the public discussion because of their toxicity and environmental problems (cadmium sulfide and selenide, lead molybdate). Focus of the development of cerium sulfide pigments was the application of these colorants in plastics [1].

The following pigments belong to the family of cerium sulfide pigments [2–4]:

Ce$_2$S$_3$/La$_2$S$_3$ (cerium lanthanum sulfide): pigment cerium sulfide light orange C.I. Pigment Orange 78, light orange.

Ce$_2$S$_3$ (cerium sulfide): pigment cerium sulfide orange C.I. Pigment Orange 75, orange.

Ce$_2$S$_3$ (cerium sulfide): pigment cerium sulfide red C.I. Pigment Red 265, red.

Ce$_2$S$_3$ (cerium sulfide): pigment cerium sulfide burgundy C.I. Pigment Red 275, burgundy red.

Cerium(III) sulfide exists in three allotropic modifications: $\alpha$, $\beta$ and $\gamma$; all forms have a different thermal stability and differ with respect to the color properties.

This article has previously been published in the journal Physical Sciences Reviews. Please cite as: G. Pfaff, Cerium Sulfide Pigments *Physical Sciences Reviews* [Online] 2021, 6. DOI: 10.1515/psr-2020-0157

https://doi.org/10.1515/9783110588071-012

γ-Ce$_2$S$_3$, which is colored dark red, is probably the only phase suitable for the use as a pigment. It is formed at temperatures above 1100 °C. The production process of this pigment is exceptional complicate. γ-Ce$_2$S$_3$ is isomorphic with Ce$_3$S$_4$ and has the ability to accommodate other cations at cerium vacancies in the lattice, preferably alkaline earth metal ions or non-cerium lanthanides [5–7]. The formula, which describes the composition is Ce$_{3-x}$S$_4$, where $x$ stands for cationic vacancies.

The dark red color of some cerium sulfide pigments can be explained by electronic transitions from the Ce 4f level into the Ce 5d conduction band corresponding to an energy gap of about 1.9 eV [3].

## 12.2 Production of cerium sulfide pigments

Cerium sulfide cannot be produced by precipitation from cerium compounds with hydrogen sulfide in water because it is highly sensitive for hydrolysis. It is also difficult to stabilize cerium in the oxidation state + 3. Almost exclusively solid–gas reactions are suitable for the production of cerium sulfide pigments [3, 4]. Synthesis routes for lanthanide sulfides involve typically high-temperature solid-gas reactions between a sulfurizing agent (H$_2$S and/or CS$_2$) and lanthanide precursors, sometimes even under high pressure [8]. Various lanthanide compounds can act as precursors, e. g. oxides, salts, alkoxides, oxalates, tartrates or malonates [9, 10].

Another method for the manufacture of lanthanide sulfides is the solid-state reaction of a precursor with elemental sulfur. Lanthanide metals can act as precursors in this case. There are several disadvantages of such processes. They are difficult to perform because they require extreme temperatures and complicate pressure conditions.

The typical production route for cerium sulfide pigments starts with the synthesis of a cerium containing precursor [3, 4]. The precursor is synthesized from a solution of a cerium compound by precipitation with hydroxide ions. The cerium hydroxide containing precipitate is filtrated, dried and then transformed at temperatures between 700 and 1000 °C in a sulfurizing atmosphere into the colored cerium(III) sulfide.

$$Ce(OH)_3 + 3\ H_2S \rightarrow Ce_2S_3 + 6\ H_2O \tag{12.1}$$

The reaction product is ground in order to destroy agglomerates that may have been formed during the sulfurization process. The Ce$_2$S$_3$ powder is surface treated in most cases using a wet process. The stability of the pigment is increased in this way and the compatibility of the pigment with the application media is optimized. The surface treatment can contain inorganic and/or organic components. Treated pigment is ready for shipment after filtering, drying, and final packaging.

## 12.3 Pigment properties and uses

Cerium sulfide pigments are characterized by a density of about 5 g/cm$^3$ and a refractive index of 2.7. Typical particle sizes are in the range of 1 μm. The color strength is comparable with that of lead chromate. It reaches 50% to 70% of cadmium sulfoselenide pigments. The hiding power of the cerium sulfide pigments is comparable with that of the lead and cadmium containing pigments. The reflectance spectra of a cerium sulfide red pigment in comparison with a cadmium red and an iron oxide red are shown in Figure 12.1. Cerium sulfide has a lower reflection than cadmium red, but reflects significantly stronger than iron oxide red. The course of the reflection curves shows that the cadmium pigment exhibits the purest red color, followed by the cerium sulfide pigment (relatively steep curve bending earlier than the curve of the cadmium pigment). The iron oxide curve with the broad and humped reflection band explains why it is the less attractive one of the three red pigments regarding the color.

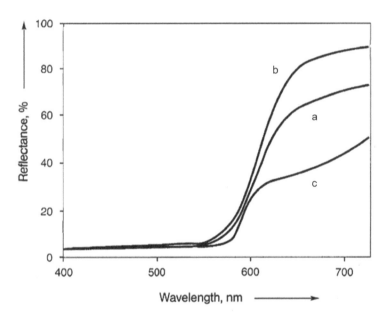

**Figure 12.1:** Reflectance spectrum of cerium sulfide red (a) in comparison with the spectra of cadmium red (b) and iron oxide red (c) [4].

The heat stability of cerium sulfide pigments in plastics is sufficient for this application material. The use is possible in polymers such as high-density polyethylene (310 °C), polypropylene, polyamide and polycarbonate (each 320 °C). Light and weather fastness were tested to be sufficient, too. Other characteristics are the easy dispersibility and the migration fastness of cerium sulfide pigments.

A disadvantage of $Ce_2S_3$ pigments is their limited stability in water containing systems and in humid atmospheres. The pigments degrade in the presence of water or humidity to form hydrogen sulfide. Coloration of plastics with $Ce_2S_3$ pigments can lead to smell problems due to reaction with residual water, especially in systems capable of swelling. This behavior limits the broader application of the pigments in various media. The use of a suitable surface treatment for stabilization can improve cerium sulfide pigments but is obviously not able to solve the problem completely.

Cerium sulfide pigments do not exhibit acute toxicity (LD50 value rat oral: >5000 mg/kg). They are not irritating to skin or mucous membranes [3, 11]. Genotoxic or carcinogenicity effects have not been found. The pigments are nearly insoluble and cerium ions show moreover effectively no toxicity (acute and chronic) [3]. As a consequence, there is practically no risk to humans or environment when using cerium sulfide pigments in accordance with the standard safety measures.

It is important for the production process that the reaction of the pigment or of preliminary stages with moisture and acids under hydrogen sulfide formation is avoided. This can be achieved by a properly-working exhaust air system for the manufacture of cerium sulfide pigments.

## References

1.  Pfaff G. Inorganic pigments. Berlin/Boston: WalterdeGruyterGmbH; 2017:155.
2.  Maestro P, Huguenin D. Industrial applications of rare earths: which way for the end of the century? J Alloys Comp. 1995;225:520.
3.  Berte J-N. In high performance pigments. Faulkner EB, Schwartz RJ, editors. Weinheim: Wiley-VCH Verlag, 2002;27.
4.  Pfaff G. In Winnacker-Küchler: Chemische Technik. In: Dittmeyer R, Keim W, Kreysa G, Oberholz A, editors. Industrieprodukte. vol. 7. 5th ed. Weinheim: Wiley-VCH Verlag, 2004:354.
5.  Julien-Pouzol M, Guittard M. Etude cristallochimique des combinaisons ternaires cuivre-terre rare soufre ou selenium, situees le long des binaires $Cu_2X$-$L_2X_3$. Ann Chim. 1972;7:253.
6.  Mauricot R, Gressier P, Evain M, Brec R. Comparative study of some rare earth sulfides: doped g-$[A]M_2S_3$ (M = La, Ce and Nd, A = Na, K and Ca) and undoped g-$M_2S_3$ (M = La, Ce and Nd). J Alloys Comp. 1995;223:130.
7.  Laronze H, Demourges A, Tressaud A, Lozano L, Grannec J, Guillen F, Macaudière P, Maestro P. Preparation and characterization of alkali- and alkaline earth-based rare earth sulfides. J Alloys Comp. 1998;275:113.
8.  Cutler M, Leavy J-F. Electronic transport in high-resistivity cerium sulfide. Phys Rev. 1964;133:1153.
9.  Henderson JR, Muramoto M, Loh M, Gruber JB. Electronic structure of rare-earth sesquisulfide crystals. J Chem Phys. 1967;47:3347.
10. Kumta PN, Risbud SH. Rare-earth chalcogenides – an emerging class of optical materials. J Mater Sci. 1994;29:1135.
11. Endriss H. Aktuelle anorganische Buntpigmente. Hannover: Vincentz Verlag, 1997:150.

Gerhard Pfaff

# 13 Chromate and molybdate pigments

**Abstract:** Chromate and molybdate pigments are representatives of the inorganic yellow, orange and red pigments. They are characterized by excellent optical and application properties in particular regarding brightness of shade, hiding power, tinting strength, and weather fastness. The declining use of lead- and chromate-containing materials in the last decades is a result of the environmental discussion and the development of less problematic substitute products, especially of bismuth vanadate and high-value organic, temperature-stable yellow and red pigments.

**Keywords:** chrome green, chrome orange, chrome yellow, lead chromate pigments, lead molybdate pigments, molybdate orange, molybdate red

## 13.1 Fundamentals and properties

Chromate and molybdate pigments belong to the inorganic yellow, orange and red pigments. Lead chromate (chrome yellow) and lead molybdate (molybdate orange and molybdate red) are characterized by colors ranging from light lemon yellow to reds with a blue hue. They are of the highest importance amongst the chromate and molybdate pigments [1–4]. Chrome orange as a further pigment belonging to this group has lost its technical importance. Mixtures of chromate pigments with iron blue or phthalocyanine pigments are marketed as green colorants.

Chrome yellow pigments with the compositions $PbCrO_4$ and $Pb(Cr,S)O_4$ (C.I. Pigment Yellow 34) consist of pure lead chromate respectively of lead sulfochromate (mixed-phase pigment) (refractive index 2.30–2.65, density ca. 6 g/cm$^3$) [5]. These lead-containing compositions can crystallize in an orthorhombic or a monoclinic structure. The monoclinic structure shows the higher stability [6]. The greenish yellow orthorhombic modification of lead chromate is metastable at room temperature. It is readily transformed to the monoclinic modification under certain temperature conditions. The monoclinic modification is found in natural sources as crocoite. Successive substitution of chromate by sulfate in the mixed-phase crystals leads to the gradual reduction of tinting strength and hiding power, but also to chrome yellow pigments with a greenish yellow hue.

Molybdate red and molybdate orange pigments with the composition $Pb(Cr,S,Mo)O_4$ (C.I. Pigment Red 104) are mixed-phase colorants, in which chromium is

This article has previously been published in the journal Physical Sciences Reviews. Please cite as: G. Pfaff, Chromate and Molybdate Pigments *Physical Sciences Reviews* [Online] 2021, 6. DOI: 10.1515/psr-2020-0158

https://doi.org/10.1515/9783110588071-013

partially replaced by sulfur and molybdenum [5]. Typical commercially available products show a $MoO_3$ content of 4–6%, a refractive index of 2.3–2.65 and densities of about 5.4–6.3 $g/cm^3$. The hue of these pigments depends on the amount of molybdate, the crystal form and the particle size. Pure tetragonal lead molybdate is colorless. It forms orange to red tetragonal mixed-phases with lead sulfochromate. The required coloristic properties are achieved by variation of the ratio of chromium, molybdenum and sulfur in the pigment compositions. Most of such commercially available pigments contain ca. 10% molybdate. Lead molybdate pigments have a thermodynamically unstable tetragonal crystal modification. This modification must be stabilized in the manufacturing process by a suitable treatment [7–9]. It is common practice to combine molybdate red and molybdate orange with red organic pigments. A considerable extension of the color range can such be reached.

Chrome orange with the composition $PbCrO_4 \cdot PbO$ (C.I. Pigment Orange 21) is a basic lead chromate. It can also be described as $PbCrO_4 \cdot Pb(OH)_2$.

Chrome greens with the composition $Pb(S,Cr)O_4 + Fe_4^{III}[Fe^{II}(CN)_6]_3 \cdot x\ H_2O$ (C.I. Pigment Green 15) are combinations consisting of chrome yellow and iron blue pigments.

Fast chrome greens with the composition $Pb(S,Cr)O_4$ + phthalocyanine (C.I. Pigment Green 48) are mixtures of chrome yellow and phthalocyanine blue or phthalocyanine green. High-grade fast chrome greens need a stabilization before they can be applied. Density and refractive index of chrome greens and fast chrome greens depend on concrete ratio of the components used. This is also true for the hues of the pigments, which can vary from light green to dark blue-green [3].

## 13.2 Production of chromate and molybdate pigments

Most of the production routes for chrome yellow pigments start from metallic lead which reacts with nitric acid to give lead nitrate solution. Sodium dichromate solution is added to the lead nitrate solution leading to a yellow lead chromate precipitate.

The adjustment of the pH value is important for the lead chromate precipitation to get the chromate dichromate equilibrium on the side of the chromate.

$$Cr_2O_7^{2-} + H_2O \rightleftharpoons 2\,CrO_4^{2-} + 2\,H^+ \tag{13.1}$$

$$Pb(NO_3)_2 + K_2CrO_4 \rightarrow PbCrO_4 + 2\,KNO_3. \tag{13.2}$$

In the presence of sulfate in the reaction solution, lead sulfochromate is formed as a mixed-phase precipitate. Eq. (13.3) shows an example for the formation of a mixed precipitate containing chromate and sulfate:

$$Pb(NO_3)_2 + 0.5\,K_2CrO_4 + 0.5\,K_2SO_4 \rightarrow Pb(Cr_{0.5}S_{0.5})O_4 + 2\,KNO_3. \tag{13.3}$$

The obtained precipitates are filtered, washed, dried and ground. The color of the so-obtained pigments depends mainly on the ratio of the starting components and on the precipitation conditions. Orthorhombic crystals are formed by using this synthesis route. These are transformed readily into the more stable monoclinic crystal phase by thermal treatment [10].

Chrome yellow pigments have the disadvantage of poor lightfastness. They tend to darken due to redox reactions under the action of light. The fastness properties can be improved by coating the pigment particles with a surface treatment. Suitable treatments contain mostly the oxides of titanium, cerium, aluminum, antimony and silicon [11–19].

Two different methods are used for the production of molybdate red and molybdate orange pigments, the Sherwin-Williams process and the Bayer process.

Main step of the Sherwin-Williams process is the reaction of an aqueous lead nitrate solution with a solution of sodium dichromate, ammonium molybdate and sulfuric acid [20]. It is possible to replace ammonium molybdate by the corresponding tungsten salt. A pigment based on lead tungstate can be obtained in this way. Stabilization of the pigment takes place by adding of sodium silicate and aluminum sulfate to the suspension of the precipitate followed by neutralization with sodium hydroxide or sodium carbonate. The hydrated oxides of silicon and aluminum are formed under these conditions. The precipitate is filtered, washed, dried and finally ground.

The stabilization of lead molybdate pigments regarding their interaction with light, weathering, chemical attack and temperature takes place using the same methods as described for chrome yellow pigments [11–19].

Chrome orange pigments are possible to produce by the precipitation of lead salts with alkali chromates in the alkaline pH range. The hue of the pigments can be varied between orange and red by adjustment of the particle size. Important reaction parameters to adjust the size are the pH value and the reaction temperature.

Chrome green and fast chrome green mixed pigments are obtained by mixing chrome yellow pigments with iron blue or phthalocyanine blue. Mixing is done using dry or wet processes. Dry mixing is executed in edge runner mills, high performance mixers or mills allowing intimate contact between the particles of the inserted chrome yellow and the iron blue respectively phthalocyanine pigments. In order to avoid segregation and floating of the two pigment components during application in coating systems, wetting agents are added [21].

The wet mixing process delivers chrome green and fast chrome green pigments with brilliant colors, high color stability, very good hiding power and good resistance to floating and flocculation. One pigment component is precipitated in this case onto the other one. Final adding of solutions of sodium silicate and aluminum sulfate or magnesium sulfate leads to further stabilization [20].

It is likewise possible to mill or mix the pigment components in a wet state or in an aqueous suspension followed by filtration. So-obtained pigment slurries are dried and ground to achieve the desired final properties [3].

## 13.3 Pigment properties and uses

Chrome yellow, molybdate orange, molybdate red and chrome green pigments are mainly used in paints, coatings and plastics. Brilliant hues, good tinting strength and good hiding power are the prominent characteristics of these pigments. Figure 13.1 shows typical reflectance spectra of chrome yellow and molybdate red pigments with their striking steep curves.

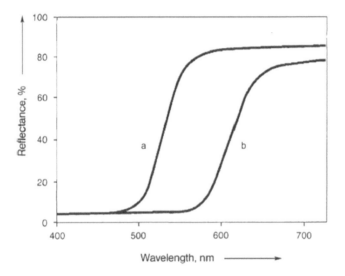

**Figure 13.1:** Reflectance spectra of a) chrome yellow and b) molybdate red [4].

The pigments can be improved regarding their resistance to light, weathering, chemicals and temperature by suitable surface treatments. They are supplied as pigment powders, as low-dust and dust-free preparations and as pastes.

Chrome yellow pigments with exceptional fastness to light and weathering, and very high resistance to chemical attack and temperature are obtained by controlled precipitation and stabilization. Four qualities of chrome yellow pigments are commercially available: unstabilized chrome yellows (limit the importance), stabilized chrome yellows with higher color brilliance, stable to light and weathering, highly stabilized chrome yellow pigments, and low-dust products (pastes or powders) [3].

Main applications of chrome yellow pigments are paints, coil coatings and plastics. Low binder demand and favorable dispersibility as well as good hiding power, tinting strength, gloss and gloss stability are characteristic for chrome yellow pigments. Stabilized pigment qualities with a silicate layer at the surface have a major importance in the production of colored plastics (PVC, polyethylene, or polyesters) with high temperature resistance.

The fastness properties of molybdate red and molybdate orange pigments are comparable with those of the chrome yellows. The pigment particles can be coated with metal oxides, metal phosphates or silicates to obtain stabilized pigments similar to the lead chromates and sulfochromates.

Lead molybdate pigments vary regarding their color from red with a yellow hue to red with a blue hue. Molybdate orange gained a much higher importance since the production of chrome orange pigments has ceased. Molybdate orange and molybdate red find their main application in paints, coil coatings and plastics (polyethylene, polyesters, polystyrene). Specific temperature-stable grades were developed for the use in coil coatings and plastics. Advantages of molybdate orange and molybdate red pigments are low binder demand, good dispersibility, hiding power and tinting strength, very high lightfastness and weather resistance.

Molybdate reds are used like chrome yellows to produce mixed pigments. Organic red pigments are used in combination with molybdate reds to provide a considerably extended color range. The pigment combinations are characterized by very good stability properties because lightfastness and weather resistance of many organic reds are positively affected by molybdates.

Chrome green pigments have an excellent dispersion behavior. Their resistance to flocculation, bleeding and floating is high, the fastness properties are suitable for all relevant demands. In particular, fast chrome greens that are based on high-grade phthalocyanine and highly stabilized chrome yellows, combine all these characteristics. They are therefore used in similar applications as chrome yellow and molybdate red pigments.

Chromate and molybdate pigments have to be handled carefully due to their not unproblematic components. Dealing with these pigments requires the compliance with the corresponding regulations during manufacture and application. Chromate and molybdate pigments do not exhibit acute toxicity ($LD_{50}$ value rat oral: >5000 mg/kg). They are not irritating to skin or mucous membranes [1].

Although lead chromate and molybdate pigments are hardly soluble, it is nevertheless possible that at hydrochloric acid concentrations as available in gastric acid a partial liberation of lead can occur. Dissolved lead can be accumulated in the organism. Animal feeding studies with rats and dogs have confirmed such an accumulation. Enzyme inhibitions and disturbances of the hemoglobin synthesis can be the consequence [1]. Lead chromate and molybdate pigments are therefore classified as chronically toxic in the relevant regulations. They are also qualified as carcinogenic and toxic for reproduction [22, 23].

A specific dust and water management is necessary for the handling of lead chromate and molybdate pigments. The limit for dust emissions from approved manufacturing plants is decided to be for the total mass flow 5 g/h or for the mass concentration 1 mg/m$^3$ (for the sum of lead and chromium) [24]. Comparable regulations exist for the wastewater management of production units for chromate and molybdate pigments [4]. Waste containing the pigments that cannot be recycled must be transported to special waste disposal sites.

## References

1. Endriss H. Aktuelle anorganische Buntpigmente. Zorll U, editor. Hannover: Vincentz Verlag, 1997:57.
2. Etzrodt G. in Kittel - Lehrbuch der Lacke und Beschichtungen, Vol. 5, 2$^{nd}$ edition, ed. J. Spille. Stuttgart/Leipzig: S. Hirzel Verlag, 2003:118.
3. Brandt K. In industrial inorganic pigments. Buxbaum G, Pfaff G, editors. 3rd ed. Weinheim: Wiley-VCH Verlag, 2005:128.
4. Pfaff G. Inorganic pigments. Berlin/Boston: WalterdeGruyterGmbH, 2017:127.
5. Hund F. Anorganische Pigmente durch iso-, homöo- und heterotype Mischphasenbildung. Farbe + Lack. 1967;73:111.
6. Algra GP, Erkens LJH, Kok DM. Lead chromates: the state of the art in 1988. J Oil Colour Chem Assoc. 1988;71:71.
7. Schäfer H. Der Einfluß der Dispergierung auf die coloristischen Eigenschaften und die Echtheitseigenschaften von Molybdatrotpigmenten. Farbe + Lack. 1971;77:1081.
8. Patent DE. 21 27 279 (Hoechst). 1971.
9. Patent DE. 20 62 775 (Hoechst). 1970.
10. Wagner H, Haug R, Zipfel M. Z. Die Modifikationen des Bleichromats. Anorg. Allg. Chem. 1932;208:249.
11. Lesche H. Über die Herstellung von Bleichromaten. Farbe + Lack. 1959;65:79.
12. Patent US. 2,808,339 (DuPont). 1957.
13. Patent DE. 18 07 891 (DuPont). 1969.
14. Patent DE. 19 52 538 (Bayer). 1969.
15. Patent DE. 20 49 519 (ICI). 1970.
16. Patent DE. 26 00 365 (Ten Horn Pigment). 1976.
17. Patent DE. 33 23 247 (BASF). 1983.
18. Patent DE. 38 06 214 (Heubach). 1988.
19. Patent DE. 39 06 670 (Heubach). 1989.
20. Patent US. 2,237,104 (Sherwin-Williams). 1938.
21. Newkirk AE, Horning SC. Flooding of paints containing chrome greens. Ind Eng Chem Ind Ed. 1941;33:1402.
22. TRGS 900. Grenzwerte in der Luft am Arbeitsplatz – Luftgrenzwerte. October 2000. (BArbBl. 10/2000, p. 34), revised August 1, 2003 (BArbBl. 9/2003, p. 42 (48)), MAK-Werte-Liste (11/2003).
23. TRGS 905. Verzeichnis krebserzeugender, erbgutverändernder oder fortpflanzungsgefährdender Stoffe. März 2001. (BArbBl. 3/2001, p. 94 (97)), revised August 1, 2003 (BArbBl. 9/2003, p. 42 (48)).
24. 1. BImSchVwV. TA Luft - Technische Anleitung zur Reinhaltung der Luft. 24 Jul 2002. (GMBl. Nr. 25-29 vom 30.07.2002, p. 511).

Gerhard Pfaff

# 14 Chromium oxide pigments

**Abstract:** The chromium(III) oxide pigments $Cr_2O_3$ (chromium oxide green) and CrOOH (chromium oxide hydrate green) are representatives of the inorganic green and blue–green pigments. $Cr_2O_3$ pigments are very important for the coloration of paints, coatings, plastics, building materials and other application systems. CrOOH pigments have lost their technical importance because of the low thermal stability. Chromium oxide green pigments are characterized by excellent application properties. They are synthesized starting from alkali dichromates or ammonium dichromate by reduction in liquid or solid phase.

**Keywords:** chromium oxide pigments, chromium oxide green, alkali dichromates, ammonium dichromate

## 14.1 Fundamentals and properties

Chromium oxide pigments belong to the inorganic green and blue–green pigments [1–4]. They are also known under the name chromium oxide green pigments. The pigments crystallize in well-known structures and consist of chromium(III) oxide respectively chromium(III) oxide hydroxide.

$Cr_2O_3$: corundum structure, pigment chromium oxide green C.I. Pigment Green 17, olive green tint, color change with increasing particle size from brighter yellow–green to darker blue-green.

CrOOH: 3 modifications with diaspore, indium oxide hydroxide, and delafossite structure, pigment chromium oxide hydrate green C.I Pigment 18, attractive blue-green color.

Chromium(III) oxide is not only used for pigment purposes but also in form of specific technical grades for other applications. One of these is the aluminothermic production of chromium metal by the reaction of aluminum powder with $Cr_2O_3$. Other applications are thermally and chemically resistant bricks and lining materials as well as grinding and polishing agents. The hardness of chromium(III) oxide is with ca. 9 on the Mohs scale very high. This leads to considerable abrasion properties of $Cr_2O_3$ pigments, which have to be considered in the applications [5]. CrOOH pigments have lost their technical importance nearly completely because of insufficient thermal stability.

This article has previously been published in the journal Physical Sciences Reviews. Please cite as: G. Pfaff, Chromium Oxide Pigments *Physical Sciences Reviews* [Online] 2021, 6. DOI: 10.1515/psr-2020-0159

https://doi.org/10.1515/9783110588071-014

## 14.2 Production of chromium oxide pigments

Chromium oxide pigments are synthesized starting from dichromates with chromium in the oxidation state +6. Dichromates are available as industrial chemicals in the required purity.

### 14.2.1 Reduction of alkali dichromates

The reaction of alkali dichromates with suitable reducing agents under formation of chromium(III) oxide may be carried out in solid mixtures or in water. Reducing agents used for the reduction of Cr(VI) to Cr(III) are sulfur, charcoal, carbon compounds or thiosulfates.

Solid alkali dichromates react with sulfur or carbon compounds to form chromium(III) oxide and sodium sulfate. The reaction is strongly exothermic and proceeds as follows:

$$Na_2Cr_2O_7 + S \rightarrow Cr_2O_3 + Na_2SO_4. \tag{14.1}$$

The formed sodium sulfate is separated by washing the resulting product mixture after the reaction. Sodium carbonate is formed as byproduct if charcoal is used as reducing agent [6–8].

$Cr_2O_3$ pigments with the desired properties are obtained when finely divided sodium dichromate dihydrate is mixed homogeneously with sulfur at the beginning of the process. The reaction of this mixture is performed at 750–900 °C in a furnace lined with refractory bricks. Sulfur is used in an excess to ensure the completion of the reaction. The water-soluble components (sodium sulfate and remaining sodium dichromate) are leached out with water after cooling down the reaction mixture. The filtrated $Cr_2O_3$ is dried, and ground. The use of $K_2Cr_2O_7$ instead of $Na_2Cr_2O_7$ leads to pigments with a more bluish hue. Jet milling is often the final process step in order to achieve the required properties necessary for the use of the pigments in paints, lacquers and other application systems.

It is likewise possible to perform the reduction of alkali dichromates in solution. Reducing agents for this manufacturing route are sulfur or thiosulfates. In case of the use of sulfur, the reaction takes place in sodium hydroxide solution at boiling temperature.

$$4\ Na_2Cr_2O_7 + 12\ S + 4\ NaOH + 10\ H_2O \rightarrow 8\ Cr(OH)_3 + 6\ Na_2S_2O_3. \tag{14.2}$$

Chromium(III) hydroxide and sodium thiosulfate are formed under these conditions. After neutralization of the alkaline suspension and addition of further sodium dichromate further $Cr(OH)_3$ is formed. The thiosulfate formed before is transformed to sulfite during this reaction.

$$Na_2Cr_2O_7 + 3\ Na_2S_2O_3 + 3\ H_2O \rightarrow 2\ Cr(OH)_3 + 6\ Na_2SO_3. \tag{14.3}$$

The so-obtained suspension is filtered. The filter cake is calcined and then washed with water to remove soluble components. $Cr(OH)_3$ reacts to $Cr_2O_3$ during the calcination at 900–1100 °C [9, 10].

$$2\ Cr(OH)_3 \rightarrow Cr_2O_3 + 3\ H_2O. \tag{14.4}$$

Grinding and sieving steps are used to adjust the final particle size of the $Cr_2O_3$ pigments.

Special treatments may be used to achieve specific properties for the chromium oxide pigments. Interesting in this regard is the deposition of titanium or aluminum containing precipitates at the surface of the particles. Such treatments have an influence on the surface properties of the pigments with an impact on the application behavior, e. g., on the flocculation tendency. They can also change the color to yellow–green. Organic compounds can also be used for the surface treatment of chromium oxide pigments [3].

## 14.2.2 Reduction of ammonium dichromate

An alternative route for the synthesis of $Cr_2O_3$ powders is the decomposition of ammonium dichromate at temperatures above 200 °C. An exothermic reaction is initialized in this temperature range leading to the formation of highly voluminous $Cr_2O_3$ under liberation of nitrogen.

$$(NH4)_2Cr_2O_7 \rightarrow Cr_2O_3 + N_2 + 4\ H_2O. \tag{14.5}$$

The $Cr_2O_3$ powder formed during this reaction does not yet have pigment quality. To obtain suitable $Cr_2O_3$ pigments, alkali salt, such as sodium sulfate, is added followed by a calcination step [11].

More suitable is an industrial process starting from a mixture of ammonium sulfate or chloride and sodium dichromate. Sodium sulfate is formed as a soluble byproduct, when ammonium sulfate is used as reducing agent [12]:

$$Na_2Cr_2O_7 + (NH_4)_2SO_4 \rightarrow Cr_2O_3 + N_2 + Na_2SO_4 + 4\ H_2O. \tag{14.6}$$

Chromium(III) oxide of pigment quality is isolated after washing the reaction products with water, drying and grinding.

### 14.2.3 Other production processes

Other routes for the synthesis of $Cr_2O_3$ powders involve the reaction of sodium dichromate with heating oil at 300 °C followed by calcination at 800 °C and shock heating of sodium dichromate in a flame at 900–1600 °C in the presence of hydrogen excess and chlorine [13, 14].

CrOOH pigments can be produced by the reaction of alkali dichromate with boric acid at 500 °C. $Cr_2(B_4O_7)_3$ and oxygen are the products formed during this reaction. The subsequent hydrolysis of $Cr_2(B_4O_7)_3$ leads to the formation of CrOOH and boric acid. Chromium(III) oxide hydrate can alternatively synthesized by the reaction of alkali dichromates with formiate under pressure [1].

## 14.3 Pigment properties and uses

Chromium(III) oxide is characterized by a very high hardness and a high thermal resistance. Pigments have typically a $Cr_2O_3$ content of 99.0% to 99.5%. The residual components are mostly silicon dioxide, aluminum oxide and iron oxide. CrOOH is thermally not resistant. Heating leads to the release of water and thus to the loss of the pigment properties.

The particle size distribution of $Cr_2O_3$ pigments is in the range of 0.1 to 3 μm with mean diameters of 0.3 to 0.6 μm. Coarser chromium oxides powders are produced for the refractory industry, as well as for grinding and polishing.

The color of chromium(III) oxide pigments can be described as olive green. The refractive index of 2.5 belongs to the highest among the inorganic pigments. Pigments with small particle diameters are lighter green with yellowish hues, larger sizes lead to a darker green with bluish tints. Chromium(III) oxide pigments are characterized by an excellent hiding power due to the combination of a high refractive index with a strong light absorption. The UV-absorption properties are excellent and contribute in organic binders significantly to an improvement of the weatherability for the overall system. Reflectance spectra of $Cr_2O_3$ pigments show a maximum in the green region (ca. 535 nm) and a second weaker maximum in the violet region (ca. 410 nm), which is caused by Cr-Cr interactions in the crystal lattice [3, 4]. The pigments have also a relatively high reflectance in the near infrared region. They can therefore be used in IR-reflecting camouflage coatings based on this reflection properties.

$Cr_2O_3$ pigments are insoluble in water, acids and bases. They are extremely stable to sulfur dioxide and in concrete. Coatings and plastics with chromium(III) oxide pigmentation are lightfast and fast to weathering. The pigments are temperature resistant up more than 1000 °C.

CrOOH pigments show a similar fastness as $Cr_2O_3$ pigments but they are instable against strong acids. The main reason for the low technical importance of CrOOH pigments is, however, the weak temperature stability.

A large part of the chromium oxide green pigments is used in the coatings industry. Excellent fastness, strong hiding power as well as very good light and weather stability are the main reasons for the broad application in this segment. The pigments are often applied in dispersion and silicate paints. Steel constructions (coil coating), facade coatings (emulsion paints) and automotive coatings belong are examples for the use in coatings. Chromium(III) oxide fulfill the high color stability requirements for building materials based on lime and cement besides the expensive cobalt green. They are therefore widely used in this application field [15].

Chromium oxide green pigments are suitable for the use in nearly all types of plastics. Good temperature and migration stability as well as excellent light and weather fastness are the main advantages of the pigments in polymers. The most attractive colors are generated when $Cr_2O_3$ pigments are blended with other colorants. Brilliant green colors are obtained, for example, when the olive green chromium(III) oxide pigments are blended with yellow pigments.

Chromium(III) oxide pigments do not exhibit acute toxicity ($LD_{50}$ value rat oral: >5000 mg/kg). They are not irritating to skin or mucous membranes. The pigments are not classified as hazardous and not subject to international transport regulations. $Cr_2O_3$ is not included in the MAK list (Germany), the TLV list (USA), or in the list of hazardous occupational materials of the EC [16]. The application of chromium oxide pigments in toys, cosmetics, plastics and paints that come in contact with food is permitted corresponding to national and international regulations [1, 3].

The use of dichromates for the manufacture of chromium(III) oxide or chromium(III) oxide hydrate pigments requires the compliance of occupational health regulations for the handling of hexavalent chromium compounds [17]. Sulfur dioxide formed during the reduction of dichromates must be removed from the flue gases. One suitable method therefore is the oxidation of the $SO_2$ to $SO_3$ followed by conversion to $H_2SO_4$.

Unreacted dichromates or chromates in the wastewater are reduced with $SO_2$ or $NaHSO_3$ to chromium III, which is precipitated as chromium(III) hydroxide and disposed after filtration [18].

# References

1.  Endriss H. Aktuelle anorganische Buntpigmente. Zorll U, editor. Hannover: Vincentz Verlag, 1997:77.
2.  Räde D. Kittel - Lehrbuch der Lacke und Beschichtungen. Spille J, editor. vol. 5. 2nd ed. Stuttgart/Leipzig: S. Hirzel Verlag, 2003:89.
3.  Rieck H. Industrial Inorganic Pigments. Buxbaum G, Pfaff G, editor. 3rd ed. Weinheim: Wiley-VCH Verlag, 2005:111.
4.  Pfaff G. Inorganic pigments. Berlin/Boston: WalterdeGruyterGmbH, 2017:110.
5.  Keifer S, Wingen A. Bestimmung der Abrasivität von Pigmenten nach einer neuen Stahlkugelmethode. Farbe + Lack. 1973;79:866.
6.  Patent US. 4,127,643 (PPG Industries). 1977.

7. Aghababazadeh R, Mirhabibi AR, Moztarzadeh F, Salehpour Z. Synthesis and characterisation of chromium oxide as a pigment for high temperature application. Pigments & Resins Technology 2003;32:160.

8. Patent US. 1,728,510 (H. C. Roth). 1927.

9. Patent US. 2,560,338 (C. K. Williams & Co). 1950.

10. Patent US. 2,695,215 (C. K. Williams & Co). 1950.

11. Patent EP. 0 068 787 (Pfizer Inc.). 1982.

12. Patent US. 4,040,860 (Bayer). 1976.

13. Patent ES. 438129 (Colores Hispania S. A.). 1975.

14. Patent US. 3,723,611 (Bayer). 1971.

15. Püttbach E. Pigmente für die Einfärbung von Beton. Betonwerk + Fertigteiltechnik. 1987;53:124.

16. EG-Guidelines. 67/548/EEC 1967 and Suppl.37. 1967.

17. Regulation (EC) No 1272/2008 on the classification. labelling and packaging of substances and mixtures. 2008.

18. Patent DE. 31 23 361 (Bayer). 1981.

Andrew Towns

# 15 Colorants: general survey

**Abstract:** This survey shows how colorants are not only ubiquitous in modern life, but also pervade many fields of science and technology. It reveals their diversity with respect to properties, origin of color, applications, composition and usage. After opening with a technical definition of "colorant", this survey exemplifies the huge assortment of uses to which colorants are put. It highlights the breadth of characteristics, other than color, that are of importance to the roles of "functional colorants". Following a description of the technical distinction between the terms "dye" and "pigment", the survey discusses the differences between these two colorant types from the perspective of how their color arises and the ways in which they are exploited, then goes on to demonstrate that particular colorants may be applied in certain instances as a dye and in others as a pigment. It outlines the wide ranges of composition, properties and economics of commercial colorants. This survey closes with a look at the nomenclature of colorants and the means by which they are classified into practically useful categories. It is intended to prepare the reader for the enormous technical and commercial variety that one encounters when dealing generally with colorants.

**Keywords:** colorant, functional colorant, dye, pigment

Colorants are ubiquitous in today's human experience. If you are reading this text from a printed page or an electronic display, then you are viewing an item that has been constructed in part with them. Look around. It is very likely that you will catch sight of at least several things which contain colorants unless your vision is impaired. In our modern world, it is difficult, if not practically impossible, to escape their reach. Colorants enhance life – they please, they warn, they persuade. They also have their detractors, who argue that they do harm. While this assertion is true under certain circumstances, trying to live without them would pose a near insurmountable challenge. But despite the pervasiveness of colorants, the majority of people do not have an inkling of just how much ingenuity and technical effort go into their creation and application. Nor perhaps do they fully appreciate the vast range of spheres of human endeavor in which colorants play a role. This chapter provides an introductory glimpse.

This article has previously been published in the journal Physical Sciences Reviews. Please cite as: A. Towns, Colorants: General Survey *Physical Sciences Reviews* [Online] 2019, 4. DOI: 10.1515/psr-2019-0008

https://doi.org/10.1515/9783110588071-015

## 15.1 A deceptively simple definition

The simple nature of the definition of "colorant" (see Box 1) disguises a subject of great complexity. Even traditional colorants span a diverse array of chemical types, covering a wide spectrum of properties. They are applied to a huge range of materials by means of a bewildering number of techniques. Be warned: the domain of colorants encompasses such a wide gamut of substances, substrates and application methods that gray contentious areas lurk within it and exceptions to rules abound. The field is so wide that it can be difficult to generalize without someone, somewhere, raising an objection. In addition, misapprehensions about colorants continue to be perpetuated, even in the scientific literature. Many papers feature inaccuracies which others repeat. This chapter will therefore help the reader navigate those basic concepts and aspects relating to colorants that are of greatest importance, defining key concepts along the way.

---

**Box 1**
**Colorant**: *A substance which is utilized to impart color to another material.*

---

Colorants are applied directly to an existing object, such as a garment, or during its creation, for example in the case of thermoplastics or glass. Alternatively, they might be incorporated into a medium to give an ink or a paint that is subsequently printed or coated onto an object. In all cases, the presence of a colorant alters the way in which that object interacts with light. As a consequence, when that object is viewed through the lens of a fully-functioning human visual system, it is perceived as having color. The word "object" is used in a very loose sense here since not only are colorants employed in the coloration of smokes and liquids, but they are also used in the production of individual elements of a more complex whole, such as the halftone dots in a printed image or the pixels of a display screen.

Coloration is such a broad subject that during this general survey, we shall encounter instances of colorants that are not themselves colored prior to, or during, their application. In addition, the next section will reveal that certain colorants are not used primarily to alter the color of materials, but instead to effect change in other ways, for example, by giving them specific physical properties. Nonetheless, just as was the case in times long past, the coloration industry predominantly utilizes substances which are initially highly colored purely to modify the appearance of materials.

## 15.2 Colorants old and new

Aesthetic usage of colorants by *Homo sapiens* stretches back tens of thousands of years. During this time, colorants shaped cultures and languages. While it is evident that our prehistoric ancestors created images and decorated themselves with colored materials, some archaeologists maintain that the employment of colorants antedates the emergence of our species [1]. Linguists argue that the growth in the palette of colorants accessible to artists and technologists during the last few millennia led to the much bigger vocabularies of color-related words routinely used in modern-day societies compared to those of ancient civilizations [2]. Over the past two centuries, our ability to manipulate and fine-tune color has increased hugely in terms of both sophistication and color gamut owing to an explosion in the number of commercially available colorants. The last few decades have also seen a rise in the importance of colored substances that transcend aesthetics by conferring physical properties or facilitating the occurrence of processes which are unrelated to color [3]: functional colorants, for which a definition is given in Box 2.

---

**Box 2**
**Functional colorant**: *A substance that does not depend entirely on its capability to impart specific color for its utility.*

---

This is another definition that covers a lot of ground. A functional colorant possesses one or more characteristics which make it useful for purpose(s) other than imparting specific color properties. The latter may be key to an intended application, but they are usually of lesser importance than these other characteristics. For some uses, color is irrelevant or even disadvantageous. While there is some overlap between the sets of industrially useful aesthetic colorants and commercially exploited functional colorants, the intersection is relatively small. The second of these two sets consists of many colorants whose chemical composition has been designed to promote functional characteristics without any consideration for aesthetic use. Consequently, their performance and/or economics rule out exploitation as conventional colorants.

The broad sweep of the definition in Box 2 lies in the wide range of functional colorant characteristics. Table 15.1 lists a few examples to give a flavor of their diversity. Some applications rely merely on the presence of color or a change in optical properties: the actual colors involved may not matter. In these circumstances, colorants act as signals to indicate, for example, the direction of an escape route in darkness, whether currency is genuine, or if a soil sample is acidic or alkaline. In other cases, they can be regarded as processors of light: they modify the nature of the incident radiation or transmute its energy into a different form. Instances of such use include tuning the color of laser light or absorbing such radiation to facilitate localized heating on a polymer disk for recording of data. Other technologies depend on them to convert light into latent electrical energy as part of image creation or the harvesting of daylight in solar power generation. For many applications, the color of the substance has no bearing on its utility or is even undesirable. This is true of certain clinical uses, such as cancer therapy, treatment of infectious disease, and pathogen removal from blood stock.

**Table 15.1:** Some examples of functional colorant characteristics and usage.

| Characteristic | Application |
| --- | --- |
| Biological activity | Chemotherapy [4], antibiotics [5] |
| Corrosion inhibition | Anticorrosion coatings [6] |
| Electrical conductivity | Antistatic coatings [7], electrostatic spraying |
| Infrared absorption | Laser marking [8], laser welding [9], optical data storage [10] |
| Infrared reflection | Heat regulating agricultural materials [6, 11], camouflage [12] |
| Magnetism | Information storage media [6, 13, 14] |
| Nonlinear optical activity | Telecommunications, photonics, holography [15] |
| pH sensitivity | Indicators [16], fuel marking [17] |
| Photoconduction | Printing [18], photocopying [19], solar power generation [20] |
| Photoluminescence | Security printing, laser optics [21], biolabelling, safety marking |
| Photosensitization | Photodynamic therapy [22], pathogen inactivation [23], photocuring [18][24] |
| Semiconduction | Electronic display technology [15] |
| UV blocking | Sunscreen products [25], cosmetics [26] |

The swelling in the number of colorants exploited for aesthetic and/or functional applications has been accompanied by an increase in the amount and complexity of chemical types produced. Well over a century ago, those involved in the coloration industry saw merit in trying to make sense of the already large number and diversity of commercial colorants by categorizing them. Various schemes, some systematic and some not, have therefore been created with the aim of aiding the colorant user to navigate the maze of brands, chemical types and uses. The remainder of this chapter will therefore discuss the many aspects in which colorants vary in character as well as outlining some notable ways of classifying them.

## 15.3 Dye or pigment?

One of the most technically important distinctions pertaining to a colorant is its classification as a "dye" or a "pigment". Simplified versions of internationally-accepted definitions for these two English-language terms are given in Box 3. (This discussion does not extend to analogous colorant-related words of other languages.) Although the demarcation between them centers on solubility alone, it has profound implications concerning how colorants interact with light, which methods of applying them are appropriate, and to what uses they are suited. The word "substrate" in the definitions refers to the object either into which the colorant is incorporated (such as a textile fiber and cake icing) or onto which colorant-containing medium is coated (like the plaster of a wall and the paper of a banknote).

---

**Box 3**

**Dye**: *A colorant that is in solution during some or all stages of its application to a substrate.*

**Pigment**: *A colorant that is not in solution at any point during its application to a substrate and whose particle structure remains unaltered throughout.*

---

Technologists categorize colorants as either dyes or pigments because the difference between the two is inextricably linked to technical properties and techniques of application. The crystal structure and size distribution of pigment particles are crucial to the color properties of the pigment itself. Dissolution during application must not occur to prevent changes in, or even destruction of, these particles. In contrast, the initial physical state of a dye is not usually preserved upon application to a substrate, because the colorant is (or its precursors are) in solution at one or more stages in the dyeing process. The starting physical form may be key to successful coloration (and thus may indirectly affect the color of a dyed object), but generally it has no bearing on the final state of the dye. The goal of most dyeing techniques is to achieve a uniform distribution of colorant throughout the area or volume of substrate at a molecular level. However, some aim for nonuniformity: an example is the visualization of different biological tissue types in histology, where use has long been made of the tendency of certain dyes to color only specific parts of specimens.

In a nontechnical setting, the terms dye and pigment are often employed interchangeably and as synonyms for colorant, perhaps because the distinction is neither appreciated nor considered important. Certain fields, such as botany or medical biology, refer to pigments in a different sense to that of the coloration industry. The word is used to denote a colorant of biological origin irrespective of its solubility properties. In this context, the phrase "biological pigment" is preferable to avoid confusion (see Box 4).

---

**Box 4**
**Biological pigment:** *A colorant originating from a living organism, which may have the character of either a dye or a pigment in terms of its solubility properties.*

---

The definition in Box 1 implies that a colorant is used deliberately and with purpose. As far as biological pigments are concerned, in many cases, explanation(s) for the existence of particular examples have been put forward – some scientific and some not. However, rather than dwell on the philosophical question of design in connection with such colorants, the remainder of this chapter will concentrate on the intentional use of coloring matter by humans. Biological pigments as defined in Box 4 will thus not be discussed further apart from instances where they are exploited by industry, for example in the dyeing of textiles or as additives to foods.

The definitions of dye and pigment in Box 3 say nothing explicit about their solubility in the substrate. One can infer, correctly, that pigments are essentially insoluble in the substrate to which they are applied. However, to define dyes as colorants that are soluble in substrates, which still sometimes happens in scientific literature, is to oversimplify. The generalization holds where dyes end up dissolved in liquid fuels and solvents as markers and tracers. It is also consistent with the outcome of injection molding processes that produce solid solutions of dye after dissolution of colorant in molten polymers. In such cases, dye is largely present in a monomolecular state, although depending upon the colorant concentration, application technique and substrate structure, loosely associated aggregates of a few molecules may also exist [27]. In contrast, conventional pigment particles with a typical mean diameter of 0.1–1 μm consist of $>10^6$ molecules. Nevertheless, the post-application physical form of certain dye types in a substrate is more akin to that of pigments. For example, the techniques of "vat dyeing" (see Box 5) and "azoic dyeing" both involve the application of aqueous solutions of water-soluble precursors to textiles, followed by their chemical conversion to water-insoluble colorants. In certain instances, washing of the dyeings with hot aqueous detergent solution transforms the dye into particles of sufficient size and crystallinity to be detected by X-ray diffraction and thus to influence dyeing color properties [28]. While all pigments are essentially insoluble in substrates, some dyes are too. Conversely, biological pigments are often soluble in substrates. When dissolved in beverages to give them color, lycopene (extracted from tomatoes) and anthocyanins (contained in grape skin extract) are examples of biological pigments acting as food dyes.

---

**Box 5**
*Indanthrone's molecular structure (shown below as **1**) means that it is essentially insoluble in water and has relatively low solubility in organic solvents – both properties contribute to its commercial utility as a pigment. Nevertheless, the colorant possesses structural features which enable its use as a dye too.*

1

*In the guise of a pigment, Indanthrone is not only used in the mass coloration of polymers, but is also employed the formulation of surface coatings like car paints and printing inks for banknotes. In these cases, micron-scale particles of the pigment are dispersed uniformly, but not dissolved, in the molten plastic or liquid coating formulation. The pigment particles become physically trapped within the substrate when it solidifies. However, Indanthrone is also used as a vat dye for cellulosic textile materials like cotton despite it being water insoluble. Fine dispersions of the colorant are treated with alkaline reducing agent to produce a weakly colored water-soluble form. Since this step involves colorant going into solution, Indanthrone is thus being used as a dye. It is only after this reduction stage, and the colorant has been dissolved, that it can penetrate and diffuse into the cotton. The substrate is then subjected to an oxidant, which regenerates the original strongly colored water-insoluble form of the colorant. Indanthrone is thus physically trapped within the polymeric matrix of the dyed cellulose fibers, giving dyed cotton that is highly colorfast to laundering with aqueous detergent. It is possible to apply pigment to a fabric, but to do so, one must coat it onto the surface of the fibers dispersed in a paste, for example by screen-printing. The coating is then cured, often by subjecting it to heat, leaving a solid film adhered to fiber surfaces in which the dispersed pigment particles are physically trapped. Indanthrone does not tend to be employed in pigment printing since the colorant is relatively expensive and it gives better quality coloration through vat dyeing.*

The definitions of dye and pigment presented in Box 3 might appear mutually exclusive at first glance. However, this dichotomy does not hold for certain colorants. For example, the reddish-blue compound Indanthrone has numerous applications not just as a pigment, but also as a dye (see Box 5).

Examples of pigments which are prepared from dyes abound. One important manufacturing method involves precipitation of pigments through combination of metal salts or complex inorganic acids with water-soluble dyes, which renders them insoluble [29, 30]. Such a pigment is known as a "lake". An alternative – and older – means of laking a dye is to adsorb it onto alumina: the result is a colored insoluble composite of improved durability [30, 31]. (The term "toner" is also used in connection with these two kinds of pigment, but it may cause confusion since the word also has separate meanings within the colorant industry.) A related method, specialized for the production of brilliant luminescent effects, involves the dissolution of fluorescent dyes into a molten organic resin which is then cooled and micronized. The resultant fine powder consisting of particles of a solid solution of dye in resin can then be used as a pigment [32]. The preparation of specialist light- and heat-sensitive pigments from dyes relies on another approach: microencapsulation. Their effects are dependent on the dye working in concert with other components in a mixture of specific concentrations, which must thus be preserved during application. Consequently, they are trapped within micron-scale shells of polymer, enabling the composites to be applied intact to substrates as pigments.

The manufacture of dyes by chemical transformation of pigments is also possible, but is less common. An example is the sulfonation of phthalocyanine pigments to give water-soluble dye analogs.

The above discussion concerning the division of colorants into dyes and pigments might perhaps seem pedantic, centered on semantics or hamstrung by exceptions to rules. In addition, developments in the use of colored nanoparticulate materials like quantum dots, whose scales occupy the gap between those typical of dyes and pigments, further challenge the distinction. Nevertheless, the labels of pigment and dye will continue to remain technically meaningful to many applications provided that they are carefully used in an informed manner and with the recognition that not every situation is clear-cut.

## 15.4 The origin of colorant color

The human visual system responds to light in the visible spectrum emanating from an object by producing the sensation of color (see Box 6). It is the nature of the non-uniformity in the energy distribution across the visible spectrum of this light that largely dictates which hue of "chromatic color" will be perceived, for example bright red or dull blue. A uniformly low level of light will be sensed as gray or even black, whereas high degrees of uniformity and intensity are seen as white: these sensations of color do not have a hue ("achromatic color"). The visual effect generated by a colorant is thus determined by the way in which it alters the interaction between the object to which it has been applied and the light incident on that object.

---

**Box 6**

**Visible spectrum:** *This region of the electromagnetic spectrum consists of radiation that can be perceived by the human visual system* [33]. *The range of wavelengths acting as stimuli is typically 380–780 nm* [34]. *These upper and lower limits vary* [35] *depending upon factors such as an observer's genetics and age as well as illuminant intensity. Radiation of even shorter or longer wavelength is capable of evoking a reaction, but only under special circumstances* [36]. *Because the sensitivity of the average human visual system to light of wavelengths of less than 400 nm or greater than 700 nm is lower by orders of magnitude compared to that in the middle of the visible spectrum* [37], *it is common to see narrower ranges quoted in the literature.*

---

Depending upon how the colorant itself interacts with light, the intrinsic color of the object may be modified or else hidden and replaced. A white colorant masks the object by ensuring that no visible wavelengths of the incident light are selectively attenuated, i. e. all are reflected. If the colorant attenuates all visible wavelengths so that virtually none reach the eye, then the object will appear black. Chromatic color on the other hand is generated when this attenuation is restricted to parts of the visible spectrum. Colorants introduce such nonuniformity in a variety of ways (see Box 7).

---

**Box 7**

**Absorption:** *A colorant has the potential to absorb photons when their energies match energy gaps between its ground and excited electronic states. These differences are primarily determined by the chemical structure of the colorant, but are also influenced to a lesser extent by its environment. Color results when absorbed photons possess energies corresponding to frequencies of light falling within the visible spectrum. Those frequencies of incident light, which are not selectively absorbed, are reflected and/or transmitted. For example, an ink containing a dye that absorbs only green light will appear purple in daylight when printed onto white paper as a consequence of the remaining light from the red and blue regions being reflected. If the ink instead is formulated with dyes that absorb all frequencies of incident light ("nonselective absorption"), then the lack of reflected light is perceived as black.*

**Scattering:** *A material that does not permit light to pass through in a straight line is said to scatter it. As the proportion of light scattered increases, the percentage reflected back diffusely in the direction it came from rises, which is perceived as opacity. When all frequencies of visible light are scattered completely and equally, the material will appear white ("nonselective scattering"). However, if certain visible frequencies are scattered in preference to others, then the material will take on a colored appearance ("selective scattering"). Pigment particles are capable of both types of scattering depending on their size and shape as well as the ratio of refractive indices of colorant and the surrounding medium. Pigments with large refractive indices tend to be useful for their nonselective scattering properties. For example, they can be used to create opaque coatings which hide the color of an object below. Particles of sizes approximating the wavelengths of visible light (0.38–0.78 µm) selectively scatter visible light frequencies determined in a complex manner by size, shape and refractive index differential ("Mie scattering"). Conventional pigments whose mean particle dimensions are in the region of 0.2–1.0 µm thus may have their appearance influenced by selective scattering. Often their particle size distribution as well as crystal morphology (i.e. particle shape) and structure (i.e. lattice arrangement) are engineered to obtain a favorable shade and intensity that complements color arising from absorption. While nonwhite pigments owe their color principally to selective absorption, selective scattering may also contribute in a technically significant way to their coloration* [30].

**Fluorescence:** *The phenomenon is a form of photoluminescence in which a colorant ("fluorophore") releases energy as visible light after absorbing higher energy visible light and/or ultraviolet radiation. The process of emission appears instantaneous to the human visual system as it typically occurs in a matter of nanoseconds. While the structure of the fluorophore largely governs the color of the emitted light and the efficiency of the process, the physical environment also heavily sways the latter. A common type of fluorophore is one which absorbs blue radiation, leaving red and green frequencies to be reflected, producing the sensation of yellow. Some of the absorbed energy is emitted by the colorant at lower energy yellow-green frequencies. This supplementary emission of light produces the illusion of more yellow-green light being reflected than is actually incident on the colorant, and the impression of extreme brightness.*

**Phosphorescence:** *As another form of photoluminescence, this process is sometimes confused with fluorescence, which is not helped by the term "phosphor" being applied to both fluorescent and phosphorescent substances. Whereas fluorophores do not perceptibly continue to emit light after* *the stimulating radiation is removed, phosphorescent colorants carry on doing so over the course of milliseconds to hours. Consequently, they glow in the dark; fluorophores do not.*

**Interference:** *Color is produced through partial refraction and reflection of different visible frequencies by thin layers of materials such as platelet-shaped pigment particles. The resultant variations in path length lead to angle-specific constructive or destructive interference so that color alters with viewing angle. The extent of change in color ("flop") is dependent on particle architecture, dimensions and refractive index.*

While pigments and dyes share some modes of color generation, the particulate nature of the former opens up possibilities for commercial exploitation which are not available to the latter (see Table 15.2).

**Table 15.2:** Potential origins of color in commercial pigments and dyes.

| Origin of color | Pigment | Dye |
|---|---|---|
| Selective absorption | Colored | Colored |
| Nonselective absorption | Gray or black | Gray or black |
| Selective scattering | Colored | Colored |
| Nonselective scattering | White | – |
| Fluorescence | Colored | Colored |
| Phosphorescence | Colored | – |
| Interference | Colored | – |

Conventional wisdom maintains that coloration with dyes relies purely on absorption and/or emission of visible light. This reasoning presumes that dyes do not scatter any light because they are in solution. Dyes which absorb all visible wavelengths uniformly ("nonselective absorption") appear gray or black. Those absorbing only a portion ("selective absorption") and/or emitting in regions of the visible spectrum ("fluorescence") produce chromatic color. For most types and applications of dye, these statements hold true but if, as outlined earlier, one accepts that under some circumstances dye is present on a substrate in particulate form which is capable of causing scattering, then it will be on a selective basis, hence the entry in Table 15.2. Note that a white dye is an oxymoron. Dyes cannot be white: they are incapable of scattering light uniformly when applied to a substrate and always absorb and/or emit at least a portion of the visible spectrum. In contrast, numerous types of pigment scatter all visible frequencies uniformly ("nonselective scattering") and absorb none, producing white coloration. The coloration produced by colored pigments is determined primarily by "selective absorption". However, as mentioned in Box 7, often "selective scattering" of visible light by pigment particles complements their absorption properties. While it is not decisive in determining predominant color, scattering can enhance that generated by absorption as well as subtly change its tone. Judicious manipulation of particle size and shape can thus alter shade [30].

The particulate nature of some pigments gives them additional strings to their bow: these extra means of color generation cannot be replicated with dyes. This is true of the interference effects utilized in surface coatings employed in applications like automotive paints, whose color shifts dramatically as angle of view or illumination changes (see Box 7). The reflection and refraction needed within the coating film are set up by layers of laminar pigment particles oriented in the same direction. Appearance is not reliant entirely on the color properties of their individual particles in isolation, which may not be intrinsically colored anyway. Instead, it is the regularity and anisotropy of the array of particles acting in concert. Dyes alone cannot therefore mimic interference colors.

Many fluorescent dyes are exploited industrially [38], yet commercial phosphorescence is achieved exclusively with pigment technology. The rigid arrangement of species within the crystal lattices of phosphorescent pigment particles effectively inhibit energy-loss pathways that compete with luminescence, promoting the efficiency of light emission. While dyes can be persuaded to phosphoresce, they must be cooled far below ambient temperature in order to discourage competing processes enough to obtain an appreciable effect. Currently, pigments remain the only option until research on "room temperature phosphorescent" organic materials [39] bears fruit in the form of commercial dyes.

While intense color is a characteristic of most colorants, some dyes are colorless or weakly colored. The latter count as colorants because they are utilized in coloration processes as precursors to colored species and thus impart color. Following their application to a substrate, strongly colored substances become generated through spontaneous *in situ* chemical transformation or combination of these intermediates (see Box 8).

---

**Box 8**

**Leuco dye:** *A colorant in a chemically reduced (and colorless or weakly colored) state, which is oxidized to its final intensely colored form. An example is given in Box 5, whereby the water-insoluble vat dye Indanthrone is reduced to its leuco form. In this state, it is weakly colored and soluble in aqueous alkali, enabling its diffusion into cotton fibers. Subsequent oxidation of the leuco dye regenerates Indanthrone, producing coloration.*

**Color former:** *This term applies to colorants that can exist in colorless or weakly colored states which are capable of being converted non-oxidatively into colored forms. They are employed in "latent coloration" applications, which call for a change in color to occur only upon the fulfillment a desired set of conditions. Colorless precursors are applied to a substrate, giving an object that has the potential to take on a colored (or differently colored) appearance. Introduction of a stimulus such as ultraviolet light, heat or acidity then results in the precursors forming colored substances: depending upon requirements, the process may be irreversible or fully reversible. For example, thermographic paper of* *the kind used for till receipts exploits latent coloration: localized application of heat leads to the color-less form of a colorant being brought into contact with an acidic substance, whereupon it irreversibly converts to a colored form only in those heated areas, generating an image. When the switching in color is reversible, the colorants are said to be "chromic" [15]. Numerous types of chromism are exploited commercially (see Table 15.3). Certain scientific disciplines refer to instances of permanent color change with chromic labels, but this is not correct. For example, irreversible fading of a colorant upon exposure to daylight is not photochromism, nor is any irretrievable discoloration on heating thermochromism. Any such phenomena must be fully reversible to qualify as chromism. (Note that there are some who extend the term leuco dye beyond its traditional redox sense to include the color-less states of color formers which can be converted to colored species non-oxidatively.)*

**Table 15.3:** Some examples of chromism and dependent commercial applications.

| Behavior | Stimulus | Example commercial uses |
| --- | --- | --- |
| Electrochromism [40] | Electrical potential | Automatically darkening anti-dazzle car interior rear-view mirrors |
| Halochromism [41] | Change in pH | Chemical analysis; thermographic printing paper |
| Heliochromism [42, 43] | Sunlight | Spectacle lenses that darken in strong sunshine and lighten indoors |
| Hygrochromism | Moisture | Humidity indicators for desiccants |
| Photochromism [44, 45] | Light (usually ultraviolet) | Anti-counterfeiting; novelty printing; toys |
| Thermochromism [46] | Heat | Thermometers; security printing; beverage temperature indicators |

**Precursor:** *These substances are colorants which chemically combine to create intensely colored products. As well as some kinds of textile coloration (e.g. vat or azoic dyeing as mentioned earlier) and chromogenic silver halide photography [47], this strategy is also employed for the most commercially important form of hair coloration [48]. Precursors and oxidant are applied to the head leading to the formation of colored products within hair fibers where they remain trapped.*

Irreversibly permanent destruction of color forms an integral part of certain coloration processes: discharge printing involves colored fabric being printed with reagents which degrade colorant in the areas that they are applied to produce patterns of lighter color [49].

The color imparted by colorants thus arises by one or more mechanisms. The majority of colorant types are intrinsically intensely colored. Certain types of colorant are not themselves strongly colored but have the potential to become intensely colored, either permanently or temporarily, by the influence of a physical or chemical stimulus or through being physically arranged in a regular manner. The next section examines the aspect of permanence in more detail.

## 15.5 Here today and gone tomorrow?

You might have noticed that the definition of colorant in Box 1 says nothing about the duration of coloration. Its silence on the matter is deliberate, because the degree of permanence of coloration demanded for applications varies so hugely. Whereas the expected lifetime of colorants in archival inks and building materials is measured in centuries, persistence of color is not even considered of relevance in uses like pH determination or microbiological staining where colorants are employed as one-use indicators. While the bulk of longevity requirements lie well within these extremes, robustness is often a key consideration when selecting appropriate colorants for the job, especially so given the marked variation in stability between different chemical types.

Ideally, for most applications, the color imparted is permanent: it remains essentially unchanged for the useful lifetime of the object to which the colorant has been applied. Often though technical shortcomings and economic constraints impose compromise. One must be pragmatic and accept that deterioration of coloration will occur during object use. Many sections of industry employ standardized accelerated test procedures to determine resistance to color change upon exposure to conditions deemed relevant to particular applications. These methods enable the definition of minimum acceptable performance and thus permit expectations to be managed concerning changes in hue or loss of color, which in turn aids selection of suitable colorants. For example, textile enterprises usually make use of internationally recognized protocols to determine whether customer-stipulated or industry-standard levels of "color fastness" or simply "fastness" to daylight, laundering, perspiration, abrasion, as well as other agencies, are met [50].

Note that these degrees of fastness are properties of the colored object, and not the colorant itself: one can speak of the fastness of a print or a dyeing, but not of a pigment or a dye. (Terms such as "high fastness pigments" do appear in colorant literature, but they should be treated as a form of shorthand; in this instance, the phrase refers to pigments that following application to substrates tend to produce coloration of great permanence.) Fastness is strongly influenced by the nature of the colorant and substrate, as well as application technique and exposure conditions.

Colorants can thus vary widely in terms of fastness of the coloration that they produce – as an extreme example, the photostability of coloration in certain applications is orders of magnitude lower than others. There are also plenty of instances of applications in which coloration is purposefully transient. One such kind is "wash-in wash-out" temporary hair color products. In some exceptional cases, the relatively poor fastness conferred by certain colorants has been turned into a feature, i. e. the susceptibility to fading of blue jeans.

We have so far seen that colorants vary widely in end use, application technique, color properties and longevity. It is only natural that they also differ enormously in composition. The next section takes a general look at colorant constitution from a chemical perspective.

## 15.6 Organic or inorganic?

Many colorants are organic. This statement refers to their carbon-based composition rather than production methods, sustainability or naturalness. Their molecular structures are based on conjugated frameworks of carbon atoms, usually aromatic, that are linked and/or studded with functions consisting of other elements. These features are designed to manipulate color, solubility and other technical properties. An indication of the prevalence of elements of which they are comprised is mapped out in Figure 15.1. It is intended to be only a very rough guideline and is not based on a detailed statistical analysis of organic colorants by type or volume consumed. The figure merely gives a flavor of what you might find if you were to pluck a dye or organic pigment at random from the pool of colorants that are in aesthetic and functional use today.

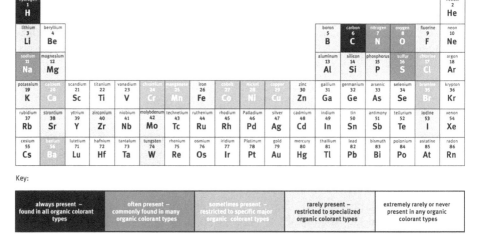

**Figure 15.1:** Summary of occurrence of elements in chemical formulae of organic colorants.

Purely hydrocarbon colorants are a select band: commercial examples exist, such as the carotenes, which are constituents of food dyes, but there are not many. Most organic colorants are made up of a combination of up to five elements in addition to carbon and hydrogen. Few commercial dyes or organic pigments do not contain at least oxygen and/or nitrogen. These two elements commonly constitute electron-rich or -deficient functional groups, as well as heteroatoms, that form part of colorants' aromatic π-systems. Such structural features are crucial to the color properties of most organic colorants because of their large influence on molecular electronic structure. Nitrogen and oxygen also participate in numerous roles that are essential to particular colorant types. For example, hydrogen-bonding centers constructed from them not only influence colorant solubility and volatility but can also impact color through modification of dye molecule conformation or pigment crystal packing. These two elements may also be used as building blocks for substituents which introduce chemistries that are crucial to certain types of colorant, e. g. acid–base sensitivity, chelation of metals, or redox properties. Figure 15.2 shows some examples drawn from typical commercial colorants.

Dye **2** is the most important blue compound for the coloration of polyester textile materials. It is a heavily substituted azobenzene (Ph-N=N-Ph), one of the most important types of organic colorant. Appropriate location of functional groups modifies its color from the yellow of the parent compound to an intense dull navy that is useful in producing black shades. Powerful nitrogen- and/or oxygen-containing electron-donating and -accepting substituents facilitate the substantial changes in electronic structure required. Others enhance intensity of color and photostability.

Compound **3** (see Figure 15.2) is another example of an intensely colored commercial dye that is reliant on nitrogen- and oxygen-based substituents for its properties. Its anthraquinonoid skeleton bears strong electron donors situated close to the electron-accepting carbonyl functions of its quinone ring. Their close proximity and orientation transforms the pale yellow of the parent anthraquinone to an intense blue hue. The pendant cationic charge confers water solubility and is the main origin of attraction between the dye and its usual intended substrate, poly(acrylonitrile).

Sulfur, and to a lesser extent chlorine and bromine, is also often present in colorants. All three are widely used to manipulate color but have other parts to play. They are found in the most exploited types of substrate-reactive groups as well as counterions of cationic dyes. Sulfur is perhaps likeliest encountered among commercial organic colorants as a constituent of sulfonate groups $(-SO_3^-)$. The presence of these functions, typically in sodium salt form, is the most common means of making dyes water soluble. (Economics dictates that the use of potassium and lithium salts occurs in dye manufacture only when required.) As was mentioned previously concerning conversion of dyes to lakes, displacement of sodium with other metals lowers solubility sufficiently to yield pigments. Alkaline earth metals, especially calcium and barium, are usually employed for this purpose [29]. Water-soluble cationic dyes are transformed into pigments through replacement of simple counterions like chloride with complex acids, which markedly lowers their solubility. A typical acid used in this way is phosphotungstomolybdic acid. The precipitation may also be performed in the presence of other elements such as aluminum.

First row transition metals feature extensively in numerous types of dye and pigment, where colorants act as ligands in the formation of complexes ("metal complex colorant") [51]. Of particular note are chromium for dyes and nickel for pigments with copper also being of importance to both. The complexation enables access

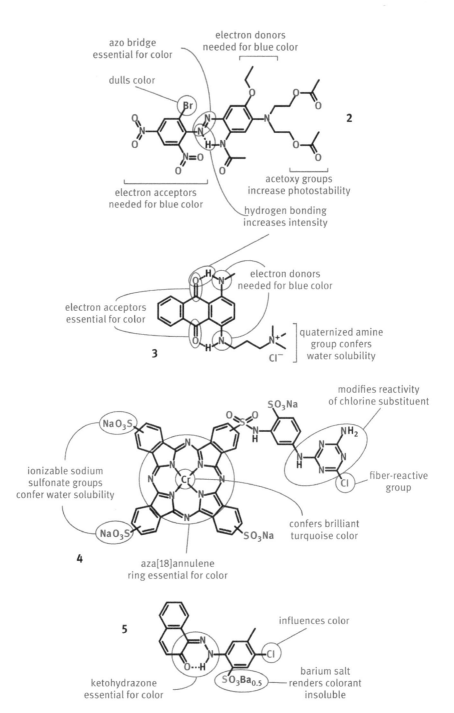

**Figure 15.2:** Some important commercial colorants and their structural features.

to colors that are otherwise difficult to obtain and/or afford substantial increases in colorant stability. Usually, the colorant is applied to the substrate as a preformed complex. However, a traditional textile coloration technique ("mordant dyeing") involves application of the metal as a salt before, after or even during dyeing of the substrate, leading to the *in situ* formation of complexes within the fiber [52].

Figure 15.2 shows examples of a dye **4** and an organic pigment **5** that incorporate sulfur, halogen and metals in their composition. Although it is a dye thanks to the presence of its solubilizing sodium sulfonate groups, turquoise compound **4**, is based on the important pigment, copper phthalocyanine, whose bright color and resilient nature owe much to it being a metal complex. The chlorotriazine fragment of **4** is a substrate-reactive group: its chlorine atom is labile enough to permit reaction between dye and cotton during application. The resultant covalent attachment leads to high wash fastness. Compound **5** is produced by laking a water-soluble dye with a barium salt. The precipitated scarlet pigment is prized commercially as a colorant for ink formulation.

The relative importance of elements making up industrial organic colorants is unlikely to change significantly, although it is possible that some previously sparingly-used elements might come further to prominence with the introduction of more exotic structures where there are breakthroughs in functional dye development.

Dyes are exclusively organic. In contrast, commercial pigments are either organic or inorganic. A form of coloration called "mineral dyeing" blurs this demarcation. While now of minor importance, this technique utilizes inorganic substances labelled "mineral dyes" [53]. Solutions of metallic salts, typically of chromium or iron, are applied to cotton whereupon they are converted to insoluble oxides by steam treatment, producing dull but fast coloration. The salts are therefore not dyes in the technical sense, despite their employment in solution form. Instead, these inorganic materials fall under the definition of precursor (see Box 8) because they are transformed into substances of a different composition from that initially applied. Nevertheless, the technique has parallels with azoic dyeing which depends upon organic precursors.

In terms of global annual consumption, inorganic colorants dwarf organic pigments and dyes (see Figure 15.3).

Depending upon which economic analyst one believes, the global coloration industry synthesizes around 10 million tonnes of colorant per year worth in the region of USD20–30 billion. Over four-fifths of the volume supplied consists of inorganic pigments. Remarkably, just one substance – titanium dioxide – accounts for over half the amount of colorant used worldwide owing to its unique combination of economy and technical properties as a white pigment. Dyes only make up around one-eighth of the total volume of the colorant market, while organic pigments constitute no more than 5 %. However, in terms of market value, these two colorant types take up far larger slices of the total sold, because inorganic pigments are generally more inexpensive. No single colorant dominates the organic sector: there are far too many different substrates and applications, all with their own particular technical and

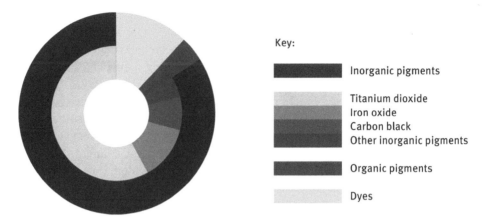

**Figure 15.3:** Proportions by volume of global colorant consumption.

economic demands, to allow that to happen. Instead, a multitude of dyes and organic pigments fulfills these needs. Commercial inorganic colorants consist of a set of only 100 or so chemically distinct types, which is only around 1% of the number of industrial organic colorant compounds [54]. Consequently, the production quantities of even the highest volume synthetic dyes at several tens of thousands of tonnes per year are orders of magnitude lower than the top inorganic pigments. Far more typical annual outputs for individual synthetic conventional organic colorants are 1–100 t. Many specialist or functional colorants are manufactured in smaller quantities. In extreme cases, just a few kg or less per year may satisfy demand. It follows that colorant unit cost varies enormously: some commodity dyes and pigments are marketed at less than USD1/kg, while niche low-volume colorants of high purity prepared by multistep syntheses may cost tens of dollars per gram or more.

A key technical property contributing to the dominance of inorganic pigments is a large refractive index. The organic or aqueous media in which they are typically dispersed possess significantly lower indices. The resultant index differences lead to high proportions of light being reflected at pigment particle–medium interfaces. Inorganic pigments thus effectively scatter light in cured coatings and plastics, producing high opacity. On a weight for weight basis, colored inorganic colorants tend to produce less intense, duller shades drawn from a more restricted color palette than their organic counterparts. The latter do not generally have large refractive indices, leading to smaller index differentials between pigment and substrate: consequently, organic pigments usually scatter light less effectively, exhibit lower hiding power and furnish greater transparency. Colored organic and white inorganic colorants are therefore often used together to produce coloration effects of high intensity, brilliance and opacity across a wide gamut. Globally, production of paints and coatings as well as coloration of plastics accounts for around two-thirds of pigment demand. Application to textiles consumes about half the production of dyes and organic pigments.

The inorganic colorant with the simplest elemental composition is perhaps carbon black, a pigment comprising ≥95 % carbon, which is used in plastics, rubber, paints and inks. Despite its structure consisting primarily of fused aromatic carbocyclic systems, it is usually classed as an inorganic pigment. It is exceptional in that the chemical formulae of all other inorganic colorants of import contain at least one of the elements oxygen and sulfur, most commonly in the form of oxides and sulfides. Nonetheless, a wide variety of elements may be found in commercial inorganic pigments today (see Figure 15.4). Despite being fewer in number than organic colorants, collectively they are more elementally diverse, although in volume terms, oxygen, titanium, iron, sulfur and carbon greatly dominate.

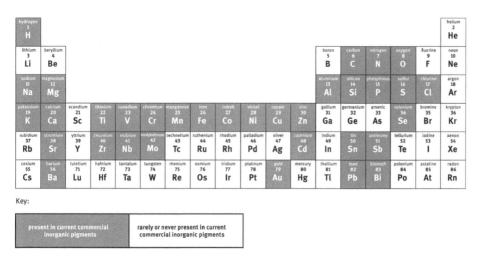

**Figure 15.4:** Summary of occurrence of elements in chemical formulae of inorganic colorants.

An important aspect of commercial colorant composition is that many products do not consist essentially of just one chemical entity accompanied by traces of impurities. A significant proportion of marketed colorants are mixtures of components, some of which are not colored but whose presence makes the imparting of color more reliable, efficient and/or effective. As we shall see in the next section, the answer to whether certain colorants are organic or inorganic is actually "both".

## 15.7 Anatomy of commercial colorants

Colorants are supplied to end-users in one of several physical forms depending upon their application. Those varieties most commonly encountered for pigments or dyes are listed in Table 15.4.

**Table 15.4:** Typical physical forms of commercial colorants.

| Physical form | Description | Pigment | Dye |
|---|---|---|---|
| Powder | Average particle size may be from the submicron domain upward | ✓ | ✓ |
| Granule | Non-dusting friable particles of several mm diameter | ✓ | ✓ |
| Presscake | Water-wet paste typically consisting of >50%w/w colorant | ✓ | ✓ |
| Solution | Liquid concentrate typically >20%w/v in water and/or solvent | ✗ | ✓ |
| Dispersion | Stabilized liquid colloidal suspension of particles in water | ✓ | ✓ |
| Masterbatch | Concentrated solid solution or dispersion of colorant in polymer | ✓ | ✓ |

Note that many commercial colorants marketed as solids do not consist wholly of a single compound [55]. For instance, "mixed pigments" are products of dry-mixing or -grinding at least two inorganic compounds together. In addition to colored compound(s), a colorant will very likely contain one or more substances whose presence is intentional or inadvertent. The inclusion of each substance may be (1) inescapable, (2) avoidable but tolerated, or (3) deliberate. Even authors of scientific papers sometimes overlook the fact that the commercial colorants at the center of their studies are unlikely to be composed of a single chemical entity. For example, textile dyes often contain substantial amounts of noncolored components in addition to one or more colored substances. Responsible manufacturers create such colorants with minimized production cost and effort in mind: they are designed to withstand industrial conditions and deliver reliable outcomes in an economic manner. Consequently, nonessential components carried through from synthesis may not be considered worth removing even if doing so was financially viable. Producers also add substances to protect and enhance colorant performance during transport, storage and use. Typically, therefore, textile dyes in powder form only comprise 30–90%w/w of colored component(s). Box 9 elaborates a little on the identity of the types of materials likely to be found in dyes and pigments marketed in a solid state.

**Box 9**

**Water:** *Low levels are usually unavoidable. Moisture may be carried through from manufacture, picked up from the atmosphere, introduced in other additives or even be present as water of crystallization. Removal of all traces is not usually practical, nor in many cases necessary. Nevertheless, manufacturers often control its presence during post-synthesis by drying below a level deemed appropriate, i.e. to ensure processability, stability and/or a particular color strength can be achieved. Certain water-soluble dyes are prone to picking up water as a consequence of exposure to atmospheric moisture during storage. An extreme example of such hygroscopic behavior is that of some cationic dyes which absorb ~10–20%w/w water and yet remain free-flowing powders.*

**Solvent:** *More than one volatile organic material may be present as a consequence of solvent use in manufacture coupled with incomplete removal during isolation and drying. The concentrations of residual solvent that are tolerated depend upon toxicological, legislative and other considerations, e.g. odor.*

**Impurities:** *These contaminants may be present in the form of residual starting materials, reagents, by-products or side products. While some specialist uses demand high purities, it is common for commercial dyes and pigments to contain percentage-level impurities, some of which are themselves colored. Economic pressure often dictates that purification to remove them is not an option given the substantial extra cost involved. In such cases, manufacturers rely on consistent operation of optimized processes to reproducibly restrict the amounts of impurities present to acceptably low levels. While their presence is usually merely tolerated, in certain cases, it is actually desirable, leading to useful modifications to shade, crystal form or solubility. In applications such as food and cosmetics, legislation or safety assessments demand that traces of particular organic compounds and heavy metals do not exceed specified limits.*

**Diluent:** *A low-cost material is added by the manufacturer during the "standardizing" stage of the production of certain colorants to reduce color strength to a consistent level.*

**Inorganic salt:** *Colorless compounds like sodium chloride and sodium sulfate are often unavoidably present in some kinds of water-soluble dye owing to their formation and/or addition during production. Substantial amounts may remain if it is neither economic nor necessary to remove them. In such cases, manufacturers usually add extra salt as a diluent to eliminate differences in colorant strength arising from inevitable batch-to-batch variation in residual salt content.*

**Dispersing agents:** *These water-soluble polymers or surfactants are intentionally blended into water-insoluble colorant powders to aid their dispersion within aqueous media and to stabilize the resultant suspensions. They are also used in production of such powders; existing levels may be topped up with further dispersant acting as a diluent, typically to contents of ~50%w/w.*

**Dust-reducing agents:** *Also called "dedustants" and "anti-dusting agents", these are predominantly high-boiling mineral or organic oils present at no more than 1–2%w/w to aid handling of powders by inhibiting aerosolization of particles during dispensing and dissolution/dispersing of colorant.*

**Buffers:** *These materials are added to chemically reactive dyes during their isolation and/or standardization in order to lengthen shelf-life by maintaining pH in a desired range during storage.*

**Shading components:** *Producers may add small quantities of other colorants to correct aberrations in color caused by deviations from a prescribed production process or to match the color of competitor products.*

**Coating:** *Manufacturers sometimes apply colorless substances to pigment particle surfaces to manipulate their hydro- or organo-philicity as well as to improve properties such as photostability.*

Some of the colorant forms listed in Table 15.4 may be preferred to powders by colorant users on the basis of convenience and/or safety. For example, liquid concentrates are readily metered by volume, save on dispersion/dissolution time and do not create dust. The liquid media employed are heavily application dependent: aqueous, water-miscible (e. g. glycol) or water-immiscible (e. g. hydrocarbon). In addition to dyes or pigments, they will contain extra components, which may include one or more of those listed in Box 9, plus other additives such as preservatives. Colorants are also supplied as concentrates in other forms. Pre-dispersed pigment "masterbatch" mixtures in pellets or slabs of thermoplastic or rubber (as well as plasticizer pastes) enable end-users to produce colored items through combination with molten virgin polymer, sparing themselves the difficulties of handling neat pigments individually.

To the above list can be added solutions and dispersions of colorant in coating formulations: inks, paints, varnishes and the like, will contain an even wider range of materials for use in conjunction with colorants, such as light stabilizers, antioxidants, rheological control agents, reactive monomers, photoinitiators, etc.

Colorant composition and form are diverse: the next section touches on distinctions in dye and pigment origin.

## 15.8 Natural or synthetic?

Many dyes and pigments are natural in origin in the sense that they are extracted from mineral, vegetable or animal sources rather than created by chemical transformation of fossil fuel [56]. Pigments refined from mineral deposits, such as iron oxides, have been known since antiquity. Dyes continue to be extracted and/or derived on a commercial basis from plant and animal sources, examples of which include curcumin and cochineal, respectively. Often these extracts or preparations consist of complex mixtures of colored compounds. By way of illustration, many literature sources refer to the principal component of the red coloring matter obtained for centuries from the plant madder (*Rubia* spp.) as alizarin (1,2-dihydroxyanthraquinone), but over 60 anthraquinonoid colorants have been identified in root extracts [57].

The vast majority of dyes and pigments are synthetic. They are manufactured using modern production methods in which well-characterized chemical reactants are combined under carefully controlled conditions, often with the aim of producing a single colored compound to defined specifications in terms of color and content [58]. These products are then utilized "as is" or in the manufacture of colorant formulations through mixture with other colored compounds and/or non-colorant additives as described in the previous section.

The advent of the synthetic dye manufacturing industry in the 1850s heralded what would become a decline in natural dye usage during the decades that followed. Today, natural dyes account for only a very small proportion of overall colorant use because synthetic materials have supplanted them in most applications. A notable exception is food coloration, for which synthetic colorants are restricted in many geographies to selection from small sets. A few natural dyes are now synthesized entirely chemically: an example is Indigotin. It is the principal blue colorant component of the natural plant extract, Indigo, which has been used for millennia in textile coloration. Worldwide production is currently ~50,000 t/year [59] of which by far the majority is synthetic. The successful commercial introduction of artificial Indigotin just before the end of the nineteenth century led to a 20-fold reduction in production of natural Indigo to less than 1,000 t at the outbreak of the First World War. The utilization of many other natural dyes also declined severely as a consequence of the commercial and technical advantages offered by synthetic dyes, the overwhelming majority of which do not occur in nature. These characteristics include a wider color gamut in addition to typically better fastness and an availability that is unaffected by seasonal and climatic variation.

In contrast to dyes, production of certain inorganic pigments from natural sources remains important: titanium dioxide is produced at a rate of a few million tonnes per year. Most of it is extracted from mined ore by chemical processing, which regenerates the compound in sufficiently pure form for pigment usage. Nevertheless, ultramarine – originally a rare and expensive natural pigment consisting of ground lapis lazuli – is an example of an inorganic pigment that has been produced synthetically since the 1820s [60]. Not only is artificial ultramarine cheaper but its color properties can be tuned by appropriate selection and control of synthesis conditions. Pigments of wholly organic composition are exclusively synthetic.

Over the past couple of decades, there has been great academic interest, especially in Asia, concerning plant-derived natural dyes. This attraction lies in their potential to be safer and more sustainable alternatives to synthetic colorants for applications including cosmetics and textiles. However, it is often presumed that "natural = safe" but the assumption that their toxicology is benign does not necessarily hold. In addition, natural colorants possess intrinsic problems connected with their production (inconsistency of quality; seasonality in availability; land-use competition with food crops) and properties (limited color gamut; tendency for poor fastness; relatively complex application techniques required). Food coloration aside [61], they will remain niche unless substantial economic or legislative shifts occur, enforcing large-scale migration away from synthetic dyes, which currently appears unlikely. Another approach to the production of naturally derived colorants that might eventually bear fruit after years of study is the attempted creation of colorants by microorganisms. Commercialization of processes in which vats of bacteria, whose metabolic pathways have been programmed genetically to secrete specific colorants, remains in its infancy.

As we have just glimpsed in this and preceding sections, colorants vary hugely in composition, origin, and end-use. Structured systems of classifying and naming colorants became important many years ago for selecting and referring to them. The marketing of dyes and pigments with supplier-specific trade names drove efforts to categorize colorants, permitting users to figure out which were equivalent and for what purposes they could be used. This chapter will close with a brief examination of the most helpful means of subdividing the colorant kingdom.

## 15.9 Colorant taxonomy

The most immediately obvious way of differentiating colorants is by color, for example, in terms of hue. One can demarcate colorant subdivisions on the basis of other properties too. Such measures include hydrophobicity and volatility. While this strategy of utilizing individual properties in isolation or combination may be appropriate when choosing between members of a narrow set of similar colorants for a specific application, it usually serves no purpose as a general method of classification at a technical level. For example, the description "red water-insoluble colorant" covers such a vast array of dyes and pigments that nothing specific can be inferred from it concerning chemical composition or end-use.

Producers and users of colorants must be able to discriminate between them in ways which they consider useful. Knowledge of colorant constitution is of prime importance to manufacturers of dyes and pigments. On the other hand, the main concerns of those applying colorants relate to suitability and properties. The two most useful general schemas are therefore based on either chemical composition or means of use. The colorant classification system which has arguably enjoyed the greatest success, that of The Colour Index™ (CI), combines these two approaches [62].

The use of full chemical nomenclature when referring to individual colorants is not practical in most cases. While it is common practice to do so for inorganic pigments with relatively simple compositions, e. g. titanium dioxide and barium sulfate, application of systematic procedures, such as that of IUPAC, to the structures of organic colorants presents several difficulties. Names are usually too inconveniently long and complex for everyday use and would be met by noncomprehension from most end-users. The approach cannot readily deal with colorants of indeterminate composition (e. g. sulfur dyes) or those made up of complex mixtures (e. g. natural colorant extracts). Other obstacles include business confidentiality: the chemical structures of many dyes and pigments remain commercially sensitive and are thus not publicly disclosed.

There is however a highly useful classification system for organic colorants that is based on structure. It centers on the key molecular features of a dye or pigment that give rise to light absorption, i. e. the motif in a colorant's molecular structure which leads to electronic transition energies coinciding with those of the visible spectrum and thus bestows on it the potential to be colored. Figure 15.2 shows four common structural classes: azo **2**, anthraquinonoid **3**, phthalocyanine **4** and hydrazone **5**. Subdivision in this manner is of value in that the same structural types share general color characteristics and chemistry. The CI not only employs this strategy but also goes further, allocating a unique numerical code ("Constitution Number", CICN) prefixed with "CI" to each dye and pigment structure (see Box 10). CICNs can often be found in ingredients listings of cosmetics products, signaling the inclusion of colorants to manipulate the appearance of formulations.

---

**Box 10**

*Indigoid colorants are allocated CICNs within the range CI 73000–CI 73999. The parent indigoid colorant, Indigotin (**6**), has a Constitution Number of CI 73000, whereas that assigned to its sulfonated derivative, Indigo Carmine (**7**), is CI 73015.*

6 : R = H

7 : R = SO$_3$Na

*As mentioned in the previous section, Indigotin has a long history as a colorant. It remains an extremely important vat dye and is listed in The Colour Index™ with a CIGN of CI Vat Blue 1. The name consists of four parts:*

- *the CI prefix;*
- *an application class (one of nineteen classes – here "Vat" signifies use as a vat dye);*
- *a hue (one of the eight divisions relevant to dyes);*
- *a numerical identifier (usually an integer).*

*Many authors of research papers incorrectly neglect to include the CI prefix when referring to colorants by CIGN.*

*Indigotin is also marketed in its leuco form as CI Reduced Vat Blue 1. The low solubility of the non-reduced form of the dye leads to its now-obsolete use as a pigment in the guise of CI Pigment Blue 66.*

*Industry also exploits the disulfonated analog **7** in various applications. It can be used as a textile dye (CI Acid Blue 74) and is an important food colorant (CI Food Blue 2). In the latter context, it has the food additive designations E132 and FD&C Blue #2 in the EU and United States, respectively. When used to color cosmetic formulations, the INCI name of the dye is the same as its CICN ("CI 73015"), but when employed as a hair colorant, the relevant INCI name is an incomplete form of CIGN ("Acid Blue 74"). After conversion to an aluminum lake, the colorant is designated CI Food Blue 1:1 and CI Pigment Blue 63.*

---

The CI also categorizes colorants in terms of use by means of CI Generic Names (CIGN). These designations consist of a combination of application class, hue and identifier prefixed with "CI". An example is CI Vat Blue 1 (see Box 10). Commercial colorants with the same CIGN are thus equivalent in the sense that they are based on the same principal dye or pigment – and so share the same CICN. Care is required though: content, additive and physical form differences between colorants of the same CIGN mean that coloration performance might not be exactly equivalent. While a CICN is often associated with just one CIGN, the versatility of certain colorants leads to their CICNs being linked with two or more CIGNs – Box 10 gives two such examples.

Another structure-based system of identification is mandated for colorants (as well as non-colorants) utilized in cosmetics products marketed within the EU and elsewhere: International Nomenclature of Cosmetics Ingredients (INCI) [63]. Its intent is that naming of ingredients on labeling is consistent between manufacturers for the benefit of consumers. Some colorants are accorded chemical names; others are referred to by names appropriated from elsewhere. For instance, the dye precursor benzene-1,4-diamine and the dye sodium 4-[(9,10-dihydro-4-hydroxy-9,10-dioxo-1-anthryl)amino]toluene-3-sulfonate, when used in hair coloration, have the INCI designations "p-phenylenediamine" and "Acid Violet 43", respectively. The latter name is an incomplete form of the dye's CIGN, CI Acid Violet 43.

Numerous other forms of colorant nomenclature exist for particular applications. The EU and the United States, respectively, originated the "E number" and "FD&C name" designations for permitted food colorants (see Box 10).

## 15.10 Summary

In comparison to colorants, there are probably few other industrial product types of which individual examples span such a wide range of cost and volume. The history of colorants is long, but perceptions of a mature unchanging technology are not correct: not only did they drive the initial growth of the modern chemical industry, but dyes and pigments lie at the heart of many new technologies being developed today. As a consequence of the breadth of colorant usage and properties, it is difficult to pin down common characteristics. We have seen that the lines between dye and pigment, aesthetic and functional, as well as natural and synthetic, are sometimes blurred. In addition, colorants vary enormously in composition and longevity. The color that they impart may be delayed, conditional or even destroyed during application.

This chapter closes with two industry-defined colorant definitions reproduced in Box 11. They are more complex and restrictive than those presented in Box 3 for dye and pigment but are nevertheless consistent with the picture presented in the preceding sections. It is hoped that this chapter not only facilitates a ready understanding of the descriptions contained in Box 11, but also leads to an appreciation of how the sphere of colorants extends beyond them.

**Box 11**

*The Ecological and Toxicological Association of Dyes and Organic Pigment Manufacturers (ETAD)
and the Color Pigment Manufacturers Association Inc. (CPMA) define dyes and pigments, respectively, in the following ways [62]:*

> *"Dyes are intensely colored or fluorescent organic substances only, which impart color to a
> substrate by selective absorption of light. They are soluble and/or go through an application
> process which, at least temporarily, destroys any crystal structure by absorption, solution, and
> mechanical retention, or by ionic or covalent chemical bonds." (ETAD)*

> *"Pigments are colored, black, white, or fluorescent particulate organic and inorganic solids
> which usually are insoluble in, and essentially physically and chemically unaffected by, the
> vehicle or substrate in which they are incorporated. They alter appearance by selective absorp-
> tion and/or by scattering of light. Pigments are usually dispersed in vehicles or substrates for
> application, as for instance in inks, paints, plastics, or other polymeric materials. Pigments
> retain a crystal or particulate structure throughout the coloration process. As a result of the
> physical and chemical characteristics of pigments, pigments and dyes differ in their applica-
> tion; when a dye is applied, it penetrates the substrate in a soluble form, after which it may or
> may not become insoluble. When a pigment is used to color or opacify a substrate, the finely
> divided insoluble solid remains throughout the coloration process." (CPMA)*

# References

1.  McBrearty S, Brooks AS. The revolution that Wasn't : a new interpretation of the origin of
    modern human behavior. J Hum Evol. 2000;39:453; Barham LS. Systematic pigment use in
    the middle pleistocene of South-Central Africa. Curr Anthropol. 2002;43:181.
2.  Deutscher G. Through the language glass. London: William Heinemann, 2010.
3.  Griffiths J. The functional dyes – definition, design, and development. Chim. 1991;45:304;
    Functional dyes, ed. S-H. Kim. Amsterdam: Elsevier, 2006.
4.  Wainwright M. Dyes in the development of drugs and pharmaceuticals. Dyes Pigm.
    2008;76:582.
5.  Wainwright M, Kristiansen JE. On the 75th Anniversary of Prontosil. Dyes Pigm. 2011;88:231.
6.  Krieg S. Anticorrosive pigments. In: Buxbaum G, Pfaff G, editor(s). Industrial inorganic
    pigments, 3rd ed. Weinheim: Wiley-VCH, 2005:207.
7.  Hocken J, Griebler W-D, Winkler J. Elektrisch leitfähige Pigmente zur antistatischen
    Ausrüstung von polymeren Beschichtungsstoffen. Farbe + Lack. 1992;98:19.
8.  Kieser M. Pearl luster pigments for laser marking. In: Pfaff G, editor(s). Special effect pigments,
    2nd ed. Hannover: Vincentz Verlag. (2008):165; Vogt R, et al. Eur Coat J. 1997;7/8:706.
9.  Troughton MJ. Handbook of plastics joining: a practical guide, 2nd ed. Norwich: William
    Andrew Inc., 2008:81; Brunnecker F, Sieben M. Laser Technik J. 2010;7:24; Jones IA, et al.
    International congress on applications of lasers & electro-optics '99 proceedings. Orlando:
    Laser Institute of America, 1999.
10. Mustroph H, Stollenwerk M, Bressau V. Current developments in optical data storage with
    organic dyes. Angew Chem Int Ed. 2006;45:2016; Wochele RE, van Houten H, Duchateau
    JPWB, Kloosterboer HJG, Verhoeven JAT, van Vlimmeren R, Legierse PEJ, Gravesteijn DJ, Wright

CD, Borg HJ, Heitmann H, Heemskerk J. Information storage materials, 2. Optical recording. In: Ullmann's encyclopedia of industrial chemistry. Weinheim: Wiley-VCH, 2011.

11. Pfaff G. Solar heat-reflecting pigments. In: Pfaff G, editor(s). Special effect pigments, 2nd ed. Hannover: Vincentz Verlag, 2008:70.

12. Burkinshaw SM, Hallas G, Towns AD. Infrared camouflage. Rev Prog Color Relat Top. 1996;26:47.

13. O'Grady K, Gilson RG, Hobby PC. Magnetic pigment dispersions (A Tutorial Review). J Magnetism Magn Mater. 1991;95:341.

14. Horiishi N. Magnetic pigments. In: Buxbaum G, Pfaff G, editor(s). Industrial inorganic pigments, 3rd ed. Weinheim: Wiley-VCH, 2005:195.

15. Bamfield P, Hutchings MG. Chromic phenomena: technological applications of colour chemistry, 3rd ed. Cambridge: RSC Publishing, 2018.

16. Sabnis RW. Handbook of acid-base indicators. Boca Raton: CRC Press, 2007.

17. Tomić T, Babić S, Biošić M, Nasipak NU, Čižmek A-M. Determination of the solvent blue 35 dye in Diesel Fuel by solid phase extraction and high-performance liquid chromatography with ultraviolet detection. Dyes and Pigments. 2018;150:216–222. DOI: 10.1016/j.dyepig.2017.12.013.

18. Verelst J, Frass W, Telser T, Hoffmann H, Bronstert B, Springstein K-A, Potts R. Imaging technology, 3. Imaging in graphic arts. Ullmann's encyclopedia of industrial chemistry. Weinheim: Wiley-VCH; Strehmel B, Brömme T, Schmitz C, Reiner K, Ernst S, Keil D. NIR-dyes for photopolymers and laser drying in the graphic industry. In: Lalevée J, Fouassier J-P, editors. Dyes and chromophores in polymer science. London-Hoboken: ISTE Ltd and John Wiley & Sons Inc., 2015:213 2015.

19. Bamfield P, Hutchings MG. Chromic phenomena: technological applications of colour chemistry, 2nd ed. Cambridge: RSC Publishing, 2010:218.

20. Mishra A, Fischer MK, Bäuerle P. Metal-free organic dyes for dye-sensitized solar cells: from structure: property relationships to design rules. Angew Chem Int Ed. 2009;48:2474.

21. Titterton DH. Dye lasers : a non-colour use of dyes. Rev Prog Color Relat Top. 2002;32:40.

22. Wainwright M. Therapeutic applications of near-infrared dyes. Color Technol. 2010;126.

23. Wainwright M. Dyes, flies, and sunny skies: photodynamic therapy and neglected tropical diseases. Color Technol. 2017;133:3.

24. Xiao P, Zhang J, Dumur F, Tehfe MA, Morlet-Savary F, Graff B, Gigmes D, Fouassier JP, Lalevée J. Visible light sensitive photoinitiating systems: recent progress in cationic and radical photopolymerization reactions under soft conditions. Progress in Polymer Science. 2015 2;41:32–66. DOI:10.1016/j.progpolymsci.2014.09.001.

25. Osterwalder U, Sohn M, Herzog B. Global state of sunscreens. Photodermatol Photoimmunol Photomed. 2014;30:62.

26. Hefford RJW. Colourants and dyes for the cosmetics industry. In: Clark M, editor(s). Handbook of textile and industrial dyeing, Vol. 2. Cambridge: Woodhead Publishing Ltd, 2011:175.

27. Wardman RH. An introduction to textile coloration: principles and practice. Chichester: John Wiley & Sons Ltd, 2018:231.

28. Kornreich E. The colour changes of dyeings of vat and azoic dyes by wet and dry heat. J Soc Dyers Colourists. 1942;58:177; Sumner HH, Vickerstaff T, Waters E. The effects of the soaping aftertreatment on vat dyeings. J Soc Dyers Colourists 1953;69:181; Valko EI. The theory of dyeing cellulosic fibers. Tex Res J. 1957;27:883.

29. Christie RM, Mackay JL. Metal salt azo pigments. Color Technol. 2008;124:133.

30. Herbst W, Hunger K. Industrial organic pigments: production, properties, applications, 3rd ed. Weinheim: Wiley-VCH, 2004.

31. Patterson D. Organic and inorganic pigments; Solvent dyes. In: Shore J, editor(s). Colorants and auxilliaries, Vol. 1. Bradford: SDC, 1990:32.
32. Ryan PJ. Daylight fluorescent pigments and colours. Pigm Resin Technol. 1972 9;1:21.; Hunger K, Herbst W. In: Ullmann's encyclopedia of chemical technology. Weinheim: Wiley-VCH, 2007:42, Chapter 17.
33. Shevell SK, editor. The science of color, 2nd ed. Oxford: Elsevier, 2003; Valberg A. Light vision color. Chichester: John Wiley & Sons Ltd., 2005.
34. Hughes HK. Suggested nomenclature in applied spectroscopy. Anal Chem. 1952;24:1349.
35. Oleari C. Standard colorimetry definitions, algorithms and software. Chichester: John Wiley & Sons Ltd, 2016:5.
36. Lynch DK, Livingstone W. Color and light in nature. Cambridge: Cambridge University Press. 2001:229; Saidman J. Sur la Visibilité de l'Ultraviolet jusqu'à la Longeur d'Onde 3130. Comptes Rendus 1933;196:1537; Gaydon AG. Colour sensations produced by Ultra-violet Light. Proc Physical Soc. 1938;50:714.
37. Klein GA. Industrial color physics, springer series in optical sciences 154. New York: Springer, 2010:11.
38. Christie R. Fluorescent dyes. In: Clark M, editor(s). Handbook of textile and industrial dyeing, Vol. 1. Cambridge: Woodhead Publishing Ltd., 2011:562.
39. Mukherjee S, Thilagar P. Recent advances in purely organic phosphorescent materials. Chem Commun. 2015;51:10988.
40. Mortimer RJ. Switching colors with electricity. Am Scientist. 2013;101:38.
41. Bamfield P, Hutchings MG. Chromic phenomena: technological applications of colour chemistry, 2nd ed. Cambridge: RSC Publishing, 2010:59, 393.
42. Corns SN, Partington SM, Towns AD. Industrial organic photochromic dyes. Color Technol. 2009;125:249.
43. Towns AD. Industrial photochromism. In: Bergamini G, Silvi S, editor(s). Applied photochemistry, lecture notes in chemistry 92. Switzerland: Springer International Publishing, 2016:227.
44. Towns A. Colorant, photochromic. In: Luo MR, editor. Encyclopedia of color science and technology. New York: Springer, 2016:447.
45. Nakatani K, Piard J, Yu P, Métiviér R. Introduction: organic photochromic molecules. In: Tian H, Zhang J, editor(s). Photochromic materials: preparation, properties and applications. Weinheim: Wiley-VCH, 2016:1.
46. White MA, Bourque A. Colorant, thermochromic. In: Luo MR, editor. Encyclopedia of color science and technology. New York: Springer, 2016:463.
47. Fujita S. Organic chemistry of photography. Berlin: Springer-Verlag, 2004:135.
48. Morel OJX, Christie RM. Current trends in the chemistry of permanent hair dyeing. Chem Rev. 2011;111:2537.
49. Berry C, Ferguson JG. Discharge, resist and special styles. In: Miles LWC, Textile printing, 2nd ed. Bradford: SDC, 2003:196.
50. Bide M. Coloration, fastness. In: Luo MR, editor. Encyclopedia of color science and technology. New York: Springer, 2016:474.
51. Jones F. The chemistry and properties of metal-complex dyes. In: Shore J, editor(s). Colorants and auxilliaries, Vol. 1. Bradford: SDC, 1990:196.
52. Bide M. Coloration, mordant dyes. In: Luo MR, editor. Encyclopedia of color science and technology. New York: Springer, 2016:482.
53. Leube H. In: Hunger K, editor. Industrial dyes chemistry, properties, applications. Weinheim: Wiley-VCH, 2003:380.

54. Zollinger H. Color: a multidisciplinary approach. Zürich-Weinheim: Wiley-VCH & VCHA, 1999:41.
55. Lacasse K, Baumann W, editors. Textile chemicals: environmental data and facts. Berlin: Springer, 2004.
56. Hofenk de Graaf JH. The colourful past. Riggisberg-London: Abegg-Stiftung and Archetype Publications Ltd, 2004.
57. Blackburn RS. Natural dyes in madder (*Rubia* spp.) and their extraction and analysis in historical textiles. Color Technol. 2017;133:449.
58. Booth G. The manufacture of organic colorants and intermediates. Bradford: SDC, 1988.
59. In the Jeans. Nature. 2018;553:128.
60. Seel F, Schaefer G, Guettler H-J, Simon G. Das Geheimnis des Lapis lazuli. Chem Unserer Zeit. 1974;8:65; Calvert D. Ultramarine pigments. In: Buxbaum G, Pfaff G, editors. Industrial inorganic pigments, 3rd ed. Weinheim: Wiley-VCH, 2005:136.
61. Coultate T, Blackburn RS. Food colorants: their past, present and future. Color Technol. 2018;134:165.
62. http://colour-index.com.
63. International cosmetic ingredient dictionary and handbook, 16th ed. Washington D. C.: Personal Care Products Council, 2016.

Gerhard Pfaff

# 16 Colorants in building materials

**Abstract:** This review article is a summary of the current knowledge on colorants in building materials, respectively, construction materials. Building materials belong as well as paints, coatings, plastics, printing inks, and cosmetic formulations to the most important application systems for colorants. The only relevant colorants used in building materials are inorganic pigments. These have to meet high demands with regard to color and stability in the application system, especially in concretes. Different processing methods are used for coloring of cement, respectively, concrete. The coloring processes need to be coordinated in accordance with the steps of the processing leading to the final building materials.

**Keywords:** colorant, pigment, building material, cement, concrete

## 16.1 General aspects

The choice of the colorant is of major importance for the appearance and the properties of an end product. In case of building materials, only inorganic pigments play an important role for coloring. Years of tests on colored concrete products exposed to different climate conditions all over the world have shown that particularly inorganic pigments on the basis of metal oxides have the required fastness properties, which are essential for the use in building materials [1].

Pigments used in building materials must withstand the aggressive influence of the strongly alkaline cement pastes. They must also be lightfast, weather-stable, and insoluble in the water used during processing of the material. The pigments must be firmly integrated in the concrete matrix after the processing [2–4].

The oxide pigments commonly used in building materials cover all the most popular colors. These include various red, brown, yellow, and anthracite color shades, but also green, blue, and white colors. Most important pigments for red, brown and yellow colors are iron oxides (iron oxide red, iron oxide yellow, iron oxide black). To a lesser extent, mixed metal oxide pigments based on the rutile structure (so-called buff rutile pigments, e. g. chromium antimony titanium oxide and chromium tungsten titanium oxide) are used in the yellow color segment. Green shades are mostly produced with chromium(III) oxide (chromium oxide green) and blue shades with cobalt aluminum oxide (cobalt blue). Titanium dioxide (titanium white) is used for white and carbon

This article has previously been published in the journal Physical Sciences Reviews. Please cite as: G. Pfaff, Colorants in Building Materials *Physical Sciences Reviews* [Online] 2020, 5. DOI: 10.1515/psr-2020-0041

https://doi.org/10.1515/9783110588071-016

black for black tones. The different pigments can be mixed together. An infinite number of different shades is possible in this way. The shades are often oriented to the opaque colors as they occur in nature, and thus fit harmoniously in with the environment.

Green and blue color shades are of less importance compared with red, brown, and yellow. Brilliant, glossy shades, like those achievable for plastics and coatings, are very difficult to produce for building materials mainly due to the composition and the structure of the different types of concrete. Figure 16.1 shows examples of concrete paving slabs with and without coloration.

**Figure 16.1:** Concrete paving slabs, from left to right without pigmentation, pigmented with an iron oxide red pigment, pigmented with an iron oxide black pigment.

Tinting strength of a pigment in a specific system used as a building material is an important quality characteristic. It is defined for building materials as the ability of a pigment to impart its natural color to the medium being colored. The laboratory test for the determination of the tinting strength of iron oxide pigments, as an example, is carried out by mixing a defined quantity of the pigment with a defined quantity of barium sulfate. The result of the test is the basis for the tinting strength tolerances given in the product specification for iron oxide pigments.

Typical pigment concentrations used in concrete are 3–6% with regard to the content of cement. In special cases, higher pigment quantities of up to 10% are possible.

The decision, which pigment quality is used in a specific case of coloring is not only dependent on the tinting strength of the possible pigments but also on other factors, in particular on the price for the pigment finally used.

The requirements concerning the processability of pigments for building materials have changed over the years. In earlier years, only pigment powders were used in the building industry. Nowadays, aqueous pigment slurries are also in use in this application field. Apart from the fact that the processor does not have any dust problems, there is also the advantage of simple handling and metering. On the other hand, such slurries contain a relatively high proportion of water, which means higher transport costs than for pigments in powder form. In addition, the pigment may settle at the bottom if the pigment suspensions have not been well stabilized for a long time. For these reasons, it is in certain cases worthwhile to purchase ready-to-use pigment suspensions if the producer is in the vicinity of the producer for building materials.

A more recent development in this field is the supply of the pigments in form of free-flowing dry pigment preparations. These have been developed specifically for the use in the building industry and allow easy emptying of silos, sacks, and bulk bags. A big advantage of working with such preparations is the avoidance of dust formation.

## 16.2 Metering and dispersion of pigments for building materials

Accurate metering and homogeneous dispersion of the pigments are essential factors in the manufacture of high-quality colored concrete products. The building industry has established in this regard a system of parameters for the dispersion of pigments in compositions for building materials [5].

Each type of mixer used for the dispersion process has an optimum mixing time. Too short mixing times lead to inhomogeneities, even if the mixing times of the individual ingredients or the order in which these are added are changed. The necessary mixing times are highly dependent on the performance of the concrete mixer. It is important for the generation of optimized pigment dispersions to know when the pigment should be added to the mixer. Experience has shown that the ideal sequence is to mix the pigment with the suitable amounts of gravel and sand for about 15 s before the cement is added. From that point on, the mixing process is the same as with unpigmented concrete. Something, which must be avoided at all costs, is the adding of all the components simultaneously, or that cement is mixed in immediately after the sand. The mixing time naturally also plays an important role in whether or not a homogeneous dispersion of the pigment is achieved.

Each mixer needs a minimum of mixing time, which is in the range of 1.5–2 min for forced circulation mixers. If the time is reduced, it will not be possible to produce a homogenous mixture, even by altering the individual mixing time or the adding sequence for the individual components.

The concrete producer has always the possibility to produce a large variety of color shades by himself. Iron oxide pigments, for example, are manufactured in the basic colors red, black, and yellow. Various shades are also available within each of these color ranges. By combining two or three iron oxide pigments, it is possible to produce an almost unlimited number of shades. In order to achieve a brown shade, an iron oxide red pigment can be combined with an iron oxide black. Figure 16.2 gives an impression of concrete stones colored with different iron oxide pigments.

Present-day concrete mixing technology allows several individual pigments to be added simultaneously to the concrete mixer. There is no need to premix the pigments.

**Figure 16.2:** Concrete stones colored with different iron oxide pigments respectively pigment mixtures.

## 16.3 Influence of the cement color on the color of the concrete

Cement is defined as a binder in the area of construction materials, and more specific as a substance used in this application field that sets, hardens, and adheres to other materials to bind them together. Cement is typically not used on its own, but rather to bind sand and gravel (aggregate) together. Cement mixed with fine aggregate produces mortar for masonry. Mixtures of cement with sand and gravel are the basis for concrete. Cements used in the construction sector are usually inorganic, mostly based on calcined lime (calcium carbonate) or calcium silicate.

Concrete is defined as a composite material composed of fine and coarse aggregate bonded together with a fluid cement (cement paste) that hardens over time. Different types of cement can be used for the production of concretes. Well known are lime-based cement binders, such as lime putty. Other cements used are calcium aluminate cement and Portland cement. Figure 16.3 shows component parts consisting of pure uncolored concrete.

**Figure 16.3:** Panels consisting of pure uncolored concrete.

The quality of the pigments used for coloring of the cement, respectively, the concrete is of fundamental importance for the appearance of these building materials. The requirements for these pigments are regulated among the countries of the European Union and some other European countries in the EN 12878 [6]. This standard describes that concrete colors must be based on inorganic synthetic pigments. Before concrete colors are allowed to be used, their conformity corresponding with the requirements of the standard must be proved.

### 16.3.1 Color of the cement

Concrete manufactured with normal gray Portland cement is not suitable for bright colors. These are typically achieved by the use of white cement. The gain in color clarity obtained by using white cement depends significantly on the properties of the applied pigments. In case of a black pigment, there is virtually no difference between concrete made by using of white or grey cement. With dark brown and red pigments, the difference is slight, and with yellow and blue pigments, it is very pronounced. The brighter and cleaner the desired shade, the greater the dependency on the type of cement.

Producers and users have to take into the consideration the fact that grey cement can vary appreciably in color from light to dark gray. The switch of a cement grade or of a cement supplier can lead to undesired deviations in the natural color of the cement, which can have an appreciable influence on the final color [3].

### 16.3.2 Color of the concrete

Coloring of concrete means, that the cement paste is pigmented, not the aggregate. The cement paste then forms a layer on the individual aggregate particles. The influence on the color of the pigmented concrete by the cement content can simplified be explained by using a mixture of two substances, the colored cement paste and the aggregate. The more the colored cement paste is "diluted" with the aggregate, the less intensive is the color of the concrete. This simplified consideration is confirmed by practical experience. A concrete with a high cement content has at an equal pigment concentration (typically a percentage based on the weight of cement) a much stronger color than a concrete with a lower cement content.

### 16.3.3 Role of the aggregate

The production of colored concrete includes that the aggregate particles are covered by the pigmented cement paste. It can happen that the grains of an intensively colored aggregate are not completely covered. In such a case, the resulting color of the final concrete is affected by the natural color of the aggregate. While this effect can be apparent even during the production of the colored concrete, it becomes particularly evident when the end product is exposed to outdoor conditions. Under these circumstances, the aggregate particles may become visible through weathering of the surface. A mixed shade resulting from the color of the cement paste and the exposed aggregate can thus be seen.

## 16.4 Influence of the formulation

### 16.4.1 Pigment concentration

It is common knowledge that the optimum pigment concentration in an application system is of high economic interest. This is also true for the production of concrete, where not more pigment is used than necessary to achieve a certain effect. If increasing amounts of pigments are added to a concrete mix, the color intensity initially rises linearly with the pigment concentration. Continued addition of pigment leads at a certain point to a pigmentation level at which further pigment does not significantly deepen the shade. Consequently, further addition of pigment is uneconomical above this level. Establishing the saturation range is dependent among other things on the system parameters of the concrete. In case of pigments with less intense color, the saturation range is often not reached until very large amounts of pigment are added. The amount of pigment required to produce a given shade can sometimes be so large that this quantity can have a negative influence on the technical properties of the concrete.

### 16.4.2 Water–cement ratio

The water–cement ratio has a significant influence on the color of a concrete. Foam is formed during the process in that moment, when the cement is incorporated in the water by stirring. The foam consists of many tiny air bubbles, which scatter the light in the same way that white powder particles do, for example pigments or fillers. The excess mixing water evaporates from the concrete and leaves behind cavities in the form of fine pores. They scatter the incident light and make in this way the concrete lighter. The concrete appears the lighter, the higher the water–cement ratio is.

### 16.4.3 Practical applications

Mixing water with its enormous influence on the technical parameters and on the color of the final concrete must be monitored very carefully during the entire process. Differences in color shade can be appreciable in cases in which the excessive moisture in the concrete results in the formation of a surface sludge on the concrete. This sludge contains the very fine components of the concrete such as cement, fines and an above-average accumulation of pigment, so that the concrete may have a different appearance in comparison with the situation when little or no sludge is on the surface.

## 16.5 Influence of the hardening conditions on the shade of concretes

The hardened cement paste formed during the reaction between the mixing water and the cement, produces crystals of varying size, depending on the temperature at which the concrete is hardened [7, 8]. The size of these crystals plays an important role for the interaction of light with the surface of the concrete. The often needle-shaped crystals act as scattering centers for the incident light. Higher hardening temperatures lead to the formation of finer crystal needles. The more distinct light scattering of a concrete with fine crystal needles generates a brighter color compared with the same concrete with larger crystals formed at lower temperatures. It should be noted, however, that this phenomenon becomes generally only recognizable, when the difference in temperature reaches a certain order of magnitude, for example, when steam-hardened concrete is compared with concrete hardened at normal room temperature [9].

## 16.6 Technical properties of pigmented concretes

Pigments used for coloring of concrete consist typically of particles with grain sizes in the range from 0.1 to 1.0 µm. Pigments with smaller or larger particles are applied in building materials only in very rare cases. A comparison shows that the pigment particles used for the coloring of concrete are about 10–20 times finer than the cement. It is to be expected that the addition of the fine pigment particles has a noticeable influence on the water demand for the production of the concrete.

Practical experience shows, however, that the addition of iron oxide red and iron oxide black pigments to water and cement has virtually no effect on the necessary water amount. The behavior of iron oxide yellow pigments under comparable conditions is different. The particles of these pigments have a specific needle-shaped structure and can therefore adsorb more water on their surface. This effect, however, becomes only noticeable at pigment concentrations above 5%, which are rather rare in practical coloration of concrete.

## 16.7 Weathering behavior of pigmented concretes

Concretes produced under optimum conditions have an extremely long steadiness. The color of such concretes is very stable as well. A precondition for this is the use of stable inorganic pigments such as iron oxides or chromium oxide green. Changes of the color are, if any, quite small if one compares the original shade with the shade at the time of consideration. Changes of the shade can also be observed on uncolored concrete. They can have various causes and may be of temporary nature as in the case of efflorescence or of permanent nature as in the case of surface exposure of the aggregates.

### 16.7.1 Efflorescence

Efflorescence is defined as the migration of a salt to the surface of a porous material, where the salt forms a coating. The process involves the dissolving of an internally held salt in a solvent, typically in water. Such formed aqueous salt solution migrates to the surface, then evaporates, leaving a coating of the salt. Efflorescence plays a not negligible role in the life course of concretes. There is no influence on efflorescence originating from the pigments used for coloring of a concrete, neither from iron oxides and chromium oxide green nor from other inorganic colorants. Deposits of white lime, on the other hand, which are formed during the process of efflorescence, have a significant color-changing effect for the concrete. Deposited lime has its origin in a variety of ions, mainly of calcium and carbonate ions, that occur in the mixing water (primary efflorescence) or in rain or dew (secondary efflorescence). The deposition of low-soluble compounds like calcium carbonate on the surface of the concrete happens when calcium ions react with carbonate ions formed from available carbon dioxide. An example of efflorescence as it can occur in the case of concrete paving slabs is shown in Figure 16.4.

**Figure 16.4:** Paving slab showing the effect of efflorescence.

The porosity of a concrete plays an important role for efflorescence. The more compact the concrete is the lower is the tendency to efflorescence. Calcium carbonate formed on the surface of the concrete is able to react with carbon dioxide dissolved in water to form calcium hydrogen carbonate, which is soluble in water. The efflorescence can therefore be washed off again by available water. Acid components in the atmosphere have also a dissolving effect on calcium carbonate efflorescence on the concrete surface [10].

### 16.7.2 Weathering of concretes

The surface of a concrete is usually covered by a sludge layer consisting of the aggregate fines and the cement. The thickness of such a layer depends mainly on its composition and on the way of formation. In the course of the years, the sludge layer is gradually eroded under the influence of atmospheric constituents. The aggregate particles in the surface are exposed during this process and contribute therefore more to the overall color of a concrete system compared with the initial state.

Color changes in pigmented building materials are normally relatively small. Construction units consisting of concrete can be exposed to various weather conditions for many years without showing bigger changes. The most obvious effect is, on the other hand, in most cases a slight soiling of the surface of the concrete.

### 16.7.3 Tests for the weather-stability of pigmented building materials

The use of weather-stable pigments such as iron oxides and chromium oxide green for the coloring of building materials is a mandatory prerequisite for a steady color under all weather conditions. Many other colorants, especially non-oxidic pigments, are not able to fulfill the high demands in this area of application. Outdoor testing is a reliable possibility to get the necessary information for a colored concrete system before it is used in technical applications [11]. Such outdoor tests with concrete and other building materials are relatively complex and time-consuming. They have to be done under representative conditions including all possible weather phenomena, which are thinkable for the location of the future use. It must be considered for all these tests that already the pure concrete is the subject of degradation and that the pigments are only a part of the whole system. Consequently, all outdoor tests have to be performed in comparison with unpigmented test probes.

## References

1. Pfaff G. Inorganic pigments. Berlin/Boston: Walter de Gruyter GmbH, 2017.
2. Püttbach E. Pigments for the colouring of concrete. Betonwerk + Fertigteil-Technik – BFT, 8/1991.
3. Büchner G. Pigments in concrete production. Betonwerk + Fertigteil-Technik – BFT, 8/1991.
4. Egger C. Einfärbung von Baustoffen – aktueller Stand und Entwicklungen. Betonwerk International, 4/2002.
5. Plenker -H.-H. The metering and dispersion of pigments in concrete. Betonwerk + Fertigteil-Technik – BFT, 9/1991.
6. EN 12878. Pigments for the colouring of building materials based on cement and/or lime - specifications and methods of test, 2018.

7.  Piasta J. Heat deformation of cement phases and microstructure of cement paste. Mater Struct. 1989;17:415.
8.  Annerel E, Taerwe L. Revealing the temperature history in concrete after fire exposure by microscopic analysis. Cem Concr Res. 2009;39:1239.
9.  Hager I. Colour change in heated concrete. Fire Technol. 2014;50:945.
10. Kresse P. Efflorescence – mechanism of occurrence and possibilities of prevention. Betonwerk + Fertigteil-Technik – BFT, 3/1987.
11. Büchner G, Kündgen: U. 25 Years of outdoor weathering of pigmented building materials. Betonwerk + Fertigteil-Technik – BFT, 7/1996.

Frank J. Maile

# 17 Colorants in coatings

**Abstract:** The aim of this chapter is to provide a compact overview of colorants and their use in coatings including a brief introduction to paint technology and its raw materials. In addition, it will focus on individual colorants by collecting information from the available literature mainly for their use in coatings. Publications on colorants in coatings applications are in many cases standard works that cover the wider aspects of color chemistry and paint technology and are explicitly recommended for a more detailed study of the subject [1–18]. Articles or information on paint formulation using coatings which contain colorants are rare [19]. This formulation expertise is often company property as it is the result of many years of effort built up over very long series of practical "trial-and-error" optimization tests and, more recently, supported by design of experiment and laboratory process automation [20, 21]. Therefore, it is protected by rigorous secrecy agreements. Formulations are in many ways part of a paint manufacturer's capital, because of their use in automotive coatings, coil coatings, powder coatings, and specialist knowledge is indispensable to ensure their successful industrial use [22]. An important source to learn about the use of pigments in different coating formulations are guidance or starting formulations offered by pigment, additive, and resin manufacturers. These are available upon request from the technical service unit of these companies. Coating formulations can also be found scattered in books on coating and formulation technology [4–27]. This overview can in no way claim to be complete, as the literature and relevant journals in this field are far too extensive. Nevertheless, it remains the author's hope that the reader will gain a comprehensive insight into the fascinating field of colorants for coatings, including its literature and current research activities and last but not least its scientific attractiveness and industrial relevance.

**Keywords:** colorants, pigments, dyes, coatings, paints, solventborne coatings, waterborne coatings, powder coatings, radiation-curing systems, coil coatings, plastic coatings, automotive coatings, thermochromic pigments, photochromic microcapsules, electrochromic pigments, phosphorescent pigments

This article has previously been published in the journal Physical Sciences Reviews. Please cite as: F. J. Maile, Colorants in Coating Applications *Physical Sciences Reviews* [Online] 2021, 6. DOI: 10.1515/psr-2020-0160

https://doi.org/10.1515/9783110588071-017

## 17.1 General comments on the application and processing of colorants used in coatings

The paint industry today can call on a wide range of colorants, with a broad spectrum of colors, effects, and functional properties. Thus, suitable products are available to meet a wide range of demands. In practice, the strengths of the individual product families are frequently combined by using mixtures of different types of colorants. This can offer a cost advantage [15].

Before dealing with the subject in detail, it is worth considering why colorants are used in coatings [28–30]. Scientific interest in this interdisciplinary specialist area and the economic importance of their use is so high that it should at the very least receive a mention in any publication on colorants.

Freedonia Group market studies indicate global growth rates for dyes and organic pigments of at 6% per annum. A global sales growth from $14.6 billion to $19.5 billion was indicated for the years 2014 to 2019 alone [31, 32].

The colors around us have a very important impact on our daily lives. In 2000, the annual production of textile dyes was worth almost €5.4 billion, organic pigments around €4.6 billion, and inorganic pigments, mainly $TiO_2$ and iron oxide pigments, €2.9 billion. This underscores the fact that dyes and pigments represent by far the largest area of commercial activity in the world of color chemistry, whereas their use in coatings is just one of many different important areas of application [2].

Many branches of industry, including paint manufacturers, focus their attention on the end users of their products to ensure that development work meets the needs of the market [33]. Furthermore, because of direct customer influence on color and design, the "time to market" needs to be kept as short as possible. This applies across the wide area of industrial design, fashion, architecture, automobile, and branded products industries. New product development is virtually always prompted by the end user's constant search for individuality or uniqueness [34–36].

The paints and coatings industry is the largest commercial application field for pigments in general [2]. Table 17.1 shows the breakdown among end users. It is worth noting that the selection criteria for the particular pigments intended for a given use depend heavily on the chemical nature of the polymer in question (polyesters, alkyds, acrylics, and latex) or the paint technology itself (solvent- or water-based or powder coating).

In today's world, the surfaces of everyday objects are coated and thus referred to as "coated objects." The term "coating" is therefore used to mean a material (usually a liquid) that is applied to a substrate and, after a drying process, becomes a continuous or discontinuous film. It however also describes the process of application, as well as the resulting dry film [5].

A paint is made up of four basic components: one or more resins, or binders as they are often called, pigments and fillers, additives and, except in so-called powder coatings, a volatile component (solvent or water) [37, 38]. The binder forms a

**Table 17.1:** World paint production 2018 by market sector [37].

| Marketing sector | as % of total* |
|---|---|
| Architectural | 50 |
| Powder Coatings | 10 |
| General Industrial | 9 |
| Industrial Maintenance & Protective | 8 |
| Wood | 7 |
| Transport | 6 |
| Coil Coatings | 3 |
| Marine | 3 |
| Vehicle Refinish | 2 |
| Packaging | 2 |

*Total world = 54 million tonnes (67.5 billion litres),
© *by R. Adams, ARTIKOL*

continuous film on the substrate, the pigments and fillers provide opacity, protection, and color and the additives act as catalysts during drying, or are flow modifiers etc. The volatile element ensures the correct viscosity of liquid paint to ensure ease of application. Coatings are normally dried by the evaporation of a solvent, curing, i. e., crosslinking, by oxidation, stoving, or by the application of ultraviolet (UV) light.

The words "coating" and "paint" are often synonymous. Normally, paints, coatings, and transparent varnishes, and the now somewhat archaic term lacquers are today referred to as coatings [3].

In general, there are three families of coating: solventborne, waterborne, and solvent-free or powder coatings:
- in solventborne coatings, the resin, pigments, and additives are dispersed in an organic solvent,
- in waterborne, the dispersing agent is water,
- in powder coatings, there is neither solvent nor water; the pigments and additives are dispersed directly in the binder.

The property of a dried coating not only depends on the resins, pigments etc. used, but also on the substrate onto which it is applied. The end result is a combination of the type of paint, nature of substrate, method of application, and drying.

The end use determines which type of coating is chosen. So-called organic paints are favored in architecture and industry where appearance and function are important, whereas inorganic paints are predominantly chosen for their protective properties. Increasingly, hybrid coatings are used for specific industrial applications. Coatings are rarely applied in a single coat. Most are used in a multi-layer system consisting of primer and topcoat. Depending on the application, there can be many more. In the automotive industry, for instance, there can be up to six layers [24].

Liquid coatings can be applied manually with a brush or roller, but industry favors automatic application by spraying, rotated disk, or cone atomization. Other automatic applications can include dipping, tumbling, pouring, and the use of rotating drums or rolling. Powder coatings are applied by electrostatic spraying or dipping. Special equipment is needed for multi-component coatings [39].

Each layer is applied to perform certain functions, although their characteristics are influenced by the other layers in the system. The interactions between different layers and the interfacial phenomena play an important role in the overall performance of multi-coat systems [40]. Different coating properties are typically associated with specific parts of that particular system [41], these are partially listed in Figure 17.1.

In the context of coating technology, the term "functional" or "smart" coatings is now regularly used and describes systems which possess an additional functionality to the classical properties of a coating, i. e., decoration and protection [3–50]. According to Baghdachi [51], smart materials are substances which are capable of adapting their properties dynamically to an external stimulus and therefore are called responsive or smart. The term "smart coating" refers to the concept of coatings being able to sense their environment and make an appropriate response to that stimulus. The standard thinking regarding coatings has been a passive layer unresponsive to the environment. The current trend in coatings technology is to control the coating composition on a molecular level and the morphology at the nanometer scale. The idea of controlling the assembly of sequential macromolecular layers and the development that materials can form defined structures with unique properties is being explored for both pure scientific research and industrial applications. Several smart-coating systems have been developed, examined, and are currently under investigation by numerous laboratories and industries throughout the world. Examples of smart coatings include stimuli responsive [52–55], antimicrobial, antifouling, conductive, self-healing, and super hydrophobic systems. Although various mechanisms are involved, as well as numerous applications, the common feature that they offer is a particular benefit which satisfies certain users' demands. Most coatings, whether inorganic or organic, perform critical functions (Figure 17.2), the content here is restricted to coatings with organic binders. Further information can be found in reference [3].

The use of particular colorants in a number of families of coatings leads to complex scientific issues and thus explains the "high tech" label that the paint and coatings technology has rightfully earned for itself, as well as the high degree of interdisciplinary cooperation that has developed as shown in Figure 17.3. As previously stated, many coatings are based on organic polymers that serve as a backbone or matrix in

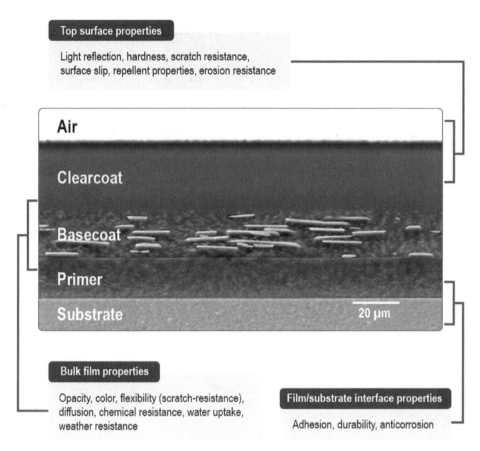

**Figure 17.1:** Left: Scanning electron micrograph of a cross-section through a typical coating layer arrangement: (metal) substrate, primer, basecoat (incl. effect pigment platelets), clear coat. Right: Topographical classification of coating properties after [3] (Copyright Wiley-VCH GmbH. Reproduced with permission).

film building, yet are, at the same time, expected to integrate discrete particles of differing morphology during the paint preparation. The surface to be coated (the substrate), can be both inorganic (metallic alloys) and organic in nature (plastic, wood, or composites). The physical–chemical properties of the final film are significantly affected by the distribution and orientation of the (sometimes anisotropic) pigment particles (percolation), the homogeneity of the distribution in the film (degree of dispersion), and the wettability brought about by the surrounding polymer matrix. For completeness, it should be mentioned that the environment during application, film building (coalescence), and treatment of the paint film (air temperature, humidity, and downdraft speed, etc.) also have a deciding effect on the properties of the resulting film. If the final paint film is subjected to external mechanical influences and/or electromagnetic radiation, the analysis of the damage

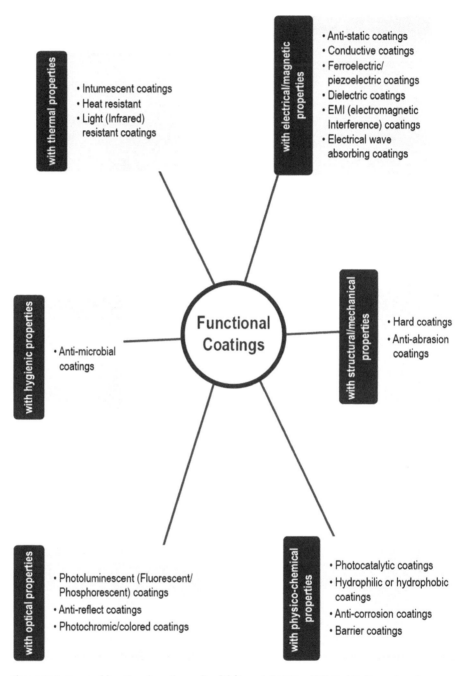

**Figure 17.2:** Types of functional coatings after [3] (Copyright Wiley-VCH GmbH. Reproduced with permission).

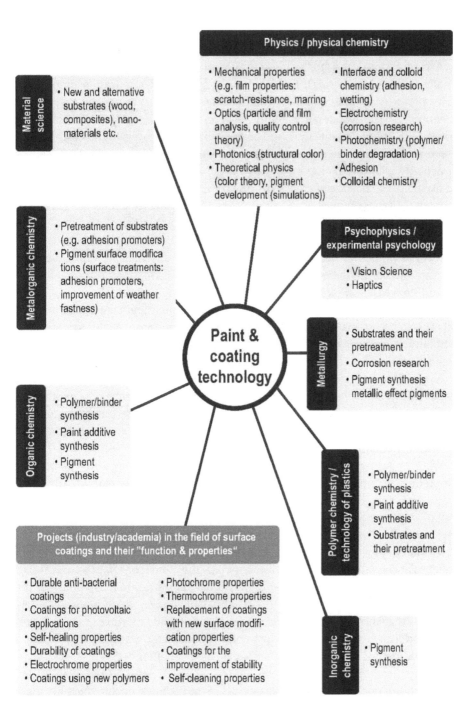

**Figure 17.3:** Specialist scientific areas and their role in relation to paint and coating technology and current research activities in the field of surface coatings (grey box bottom left).

that is incurred makes the interpretation of the degradation mechanisms that occur immensely complex.

According to [1], the most important properties of coatings are: durability (specifically exterior weathering), rheology (which affects the application properties), hiding power (coverage), pigment volume concentration (PVC) (influences stress, water uptake, water vapor permeability, weather resistance), water vapor and water permeability (influences the moisture balance of the substrate and thus the durability and protection of the substrate against weathering), adhesion to the substrate (particularly with alternating moist–dry, moist–hot, and moist–frosty conditions), and elasticity (in the case of crack-covering coatings and wood coatings).

In 1981, Kaluza published a useful "trouble shooting guideline" (Figure 17.4) describing the complexity of the physical–chemical correlations of pigment–binder interactions in the dispersion, application, and film-formation process [56].

If these properties are taken into consideration, an efficient "error analysis" (for example, during the optimization of a paint formulation [56]) can only be successful if looked at on an interdisciplinary level. Figure 17.3 lists the typical scientific disciplines and their applications in relation to paint and coating technology. Each specialist area is given the same degree of importance. Thus, for example, "metallurgy" is not classified as a subcategory of "inorganic chemistry," but as an independent field for error analysis.

Empirically obtained knowledge is carefully protected. For this reason, it is often not published (even in the form of patents). Guidance formulations now exist which can be used as starting points. Given their source (often manufacturers of additives, binders, or pigments, who seek to use these formulations to promote their products), the specialist information over time becomes less sensitive [25, 57–61].

When choosing an appropriate colorant, one should not forget that the profile of a product's properties does not depend solely on, e. g., the colorant. The form, distribution, and alignment of the particles and their interaction with the polymer matrix, as well as with additives determine color, effect, hiding power, and weathering properties. Thus, although each component is important, it is sometimes only one of the building blocks which determine the quality. For this reason, the closer one looks at the various classes of raw materials, the more it becomes apparent that they are made up of a large number of individual substances [62, 63]. Consider the range of binders available. Apart from natural materials, which are still widely used, the range increasingly comprises a large and steadily growing number of synthetic products. Here there is a clear trend towards water-based coating products that can be applied under conventional application conditions [15].

Taking into consideration all work and environment-related factors, high quality is driving the development of the raw materials in the high-performance field like automotive or coil and powder coatings for exterior applications [15]. As a result, especially the high-performance driven industry has set challenging requirements for the manufacturers of the raw materials used these coatings, this includes colorants [28, 63].

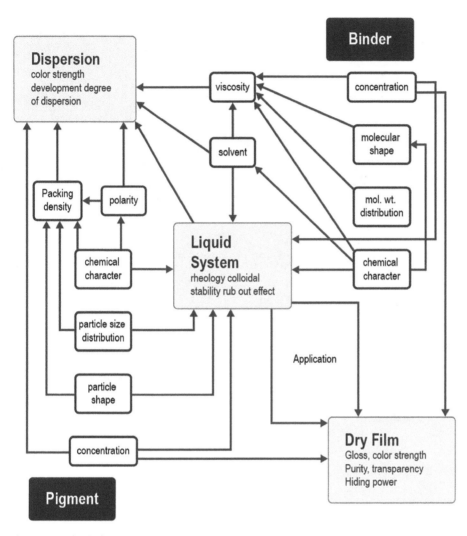

**Figure 17.4:** Physical–chemical fundamentals of pigment processing for paints and printing inks [56].

An overview of the most important requirements created by the author is shown in Figure 17.5.

The above information underlines the importance of colorants in paints and coating materials. After a short introduction to the history of paints and coatings and their composition, as well as a guide to the physical–chemical reasons for color creation, there follows a brief summary of the properties of individual colorants in paints and coatings.

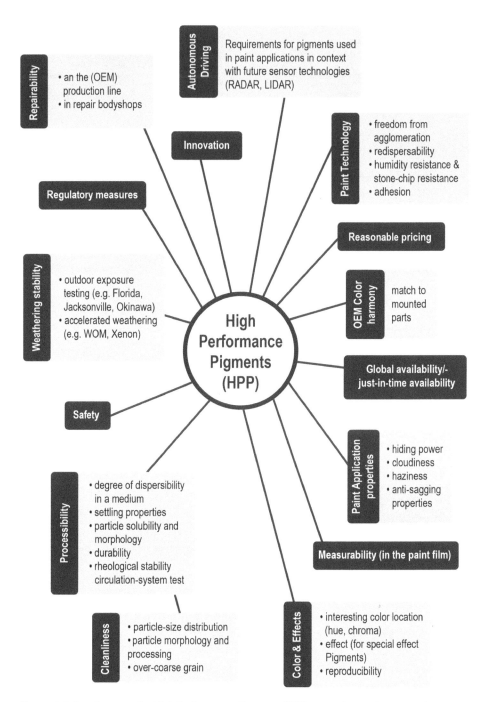

**Figure 17.5:** Requirements for High Performance Pigments (HPP) as used, e. g., in automotive coatings.

### 17.1.1 Paints and coatings – a brief history

Mankind has been using paint since 16,000 BC. The first concrete evidence of this are prehistoric paintings discovered in caves as far apart as Southern France, Spain, and South Africa. The artists of the time mixed pigments such as ocher, iron oxide, and chalk with animal fat [1, 64, 65] to produce paints which have stood the test of time. Ten thousand years later, artists in North Africa were painting on rocks. In the two millennia before Christ, painting developed in the Middle East and, in China, furniture and utensils were being covered with a layer of paint in an artistic design. First evidence of traditional activity in China dates from much later, around 200 BC. There, instead of animal fat, they used resins from the lacquer tree (*Rhus vernicifera*) and much natural black and red pigment. The Chinese even anticipated today's use of metallic powders by employing gold dust or leaf gold.

When someone finally wrote the first formulation for a paint, he chose linseed oil and another natural resin. This was at the beginning of the twelfth century and subsequently little changed in paint technology and raw materials for 800 years. It was the demand for ever faster and automated production methods that called for quick-drying paints at the start of the twentieth century. It took the technology of producing gun cotton in World War I to give the paint industry its first, fast-drying resin: nitrocellulose. Since then, a whole gamut of synthetic resins such as alkyds and acrylics has transformed the way paints can be used in industry and make it possible to customize the paint to the application method and area of use. The entire industry is dependent on the petrochemical industry, but there is increasing pressure to use more sustainable raw materials fueled by consumer demand.

For these new binders to dry quickly enough fast-drying organic solvents replaced the natural solvents. They also enabled spray application and other technologies that accelerated the application of paint on the assembly line. Beginning in the 1920s, the production of such solvents increased worldwide. Synthetic colorants also entered the market in the nineteenth century to facilitate the industrial production of pigments and dyes. As a result, almost everything became paintable. Everyday products became colorful and differentiation became possible. Until the beginning of the industrial age, the objects to be painted were naturally occurring or made of natural materials such as leather, wool, cotton, or linen, which are polymers containing sugars or peptides. This meant that the choice of dyes that could be used on them was quite limited, for example, alizarin or indigo. The goal was to replace these materials with synthetic versions. After much work, entirely new chromogens were developed, different from any found in nature, including azines, triarylmethanes, and others from arylamine oxidation. Azo dyes were prepared from the diazo reaction, and later azo-metal complexes and phthalocyanines. This led to new chromogens with excellent performance and novel pigments [6].

At the same time, the range of substrates had to be expanded. New tools were needed and created, and synthetic fibers as well as resins and new products such as plastomers and elastomers emerged. They were noticeably different from their natural counterparts [7].

As the twentieth century progressed, the need for new colorants and ways of applying coatings increased. Fibers made from esters required disperse dyes, modifications of basic dyes became necessary for acrylic polymers, and new pigments were required for mass coloration of plastics and certain fibers. Polymer chemists met the challenge and, as a spin off from their work, natural fibers benefited with developments in the field of reactive dyes for cellulosic and protein fibers, and fluorescent brighteners for use on undyed textiles and paper. Even detergents profited from this tremendous flow of developments [7].

The emergence of metal effect pigments in the 1920s meant that a completely new group of colorants had gained importance, particularly in the automotive industry [24]. The pigment particles are platelets which produce a color change from light to dark depending on the angle of observation. Their anisotropy and high reflectivity have given them the title of "miniature mirrors." Their shape and the potential range of particle size results in a, at times, coarse appearance and texture.

By the late 1960s, concerns grew about environmental pollution. The quest to reduce the level of solvents in paint led to a great deal of development work. The result was the creation of waterborne and low-solvent coatings and even solvent-free, so-called powder coats and radiation-curing processes where reactive solvents are chemically bound to the paint during the drying process. Although these environmentally friendly systems have gained significant commercial importance, in some coatings it is difficult to replace solvents and maintain an acceptable quality. To combat this, solvent recycling and solvent combustion equipment has been introduced to remove or incinerate the solvents in waste air.

The 1970s mark the beginning of the next chapter in color paint development. At this time, mica-based pigments appeared on the market for the first time. Although they have a similar platelet shape to the metal effect substances already mentioned, they differ significantly in their chemical and physical properties. The effects they produce are used to create memorable and distinctive appearances for industrial products. Detailed information on the different types of substrate-based effect pigments can be found later in this chapter.

At the turn of the twentieth century, there were also major developments in paint application. Until then, all paints were applied manually with a brush – even in industry. Today, brushes tend to be used only in DIY (Do It Yourself) and craft applications. Instead, the industry has adopted automated and mechanized methods of paint application, resulting in increased quality, efficiency, and reduced material waste and labor costs. Application methods include high-pressure or electrostatic spraying, environmentally friendly dipping, and electrophoretic processes and even roller application.

Late in the twentieth century, it became rare for new or unknown substrate types to appear that required specific treatment. Thus, there was a fall in the need for radically new colorants or auxiliary processes or products. The ever-rising costs of research and safety requirements compared unfavorably with the technical or economic benefits achievable from standard colorants and substrates which were seen as marginal. This has prompted work on dyes and pigments to concentrate on more esoteric solutions [66–68]. These areas are never going to meet the volumes seen in textile dyes, for example.

Many of these areas focus on the way the incident light is treated by the pigment particles. The ability of a dye molecule to absorb light is highly dependent on the electrical vector of the incident light (polarization or absorption). This has become critical in the field of liquid crystal displays [69]. Colorants showing high infrared radiation absorption are widely used for, e. g., solar energy traps or laser absorbers in electro-optical equipment [70].

Many innovative uses, from indicator systems, sensors, imaging technology, monitors, sunscreens to optical data recording have been found and use the chemical or photochemical activity of colorants [2, 71].

## 17.1.2 Composition of coatings

Coatings consist of a variety of components depending on the application method, the desired properties, the substrate to be coated, and ecological and economic constraints.

These components are non-volatiles/volatiles and include organic solvents, water, coalescents, binders, resins, plasticizers, coating additives, dyes, pigments, and extenders. Chemical curing involves the formation of products such as water, alcohols, and aldehydes or their acetals by condensation in some binders and their release into the atmosphere.

In this process, virtually every individual component in the liquid phase or in the cured film fulfills a specific function, with solvents, binders and pigments making up the major proportions and additives the smallest. However, important coating properties such as flow behavior, substrate or pigment wetting, catalytic acceleration of curing are significantly influenced by the latter. Solvents and pigments can also be absent from a coating formulation; this is the case in solvent- or pigment-free coatings. The binder is the backbone of every paint formulation and determines application, drying and curing behavior, adhesion to the substrate, mechanical properties, chemical resistance, and weathering resistance.

Binders and resins are macromolecular products with a high molecular mass, such as cellulose nitrate, polyacrylate and vinyl chloride copolymers, which are suitable for physical film formation. Low molecular weight products represent alkyd, phenolic, polyisocyanates, and epoxy resins. To achieve an intact film, these binders must be chemically cured to produce high molecular weight, cross-linked macromolecules after application to the substrate. Increasing the relative molecular mass of the binder in the polymer film improves properties (elasticity, hardness and impact deformation), but at the same time leads to higher solution viscosity of the binder. A balance has to be found here, since the mechanical film properties essentially determine the benefit of a coating, but at the same time a low viscosity combined with a low solvent content means easier handling.

Low-molecular weight binders are characterized by low solution viscosity and enable the production of low-emission coatings with a high solids content or even solvent-free coatings. The binders consist of a mixture of several reactive components and film formation takes place by chemical drying after coating application. Two- and multi-component systems are a special case, since their chemical curing already takes place at room temperature and the binder components are mixed with each other shortly before or even during application.

Modern binders are based on alkyd or epoxy resins, and the property profile of the coatings produced with them can be controlled particularly effectively depending on the choice of monomers used in their structure. The basis for their molecular design is an understanding of macromolecular structure–activity relationships, which are valid for both aqueous and solvent-based coating formulations.

The glass transition temperature $T_g$ is used to characterize macromolecules or polymers; it represents the inflection point in the cure curve of a coating [72]. Modern resins are designed to have their own $T_g$. In the case of acrylics, high glass transition temperatures in the copolymers are associated with high surface hardness and thus increased scratch resistance, while the adhesion properties, cracking tendency, and impact strength of the coating are affected by the glass transition temperature in exactly the opposite direction [24].

Plasticizers lower the softening and film-forming temperatures of a binder and can improve flow, flexibility, and adhesion properties. They are organic liquids of high viscosity and low volatility and can be considered chemically inert because they do not react with binder components. Prominent examples of this are dicarboxylic acid esters such as dioctyl phthalate. Their use is declining, since modern binders already possess these flexible properties ("internally plasticized").

Fillers serve to increase the solids content and influence the mechanical properties of the paint film. Furthermore, so-called cavity fillers exist which improve the hiding power in certain systems such as in architectural coatings [73]. These raw materials will be discussed again at a later stage.

Paint additives [74–77] are auxiliary products which are added to coatings in small quantities of between 0.01% and 1% in order to improve certain technical properties. These include the prevention of defects, foam bubbles, flocculation, and sedimentation, as well as the improvement of surface slip, flow properties, flame retardancy, and UV stability. An important group of coating additives are so-called humectants [78]. They are used to keep films open after coating application and to prevent skinning during storage of liquid coatings. Wetting agents form one of the largest groups of coating additives. They are surfactants that assist the wetting of pigments by binders and prevent flocculation of pigment particles, resulting in the formation of a uniform, haze-free color and a consistently high gloss of the paint film. This group also includes the dispersants, which ensure good pigment wetting and thus optimum dispersion of the pigments in the paint to prevent sedimentation, especially in the case of high-density pigments. They also ensure that the individual particles in the paint do not combine to form agglomerates. A large number of paint additives sorted by application can be found here [15]. Additives are indispensable [74, 77–79] due to stricter quality and environmental requirements in the manufacture and application of coatings. For their successful use, understanding their mode of action or the physical and chemical principles of paint manufacture and application is very helpful [56, 80–82].

Wetting and dispersing additives play a crucial role in paint formulation [4, 83, 84], which is why they will be discussed in more detail here. Many pigments are sold as solids, and the ease of dispersibility in the application medium depends largely on the size of the particles in these solids. Pigment synthesis increasingly involves the production of large agglomerates, which are subsequently broken down by grinding or comminution by processes such as dry or wet grinding in rollers or impact mills, or chemically by precipitation or thermally. Alternatively, "flush" pastes are produced by transferring pigments at the end of the manufacturing process directly from an aqueous medium into a non-aqueous medium containing the binders, etc., which are then mixed directly into the application media without drying or further abrasion.

The chemical modification of pigment surfaces represents an important tool for the successful processing and use of pigments in coatings, which can only be briefly discussed here. This know-how is largely guarded as a trade secret by the pigment industry [85], which often leads to empirical mixing and a quasi-science on the part of paint manufacturers, which continues to be practiced in the paint industry in the twenty-first century. The following is a brief description of the contents of current scientific work in this field.

In [86], the surface modification of Pigment Red PR 170 to improve thermal stability and solvent resistance using hydrous alumina by hydrolysis of $Al_2(SO_4)_3$ was investigated. The effect of pH, temperature, and its content on the structures of hydrous alumina coating was investigated by transmission electron microscopy (TEM), zeta potential analysis, and various spectroscopic techniques.

In [87], a novel coating process of hydrous alumina on organic pigment particles by direct precipitation in aqueous solution is presented. The aqueous suspensions of the organic pigment particles were prepared using cetyltrimethylammonium bromide and sodium dodecylbenzene sulfonate as additives before coating. The organic pigment particles were then coated with hydrous alumina using $Al_2(SO_4)_3$ as precursor. The morphology and surface state of the as-coated organic pigment particles were analyzed by high-resolution TEM and zeta potential.

In [88], a chemical treatment of the pigment titanium dioxide ($TiO_2$) with silicone is reported. Coatings were formulated by incorporating this functionalized $TiO_2$ into an epoxy polymer matrix and compared with untreated $TiO_2$ in terms of coating properties. The influence of the functionalized $TiO_2$ on various coating properties in terms of physical–mechanical properties, corrosion protection effect, UV resistance, and chemical resistance was studied in detail.

During paint production, the pigment particles must be finely and homogeneously distributed in the liquid phase. After wetting of the pigment agglomerates with the binder solution, which is described in detail in [89], pigment chemistry, solvent and binder are affected.

When discussing the particle size of pigments, it is useful to also mention the influence of the pigment surface, which can be particularly important when discussing "wetting" or oil absorption. Also, for surface functionalized materials and general paint systems, especially colloidal dispersions, the zeta potential is the electrical potential at the slip plane and can be used to predict the long-term stability of a liquid paint system. Optimization of the zeta potential can be brought about by controlling the pH value [90].

Flocculation is caused by the action of van der Waals forces on particles [90], although these forces act only over very short distances. However, the pigment particles suspended in the liquid are also subject to Brownian motion, which leads to collisions and thus flocculation. Therefore, to prevent this problem, an energy barrier is created that separates the particles with placeholders and succeeds, however, only if the barrier is higher than the thermal energy (kT) of the particles.

Dispersing additives act as stabilizing substances that are adsorbed onto the pigment surface with the help of pigment affinity groups (anchor or adhesive groups that provide strong, permanent adsorption onto the pigment surface) and then cause repulsion between the individual pigment particles [91]. Pigment particles are stabilized by electrostatic charge repulsion or steric hindrance due to molecular structures projecting from the pigment surface into the binder solution (Figure 17.6) [92]. Electrostatics are used, for example, in the case of aqueous emulsion systems using polyelectrolytes to generate charge on the pigment surface, and steric hindrance in solventborne coatings.

Due to their non-polar surface, organic pigments cause problems in stabilization, with polymeric wetting and dispersing additives being superior to the low molecular weight structures. The reason for this is their macromolecular structure and the large number of pigment affinity groups, which are better able to stabilize these pigments (Figure 17.6, Figure 17.7). In the case of fine, uniform pigment dispersions, however, it can be seen that very closely distributed submicron and, above all, nanodispersions are much easier to disperse and can remain stable over longer periods if dispersants with low molecular weights and high equivalent weights of the functional groups are used. The type, number, and physical arrangement of the polymers or oligomers are

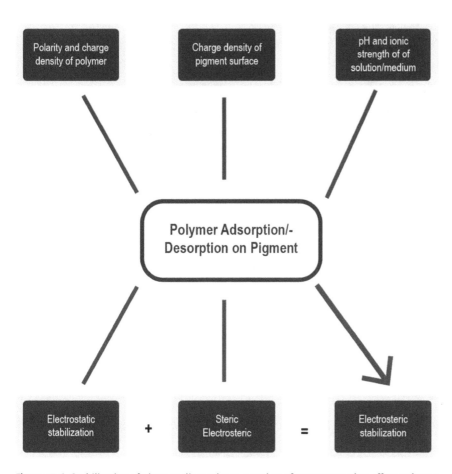

**Figure 17.6:** Stabilization of pigment dispersions: overview of parameters that affect polymer adsorption/desorption on a pigment surface according to [93–95].

crucial influencing factors to obtain stable as well as cost-effective dispersions. Further information can be found in the section on nanopigments.

In connection with wetting and dispersing additives, floating will be briefly discussed here, referring to the random instead of the desired uniform pigment distribution in the coating surface (Bénard cells and streaking). Coatings are often formulated with multiple pigments and can segregate in the paint film as it dries. Floating, on the other hand, produces a homogeneously colored surface, but there is a difference in pigment concentration perpendicular to the surface, as evidenced by the rub-out test. If, after drying of the coating, there is a clear difference between the rubbed and untouched surface, this is an indication that floating has occurred.

Floating is the result of local turbulence in the paint film during the drying process due to different mobility of the pigments. Pigment density, size, and the relative strength of their interaction with the binder molecules influence pigment

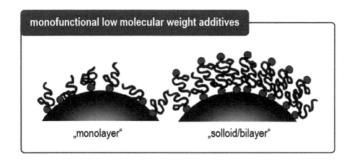

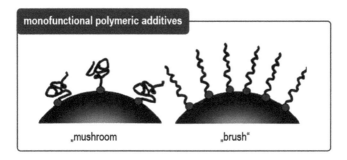

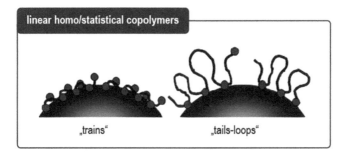

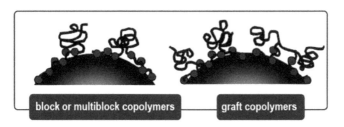

**Figure 17.7:** Visualization of possible additive conformation for low-molecular-mass and polymeric-dispersing additives of a pigment surface after [98].

mobility. Additives can compensate for the different mobility behavior of different pigments by regulating pigment–binder interactions to neutralize flooding and floating.

Inhibition of pigment mixing through co-flocculation by controlled flocculating additives is another approach. Bridging between pigment particles triggers flocculation, controlled by the size and stability of the additive. Since this also changes the rheology of the coating system, wetting and dispersing additives with controlled flocculation properties are often used synergistically with other rheological additives and counteract pigment sedimentation, for example. If an additive is directly on the pigment surface, it is able to prevent sedimentation by preventing the formation of hard, compact sediment and causing the formation of soft sediment that can be stirred up again [96, 97].

The formation of a firmly adhering film on the substrate takes place in the course of the drying process of the applied paint, whereby the film properties are determined by the substrate, its pretreatment (cleaning, degreasing, etc.) and paint composition as well as its application method.

Physical drying is defined as the evaporation of organic solvents from solvent-based paints or of water in the case of water-based paints, and of polymer melts in the case of powder paints. Chemical drying is referred to when the reaction of low molecular weight products with other low or medium molecular weight binder components results in macromolecules by polymerization or crosslinking. Chemical drying is carried out by polymerization, polyaddition, or polycondensation and is used for coatings whose binders react during the drying process to form crosslinked macromolecules. Due to the comparatively low molecular mass of the binders, their solutions can have a high solids content and low viscosity. In certain cases, this allows the production of solvent-free liquid coatings. Under higher temperatures, film formation can be forced [98].

Polymerization enables the crosslinking of reactive ingredients to form a binder, such as unsaturated polyesters with styrene or acrylate monomers. In this form of crosslinking, one component acts as a reactive solvent to the other, resulting in a low-emission coating system. Such cold or radiation curing can be performed at room temperature.

Polyaddition curing produces crosslinked macromolecules by reacting reactive polymers such as alkyd resins, saturated polyesters, or polyacrylates with polyisocyanates or epoxy resins. These are two-component coatings that already react at room temperature, so the binder components are mixed just before application. Systems of this type, so-called stoving coatings, can be kept stable at room temperature, whereby one of the polyaddition binder components (e. g., polyisocyanate) is chemically blocked in this case, and the activation of the crosslinking reaction takes place by the addition of heat (see also the section on powder coatings).

Polycondensation requires the use of catalysts or elevated temperatures, and water, low molecular weight alcohols, aldehydes, acetals, and other volatile compounds are released during this crosslinking process.

Solvent-based coatings that are dried by physical means have a low solids content because the molecular mass of the binder is relatively high. Dispersing the binder in water (dispersions, emulsions) or in organic solvents (non-aqueous dispersion systems) results in a higher solids content. The coating films formed by physical drying have poor solvent resistance properties, which also applies to powder coatings based on thermoplastic binders.

In practice, it is common to dry coatings using more than one method. For example, physical and chemical drying can also occur simultaneously, depending on the composition of the binder system, and the individual steps of chemical drying can occur simultaneously or back-to-back, depending on the binder. In either case, a thorough understanding of the binder system is critical to selecting correct drying conditions to achieve a functional coating.

To fully protect a substrate and produce the desired performance requirements (corrosion protection, weathering resistance, etc.), multiple coats are often necessary, in part because dry paint films are not always non-porous. Therefore, successive coats of different compositions are often applied, which together provide the desired properties. The first layer, usually a primer, provides good adhesion and corrosion protection followed by a topcoat, which provides the desired color, gloss, and weathering stability.

A prerequisite for such multiple coatings in the case of waterborne and powder coatings is the ability to produce defect-free film. In the case of high-performance applications, such as in the automotive industry, additional coatings are applied between the topcoat and the primer to ensure adhesion between the primer and the topcoat. Furthermore, they conceal unevenness of the substrate and thus indirectly contribute to good flow of the topcoat and a high gloss level without unevenness.

### 17.1.3 The origin of color

The various types of physical or chemical color formation and their application in a wide variety of technologies have been described in detail elsewhere [2, 10], which is why only the basic mechanisms and their terminology will be briefly discussed here (Figure 17.8). The majority of these mechanisms are physical in nature, while ligand field effects represent a borderline between chemistry and physics, and color formation in the case of molecular orbitals is purely chemical. The books of Bamfield [2] and Tilley [11] can be recommended here, which describe in detail the relationship between light and the optical properties of materials.

Classic, light-absorbing pigments and dyes are still predominantly used for coloring objects of all kinds, including coatings, printing inks, and plastics [2]. Historically,

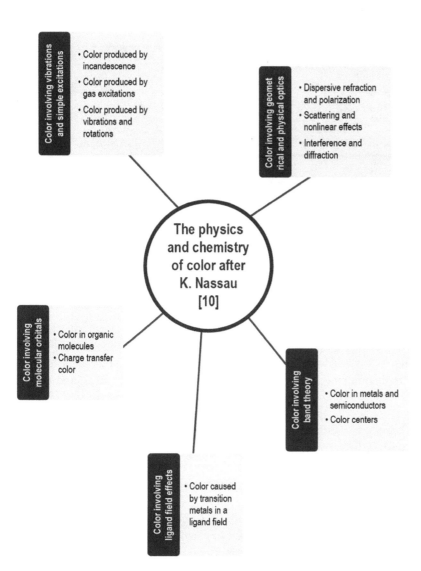

Color involving vibrations and simple excitations
• Color produced by incandescence
• Color produced by gas excitations
• Color produced by vibrations and rotations

Color involving geometrical and physical optics
• Dispersive refraction and polarization
• Scattering and nonlinear effects
• Interference and diffraction

The physics and chemistry of color after K. Nassau [10]

Color involving molecular orbitals
• Color in organic molecules
• Charge transfer color

Color involving band theory
• Color in metals and semiconductors
• Color centers

Color involving ligand field effects
• Color caused by transition metals in a ligand field

**Figure 17.8:** The physics and chemistry of color after Nassau [10].

the focus was initially on textiles, and the application then expanded to include coatings, printing inks, cosmetics, and food applications.

Modern synthetic effect pigments, which originated at the beginning of the last century [90], now include a variety of different types [99]. They are used in coatings to enhance the attractiveness of the coated object through visually appealing effects such as pearlescence, glitter or, depending on the viewing angle, the oscillation of the (interference) colors they produce. Figure 17.9 provides an overview of effect pigments.

**Effect Pigments**

- Pearlescent pigments coated with organic pigments
- Cobalt titanate coated pearlescent pigments
- Liquid crystal pigments
- Graphite
- Molybdenum disulfide
- Iron oxide coated aluminum
- Flake iron oxide (P.I.O.)
- Silicon dioxide coated with $TiO_2$
- Silicon dioxide coated with $Fe_2O_3$
- Alumina coated with $TiO_2$
- Alumina coated with $Fe_2O_3$
- Optically variable pigments (OVP) / goniochromatic pigments
- Holographic pigments

- Conventional aluminum bronzes (Cornflake)
- Elliptical aluminum bronzes (Silver Dollar)
- PVD pigments (VMP/VMF*)
- Aluminum bronzes coated with organic pigments
- Pearlescent silver pigments ($TiO_2$)
- Pearlescent interference pigments ($TiO_2$)
- Pearlescent silver-grey pigments ($FeTiO_3$)
- Iron oxide coated pearlescent pigments ($Fe_2O_3$)
- Combination pigments gold ($TiO_2$ / $Fe_2O_3$)
- Combination pigments green ($TiO_2$ / $Cr_2O_3$)
- Aluminum bronze multiple coated
- Iron oxide in platelet morphology multiple coated
- Metal interference pigments based on UTP-technology

*Remark: this list shows important representatives - for more information see references 16, 90 and 97.*

**VMP/VMF = Vacuum metallized pigments or flakes*

**Figure 17.9:** Overview of effect pigments.

Pigments which are sometimes also used in specialty coatings are based on color change phenomena such as thermochromism, electrochromism, and photochromism [2, 71].

## 17.2 Colorants in coatings

### 17.2.1 Introduction

It is important to distinguish clearly between dyes, pigments, and colorants. Sometimes such terms are incorrectly used in various major scientific terminologies, as though they were synonymous [6, 100].

All dyes and pigments are colorants: when present on a substrate they selectively modify the reflection or transmission of incident light. During application to a substrate, a dye either dissolves or passes through a state in which its crystal structure is destroyed. It is retained in the substrate by adsorption, solvation, or by ionic, coordinate, or covalent bonding. A pigment, on the other hand, is insoluble in and unaffected by the material in which it is incorporated. These inherent characteristics mean that dyes and pigments have quite different toxicological and environmental profiles [6, 101].

## 17.2.2 Definitions

There are fundamental differences between dyes and pigments which are best explained by the definitions below [102]. These definitions have been agreed internationally through various organizations.

ETAD (Ecological and Toxicological Association of Dyes and Organic Pigment Manufactures) definition of dyes [103].

Dyes are intensely colored or fluorescent organic substances only, which impart color to a substrate by selective absorption of light. They are soluble and/or go through an application process which, at least temporarily, destroys any crystal structure by absorption, solution, and mechanical retention, or by ionic or covalent chemical bonds.

CPMA (Color Pigment Manufacturers Association) definition of pigments [104].

Pigments are colored, black, white or fluorescent particulate organic or inorganic solids which usually are insoluble in, and essentially physically and chemically unaffected by, the vehicle or substrate in which they are incorporated. They alter appearance by selective absorption and/or by scattering of light. Pigments are usually dispersed in vehicles or substrates for application, as for instance in the manufacture or inks, paints, plastics, or other polymeric materials. Pigments retain a crystal or particulate structure throughout the coloration process.

## 17.2.3 Pigments and dyes

As can be seen from this definition, pigments are insoluble in the application medium and characterized by fine particles of small particle size. They are incorporated into binders to provide decorative effects (e. g., color, gloss, matte effects) or functional effects (e. g., corrosion protection) in coatings. In order to achieve an optimum expression of the desired effects, the pigments must be processed in the respective binder in a suitable manner and in appropriate quantities, whereby the distribution of the primary pigment particles in the polymer matrix of the binder should be as homogeneous as possible.

In contrast to pigments, dyes that are soluble in the application medium are used much less frequently in coatings, e. g., for wood coatings.

The particle size of pigments influences important properties, which include color strength, opacity, weather resistance, or chemical resistance as well as other performance characteristics. Typical particle sizes are in the range of 0.1–2.0 µm, with values of 5–150 µm (Ferret diameter) common for flake effect pigments. The size of the particles typically has a bell-shaped or oblique distribution. As discussed earlier, wetting of the particles can be a challenge. If this is not ensured during processing, it can negatively affect important properties of the coating, including gloss or weathering properties. Therefore, particle size reduction, e. g., by grinding, is of great importance. The pigment manufacturer has to ensure the reproducibility of the particle size distribution of the pigment by suitable production steps. The production process takes place in different media, with precipitation from water being a typical method of achieving the desired particle size in a precipitation process. For a large number of pigment types, the precipitation step is followed by surface treatment [85, 86]. Here, the precipitated pigment is first filtered and the filter cake is then dried. The drying process leads to aggregate formation of the pigment particles, which is why the dry powder of the pigment is first processed by paint manufacturers in a dispersion process to break up the aggregates and produce a stable dispersion.

The concentration of pigments in paint films is described by the PVC. This is the volume ratio of pigments and fillers to the total volume of non-volatile components. Each coating system has a critical PVC (CPVC). At this value, the binder just fills the free space between the densely packed pigment particles. At higher pigment concentrations, the pigment particles in the paint film are no longer completely wetted by the binder, which leads to a significant deterioration in the properties of the paint film. These include properties such as gloss, stability, strength, and corrosion resistance.

Pigments and dyes that are particularly important for the use in coatings are presented in more detail in the following sections.

## 17.2.4 Inorganic pigments

The history of paint technology is closely related to the development of plastics in the twentieth century, which was characterized by high innovation and the development of a variety of different types. Linseed oil was a widely used paint and pigments were mainly inorganic [6, 105–109]. Innovations in the field of plastics led to the painting of the material by direct coloring using pigmentation. It is important to mention here that the additional pigment consumption resulting from this was slightly compensated by the savings in the production method (one step) and the color uniformity across batches [6], which led to the increase in demand for pigments. One problem here was that many organic pigments did not meet the requirements for fastness and temperature stability, nor did they interfere with the molding process of the finished products. For this reason, inorganic pigments or blends with organic pigments were used, even though these posed problems in terms of dispersion. In the case of inorganic pigments, mechanical methods are used to achieve the desired particle size distribution. These include grinding of the particles in ball mills, hammer mills and jet mills, or their targeted production by spray drying. Detailed explanations of these processes can be found in the relevant literature [15].

Particle sizes of the inorganic pigments used in (top) coatings are typically 0.05–10 µm, in the case of transparent pigments 0.01–0.05 µm. Typical PVCs in gloss paints are between 10% and 35%, in matte dispersion paints also over 80% [6].

Two prominent representatives of inorganic pigments are white opaque $TiO_2$ and carbon black [106].

Inorganic pigments consisting of cadmium derivatives and lead chromates ($PbCrO_4$) entail toxicological risks, which is why their use is regulated by law in many countries.

When it comes to improvement in terms of coloristic properties and toxicological alternatives along with insights on advanced inorganic color pigment design through materials chemistry, the subjects were presented here [110–114].

A detailed overview of options with colored complex inorganic pigments when faced with elemental restrictions is available in [115].

Today, colored inorganic pigments are generally produced to offer special advantages that organic pigments cannot, for example, in high temperature applications like coatings for grills, wood stoves, car exhaust systems or engine parts or for the use in (high temperature) plastic applications.

### 17.2.5 Organic pigments and dyes

In coatings, so-called solvent dyes [6, 7] are used, which are characterized by their solubility in organic solvents, but are insoluble in water. Alcohols, ethers, esters, ketones, chlorinated solvents, hydrocarbons, oils, fats, and waxes are relevant in practice. In applications, this specific solubility is exploited [116] to achieve the desired properties. Further information including chemical structures can be found in [117].

Their application is the coloring of transparent coatings and so-called glazes, as often found in the furniture and leather industries. In this case, the surface of the manufactured product is protected by the coating, but not optically obscured.

In contrast to solvent-based dyes, organic pigments retain their particle shape in the application medium due to their insolubility, which is a parameter for their performance in coatings. After the dispersion process described above, these lead to the desired properties in the coating, which include color and/or light scattering.

The low solubility leads to low bioavailability and low toxicity [101], the relevant determination is described in [118].

To produce yellow, orange, brown, or red hues, monoazo and disazo compounds play an important role chemically.

Red, violet, and blue hues are achieved by anthraquinonoid pigments, and their high lightfastness should be emphasized at this point.

Due to their industrial relevance, other chromogens should be mentioned here, which include so-called isoindolinones (yellow–orange), quinacridones (red–violet), and phthalocyanines (blue–green).

Many organic pigments are suitable for use in coatings [1], although it is important to achieve a comprehensive property profile to achieve a functional coating. In addition to the properties already mentioned, these include their rheological properties, which have an effect on dispersion or the coating itself. Ultimately, the rheology determines how much pigment can be dispersed per batch, whereby a favorable rheology (i. e., low viscosity) allows higher pigment and binder concentrations at a given application viscosity.

As already explained by Kaluza [56], the pigment–binder–solvent interaction as shown in Figure 17.2 is complex and influences a wide range of paint properties, which also includes the surface gloss of the paint film.

When selecting a pigment for a coating, the main factors that can be mentioned are the desired final color shade, the binder used, and the resistance of the pigment.

Paint manufacturers choose not to disperse pigments themselves, but to use so-called pigment preparations for various reasons, including difficult dispersibility of the pigments or economic reasons. These preparations contain resins or dispersants, and their compatibility with the other formulation components is a prerequisite for successful use by the paint manufacturer. These preparations exist for both solventborne and waterborne systems.

At this point, so-called "high-solids" coatings should be mentioned, as they have a low solvent content and are therefore considered to be more environmentally friendly. At the same time, however, this means an increase in the solid content, which is why the rheological properties are of greater importance. Due to the effect of pigment properties on rheology, both particle sizes and chemical modification of the pigment surface are used to optimize rheological parameters.

For organic pigments, this plays a role in that their smaller particle size tends to result in higher viscosity compared to inorganic pigments. Wetting and stabilization of the pigment takes place through the dispersion process in the binder and solvent, whereas destabilization, which occurs through reagglomeration, flocculation, and flotation, must be prevented during the addition of the other coating components.

It goes without saying that the dispersion of pigments of an organic nature in an aqueous phase is a challenge due to hydrophobicity, which is why surface-active substances are used. These reduce the surface tension of the water and lead to improved wetting properties. Alternatively, this can be achieved by surface treatment of the organic particles. The high dielectric constant of water enables both steric and electrostatic stabilization of the pigment.

A large number of pigments used in solventborne coatings are also suitable for aqueous systems, although the portfolio is reduced due to limited alkali resistance. Examples include $PbCrO_4$s and molybdates, manganese colors (PR 48 and PR 52), some perylenes (PR 223), benzimidazolone yellow (PY 151), and isoindolines (PY 139).

A well-known phenomenon in waterborne coatings is the coagulation of the binder or dissolved resins, which is why pigments with a high content of water-soluble components including electrolytes precipitate. Due to this property, they lead to reduced storage stability or loss of adhesion or blistering of the coating when exposed to moisture and water. This applies to both organic and inorganic pigments.

## 17.2.6 Fillers

Fillers, also known as inert pigments, inerts, or extenders [5], are a class of raw materials in their own right and occur in a variety of forms [119–121]. Reference [73] is recommended to the interested reader as it covers their application in coatings in detail.

They are made from low-cost chemicals such as barium sulfate, calcium carbonate ($CaCO_3$), or kaolin. This makes them interesting as a partial replacement of other, more expensive, high-opacity pigments. However, cost is not the only criterion. These materials may be used to optimize properties such a film thickness or structure to give a film more bulk. Many properties are dependent on the volume of a pigment. Furthermore, extenders may be employed to adjust the rheology of a wet paint film or improve mechanical properties and gloss in the dry state. Another aspect of the presence of extenders is the degree of water uptake and permeability to gases such as carbon dioxide, sulfur dioxide, and nitrogen oxides. The list of natural minerals used for extenders is long and includes calcite, chalk, dolomite, kaolin, quartz, talc, mica, diatomaceous earth, and baryte. These are crushed, ground, elutriated, and finally dried. Another alternative is synthetic inorganic extenders. These are made by the digestion, precipitation, or annealing of inorganic products. $CaCO_3$, baryte, or blanc fixe, and silicates are the major synthetic extenders. Synthetic organic extenders are produced as polymer fibers, but the development of hollow spheres has allowed the creation of coatings with very low density [73].

$CaCO_3$ is widely used, especially as ground limestone or mixed calcium-magnesium carbonate ore (dolomite). However, the application of these raw materials is limited with their reactivation with acids, such as for exterior paints, as the coatings are susceptible to acid rain. Furthermore, they are not used in exterior latex paints due to the penetration of water and carbon dioxide, as some of the $CaCO_3$ reacts to form the water-soluble calcium bicarbonate, which escapes the coating. On the surface of the coating, the water evaporates and after reversal of the reaction, a frost of insoluble $CaCO_3$ is formed.

A wide range of clays or aluminum silicates are also used as fillers. They are available in various particle size ranges, the whiter ones being more expensive. Bentonite and attapulgite clays help modify the viscosity of coatings.

Mica (aluminum potassium silicate) has a platelet-like structure and reduces oxygen and water vapor permeability, e. g., in anti-corrosion coatings.

Talc or magnesium silicate minerals of different crystal structures are used to influence the film strength of coatings in different ways. Some exhibit anisotropic, flaky morphology and, as in the case of aluminosilicates, influence vapor permeability; modifications with fibrous morphology are very useful for strengthening the coating film.

Fillers based on silica represent an important class, with natural silica available in various particle sizes. Fine, synthetic silicon dioxides ($SiO_2$) are used for gloss (clear coats) or rheology control (shear thinning) in coatings.

Barite (barium sulfate) has proved particularly useful as fillers in the formulation of primers for automotive coatings, as its use has a positive effect on the hardness of the coating.

Organic materials are also used as fillers in coatings, with polypropylene, polystyrene for latex paints and synthetic fibers from aramid to increase mechanical strength being mentioned here as representative examples.

One aspect that makes the formulation of coatings very complex is the fact that pigments and fillers are discontinuous particles with absolute volume. Regardless of the effect they are to have on the paint, this and the interaction of the particles in the paint film must be taken into account. This is especially true for highly filled systems, where the effects of PVC and CPVC can dominate [73].

### 17.2.7 White pigments

A large proportion of coatings contain white pigments [1, 18, 122, 123]. They are used not only in white coatings, but also in a significant proportion of other pigmented coatings to produce lighter colors than would not be possible with pure colored pigments. In addition, many colored pigments produce transparent films, so white pigment is used to provide much of the hiding power. The ideal white pigment should not absorb visible light and should have a high scattering coefficient. Since an important factor controlling the scattering effect is the difference in refractive index between the pigment and the binder, this is a critical property of a white pigment.

#### 17.2.7.1 Titanium dioxide pigments

$TiO_2$ is the most frequently used white pigment in coatings [18, 122–127], whereby a distinction must be made between two crystal forms which differ in their physical properties, rutile and anatase.

A coating pigmented with rutile gives it a higher hiding power due to the higher refractive index (rutile: 2.70, anatase 2.55). In contrast to anatase, the crystal modification rutile absorbs violet light, which is why a coating pigmented with this modification appears white in contrast to anatase, which is subject to a yellow cast. Organic pigments used to correct the "white hue" of $TiO_2$s are carbazole violet or phthalocyanine. The most common pigment for tinting is carbon black, which absorbs light over the entire wavelength range, is less expensive than violet or blue pigments and achieves a higher hiding power. So-called "transparent $TiO_2$" (i. e., $TiO_2$ in very fine particle size) is used, among other things, in automotive metallic shades to achieve matte effects [18].

The crystal modification anatase behaves as advantageous in white UV-curing coatings due to its physical properties, since the lower absorption of UV radiation implies a lower interference by absorption of UV radiation by photoinitiators.

Furthermore, both crystal modifications differ in their photoreactivity. Anatase pigmented coatings are subject to faster chalking when used outdoors, which may be undesirable in the case of a highly weather-resistant automotive coating, but may be intentional in the case of self-cleaning facade paints. If erosion of the surface film occurs, accumulated dirt reaches the surface and is removed, giving it its white hue. However, formulation strategies of coatings for exterior use basically aim at preventing the effect of chalking, which is why rutile is usually resorted to. Nevertheless, this modification is sufficiently photoactive to affect durability in outdoor use. To counteract this reactivation, the surface of these particles is treated with additional oxides [18]. Typical strategies include oxides of the main group elements $SiO_2$ and/or aluminum ($Al_2O_3$) and a number of other metal oxides, which include cerium oxide. The aim of this surface treatment is to create a protective shell around the pigment particles, preventing contact between $TiO_2$ and the organic binder. It should be noted at this point that both $TiO_2$ (especially rutile) and components of the binder are highly UV absorbent, and UV absorption competition occurs here [123]. The use of $TiO_2$ grades, which positively influence the chalking, thus also reduces the UV absorption by the binder, which is why the external resistance of the coating is improved in this way.

Other pigment properties can also be regulated by surface treatments, for example, specially treated $TiO_2$ pigments enable the production of more stable dispersions in aqueous coatings. These so-called slurries are used in the production of latex coatings and are supplied by $TiO_2$ producers instead of pigment powders.

Optimization of $TiO_2$ pigments by surface modification to improve the performance of pigmented coating systems is always the subject of research and development activities [88].

In order to select suitable $TiO_2$ types, the formulator of a coating must choose from the large number of types available from $TiO_2$ manufacturers the type that is suitable for him according to his requirements profile. His choice will depend, for example, on whether he needs or prefers properties such as weathering resistance or gloss or dispersibility in aqueous or solventborne systems. In this context, it is clear that the selection of a $TiO_2$ type unsuitable for its field of application will lead to impairment of the coating performance provided and thus to increased costs.

Finally, an important aspect to mention in connection with $TiO_2$ pigments relates to regulatory matters. In 2016, France had submitted a proposal to classify $TiO_2$ as carcinogenic (Cat.1) by inhalation. The relevant scientific hazard assessment body, RAC, disagreed with France's proposal. However, it is recognized that inert dusts, regardless of their chemical composition, may well be problematic by inhalation due to general particle effects. From this, a classification for $TiO_2$ as a suspected carcinogen (Cat. 2) was derived. Based on this, a discussion arose about the replacement of $TiO_2$ by substances such as zinc sulfide (ZnS), lithopone (ZnS/barium sulfate), zinc oxide (ZnO), or $CaCO_3$ other compounds used as white pigment. At this point, it should be mentioned that, according to current knowledge, $TiO_2$ cannot be adequately replaced in many applications due to its unique, technical properties.

### 17.2.7.2 Other white pigments

ZnO, ZnS, lithopone, and basic lead carbonate may be mentioned as representatives for white pigments besides $TiO_2$ [128]. Basic lead carbonate (refractive index 2.0), however, plays only a minor role today due to its toxicity, which is coupled with water solubility. Because of its toxicity, many countries have a legal limit on the lead content in any paint sold to consumers. ZnO with a refractive index of 2.02 cannot compete with $TiO_2$ for hiding power, but is used as a fungicide in masonry paints. Its use in primers should be considered critical due to its water solubility, as osmosis can lead to bubble formation. ZnS with a refractive index of 2.37 and lithopone ($ZnS/BaSO_4$) are important white pigments for coatings applications, but also not comparable with $TiO_2$. Another possibility for achieving a white color impression is the inclusion of scattering air bubbles in particles used in paint films, which can provide additional hiding power at the air–binder interface. Furthermore, latices of high glass transition temperature with water entrapped in particles are used in latex paints as partial substitutes for $TiO_2$ [129].

### 17.2.8 Colored pigments

### 17.2.8.1 General considerations

The variety of colored pigments used in coatings is very wide [109, 119, 130, 131], so guidelines for their selection and use for specific coating applications are important and can be found elsewhere [4, 5, 19, 27]. These include criteria such as color and tinting strength, opacity or transparency, ease of dispersibility, weatherability, heat and chemical resistance, water solubility, moisture content, toxic and environmental hazards, IR reflectance, and cost.

This list explains why formulation strategies can be very complicated, as the industry may have different views and goals when formulating coatings. For example, as far as opacity or transparency is concerned, this depends on the end application. It may be desirable to use a pigment that increases opacity by both scattering and absorbing radiation, or it may be important to choose pigments that scatter little or no light in the paint film so that a transparent color can be achieved.

Also, sometimes the same pigment chemistry is found in coatings for different end uses, which in turn leads to a variety of pigment variants. For example, consider a phthalocyanine blue pigment used simultaneously in architectural, industrial, or automotive paint formulations. It is clear that particle size, surface treatment, stabilization, and light fastness will always be different for various applications due to the opacity or transparency of the formulation.

In general, pigments should be easy to disperse, therefore also a large number of colored pigments are surface treated by manufacturers to improve their dispersibility as already mentioned for $TiO_2$ pigments. Since the dispersion of pigments is a complex matter, this also explains why a wide variety of surface treatments are used. For more information, references [5, 85, 93–95, 132] can be recommended. In addition to improving dispersibility, surface treatments are also developed with the intention of improving light and weather fastness [63], as some pigments are more sensitive to photodegradation, resulting in color loss or hue changes.

### 17.2.8.2 Yellow and orange pigments

#### 17.2.8.2.1 Inorganic yellow and orange pigments
Iron oxide yellow ($\alpha$-FeOOH) is a brownish-yellow pigment that allows the formulation of opaque coatings with high weathering resistance and has excellent chemical and solvent resistance and dispersibility. Temperature stability is limited, as above 150°C they gradually change color to red due to dehydration to iron oxide red ($\alpha$-$Fe_2O_3$). The pigments are produced synthetically, and natural ocher pigments are also used. The presence of soluble iron components or metal salts affects coating stability in the case of radical-curing coating systems. As already explained, particle size exerts an influence on properties such as opacity.

Chrome yellow is a light, strongly chromatic yellow pigment, whereby these can be produced from $PbCrO_4$. So-called co-crystals of $PbCrO_4$ and lead sulfate result in greenish-yellow colorants, while co-crystals of $PbCrO_4$ with lead oxide (PbO) result in redder yellows, and this red shift is further enhanced by $PbCrO_4$ with lead molybdate ($PbMoO_4$).

Chrome yellow changes color when used outdoors, but shows an adequate range of properties for many coatings and further shows resistance to bleeding and heat. Regulatory issues have led to restrictions in some countries over the last few years.

The search for adequate substitutes for $PbCrO_4$ pigments is still ongoing. For example, inorganic yellow pigments based on niobium-tin pyrochlore, $Sn_2Nb_2O_7$, have been developed to replace $PbCrO_4$s [133]. These pigments provide a chromatic, bright red-yellow in combination with high temperature, chemical and UV stability and extend the colors for systems with high durability [134].

In [135], a series of inorganic–organic hybrid pigments with a main component of inorganic materials ($SiO_2$, $TiO_2$, and sepiolite) were prepared and used as environmentally friendly substitutes for $PbCrO_4$ pigments.

The preparation of a series of orange Cr- and Sb-containing TiO$_2$ nanopigments has been reported [136]. Although they are still too new to have commercial relevance, their properties seem to be very promising.

Inorganic, highly chromatic orange pigments based on Sn/Zn-doped TiO$_2$ orange pigments are commercially available from many sources.

Greenish-yellow shades are obtained by incorporating antimony and nickel. Reddish-yellow types contain antimony and chromium. Their resistance to external influences, chemicals, heat and solvents is excellent. However, only relatively weak yellows are possible, so the cost is high and the range of colors that can be produced is limited.

Bismuth vanadate, a yellow pigment with high brilliance and good resistance, is available on the market. A recent development uses Bi$_4$Zr$_3$O$_{12}$ as an environmentally friendly inorganic pigment a V-containing inorganic ZrO$_2$ yellow nanopigment prepared by a hydrothermal approach [137, 138].

### 17.2.8.2.2 Organic yellow and orange pigments

A selection of representative chromophores is shown in Figure 17.10, with code names taken here and in the following discussions from the Colour Index (C.I.) system, e. g., PY, PO or PR with the corresponding index numbers [117].

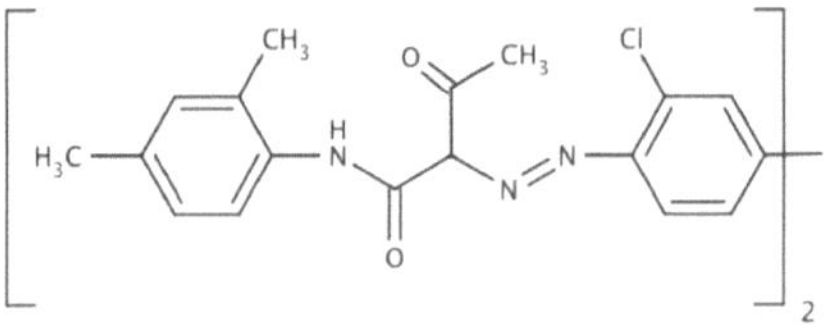

Diarylide yellow PY 13

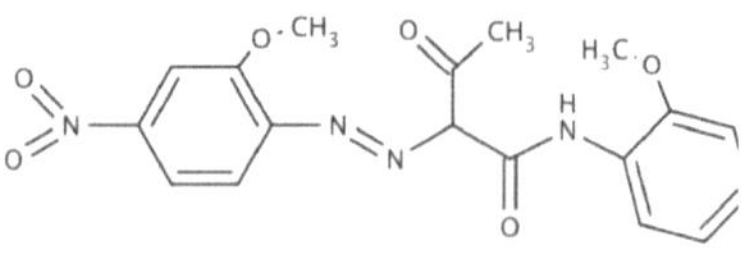

Monoarylide yellow PY 74

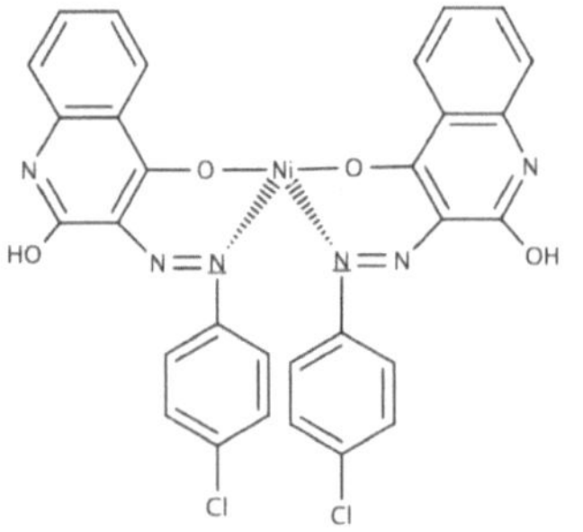

Nickel azo yellow PY 10

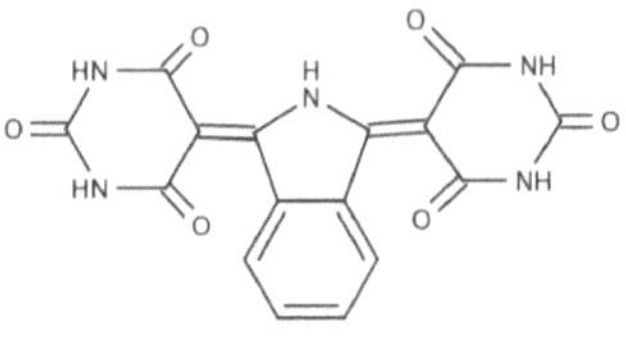

Isoindoline yellow PY 139

**Figure 17.10:** Examples of organic yellow pigments [5, 7].

Figure 17.10 shows organic yellow pigments, which are characterized by both high color strength and chroma. Diarylide yellow pigments such as PY 13 fade due to their photochemical instability, whereas their solvent, heat and chemical resistance can be described as excellent. Monoarylide yellow pigments such as PY 74 also have high color strength, but show a tendency to fade and sublimate at higher service temperatures. Although their lightfastness is better than that of diarylide yellow pigments, the level is lower than that of inorganic yellow pigments. Nevertheless, certain types are used in outdoor applications to displace chrome yellow. A pale, greenish yellow pigment with high weather and heat stability is nickel azo yellow PY 10, which is also used in automotive coatings, as is the vat yellow pigment isoindoline yellow PY 139 [24, 139, 140].

Benzimidazolone orange pigments offer excellent lightfastness and resistance to heat and solvents. They are a substitute for molybdate orange. An example of an orange benzimidazolone pigment is shown in Figure 17.11.

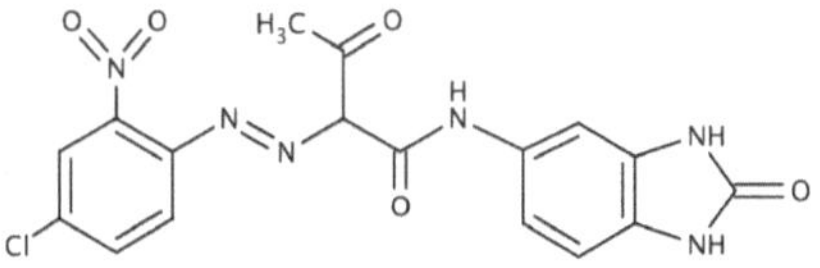

Benzimidazolone orange PO 36

**Figure 17.11:** Structure of benzimidazolone orange PO 36 [5, 7].

The commercial importance of monoazo pigments has been reduced due to alternative developments such as those based on acetoacetanilides with higher fastness properties. Although they cover almost the entire color range from green to yellow or up to deep reddish-yellow or orange shades, their property spectrum, especially fastness, solvent (bleeding), and migration fastness is responsible for this. So-called *P*-naphthol pigments are found in drying coatings, which are located in the yellow, yellow–orange to bluish-red range in terms of color, although they tend to exhibit poor fastness properties in contact with organic solvents. So-called benzimidazolone pigments are classified as high-performance pigments, whereby the yellow/orange pigments based on 5-acetoacetylaminobenzimidazalone and the red and brown pigments made from arylamide should be mentioned at this point. The benzimidazolone group imparts the properties of insolubility and improves solvent, migration, light, and weather fastness, which is why these pigments are used in coatings. Furthermore, so-called polycyclic quinones should be mentioned here, which are used as pigments due to their high fastness properties, especially weather fastness. Ring systems such as flavanthrone, the yellow pigment PY 24, and pyranthrone, whose halogenated derivatives provide very lightfast orange and red tones, are used in coatings. Yellow and red pigments based on metal complexes, mostly $Cu^{2+}Ni^{2+}$ or, more rarely, $Co^{2+}$, are also found in industrial and automotive coatings. One of the most heat-resistant organic pigments is the orange pigment PO 68, which contains a benzimidazolone moiety.

### 17.2.8.3 Red pigments

#### 17.2.8.3.1 Inorganic red pigments

$\alpha$-Iron(III)oxide ($\alpha$-$Fe_2O_3$) is a pale red pigment with an excellent property spectrum, which, in contrast to Fe-based yellow pigments, is characterized by its thermal stability and is also available as an orange–red pigment. Very fine grades are available to achieve transparent coatings and are used, among other things, in automotive shades and, because of their UV-absorbing properties, in wood stains to protect the latter from photooxidation [1, 107].

More recent developments have been concerned with optimizing the process technology, resulting in a new iron oxide pigment with a higher red value [141].

Heavy metal alternatives such as cerium sulfides played an industrial role for a time. Purple of Cassius (colloidal gold) is also still a common red for glass, ceramics, and glass enamels. From the YInMn blue research comes the magenta-colored YIn(CoTi)$O_3$, which can be used as a chromatic and low-cost glass alternative to the purple of Cassius [142].

Recently, the synthesis of a new class of environmentally friendly red pigments based on the tetragonal $\beta$-phase of $Bi_2O_3$ was reported, where doping with zirconia ($Zr^{4+}$) yielded compounds showing the most promising red color coordinates [143].

#### 17.2.8.3.2 Organic red pigments

A light red azo pigment with high tinting strength, which can be found in baking enamels due to its sufficient heat resistance, is toluidine red PR 3, which also has good properties in terms of weather and chemical resistance, but must be tested for its use in solventborne coatings due to its reduced bleed resistance (Figure 17.12). As an example of a bleed-resistant chromatic azo pigment, permanent red PR 48 represents the largest volume of organic red pigments used in coatings in this general class of grade (Figure 17.12). Only their alkali sensitivity limits their use for some latex paints. A large number of red pigments are naphthol-based azo pigments, which are equipped with various ring substituents and thus exhibit increased resistance.

Also worth mentioning are high-priced quinacridone pigments, which are used as fine grades in high-quality automotive coatings, among other applications. By chemical substitution or by variation of the crystal form, shades of orange, maroon, shagreen, magenta, and violet can be achieved [2]. The resistance of their dispersions to flocculation is also achieved with these pigments by specific surface treatment [144]. Other red pigments with high-performance properties can be found here [7, 119, 131]. The quinacridone-based yellow–red to reddish-purple colored pigments exhibit excellent stability and are therefore used in both industrial and automotive

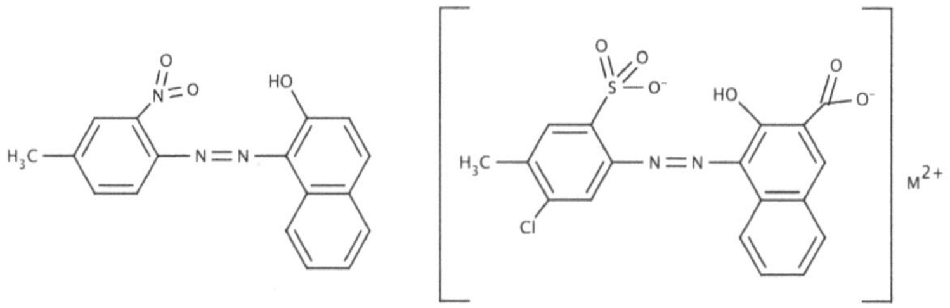

Figure 17.12 diagram labels:

Toluidine red PR 3

Pigment red PR 48 (M = Ba, Ca, Sr, Mn, Mg)

Ba salt: PR 48:1
Ca salt: PR 48:2
Sr salt: PR 48:3
Mn salt: PR 48:4
Mg salt: PR 48:5

**Figure 17.12:** Examples of organic red pigments [5, 7].

coatings as well as in exterior paints. The unsubstituted linear quinacridone exhibits polymorphism, i. e., two crystal forms exist, the reddish-purple beta form and the red gamma form, the pigment PV 19.

High-performance pigments based on isoindolinone are available in colors ranging from yellow to orange to red, and are particularly important in the greenish to reddish-yellow color range. They are used in industrial and automotive coatings, such as the yellow pigment PY 10 [2]. So-called Vat pigments are also used in these applications. These are a class of pigments that also includes perylenes and perinones.

Depending on the type of amide substituent, perylenes provide pigments with high color strength, good lightfastness, weather fastness and solvent fastness in a color spectrum ranging from red through bordeaux to violet, such as the red pigment PR 179 (R = $CH_3$). Perinones are similar to perylenes and are available in orange to burgundy shades with similar properties to perylenes. A significant but not always desirable property of perylenes and perinones is their fluorescent properties. The scarlet pigment PR 168 is one of the most lightfast and weather-resistant pigments known and is used in high-performance coatings. Other pigments worth mentioning are the yellow pigment PY 24, which is based on a flavanthrone ring, and pyranthrone, whose halogenated derivatives give very lightfast orange and red tones that are also used in coatings.

Diketopyrrolopyrrole is undoubtedly the most important chromophore of the last few decades [145], whereby the first commercially introduced red pigment PR 254 should be mentioned here as representative, which quickly established itself in the field of high-performance coatings such as automotive coatings [146].

As already mentioned in the chapter on organic yellow and orange pigments, metal complexes have also been used to produce red pigments with commercially interesting properties for industrial and automotive coatings.

### 17.2.8.4 Blue and green pigments

#### 17.2.8.4.1 Inorganic blue and green pigments

For historical reasons, the synthetically produced iron blue, $FeIII[FeIIFeIII(CN)_6]_3$ x $H_2O$ ($x = 14$–$16$), should be mentioned here, which was displaced in most coating applications by phthalocyanine blue due to higher color strength. Chromium green can be produced from iron blue by co-crystallization with chromium yellow, although the use of these pigments has declined due to their lead content and the availability of phthalocyanine green [2, 106, 107].

When it comes to improvement in terms of coloristic properties and toxicological alternatives, interesting work has been published by Smith et.al. using Mn substituted $YInO_3$ ($YIn_{1-x}Mn_xO_3$) to create an inorganic, blue chromophore with UV absorbing and NIR reflecting properties [110–113]. Industrial analysis has shown drastic improvements in the infrared reflectance of $YIn_{0.8}Mn_{0.2}O_3$ as compared to industrial $CoAl_2O_4$, Figure 17.13, while exhibiting a slightly redder hue and improved UV absorbance. The combination of these attributes and robust weatherability makes for a favorable alternative in exterior applications including automotive and coil coatings.

 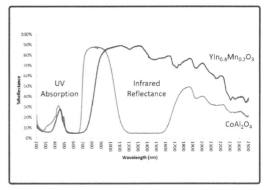

**Figure 17.13:** Spectral properties of $YIn_{0.8}Mn_{0.2}O_3$ and $CoAl_2O_4$ in polyvinylidene fluoride (PVDF)/ acrylic coatings (images provided by The Shepherd Color Company).

The following new developments have been reported recently.

[147] reports a non-toxic intense blue inorganic pigment with near-infrared reflecting properties. The pigment was developed as a viable alternative to existing cobalt-based blue colorants. [148] reports the development of a novel inorganic sky-blue pigment. [149] reports the development of a violet inorganic pigment based on $Mn^{3+}$-doped $LaAlGe_2O_7$.

A green colored nano-pigment with the formula $Y_2BaCuO_5$ was synthesized by a nano-emulsion method. The potential utility of the nano-pigments as "Cool Pigments" was demonstrated by coating on a building roofing material like cement slab and PVC coatings [150].

### 17.2.8.4.2 Organic blue and green pigments

Phthalocyanine pigments, in short phthaloblue or phthalogreen, are the most important representatives of organic blue and green pigments with impressive high-performance properties [151, 152], which are characterized by weathering, chemical and heat resistance, or bleeding combined with high color strength (Figure 17.14). The cyclic macromolecule of this pigment class contains copper as the coordinated central atom, hence the name copper phthalocyanine. These pigments are produced in large quantities, with phthalocyanine blue PB 15 being the most popular blue for use in coatings [153, 154]. Halogenation of copper phthalocyanine produces green shades, which are also the pigment of choice for coatings in this shade range. The three crystal forms of phthaloblue, alpha, beta, and the rarer epsilon, differ in their properties. The beta form has a light greenish-blue hue and is stable, whereas the more reddish alpha form has less stability, which in certain cases can cause serious problems with color and strength changes during storage or baking of coatings. More stable alpha-form pigments are available. These contain various additives that stabilize the crystal form and minimize problems with flocculation of dispersions. Some grades of phthalocyanine blue are slightly chlorinated and are characterized by a green tint in color [7].

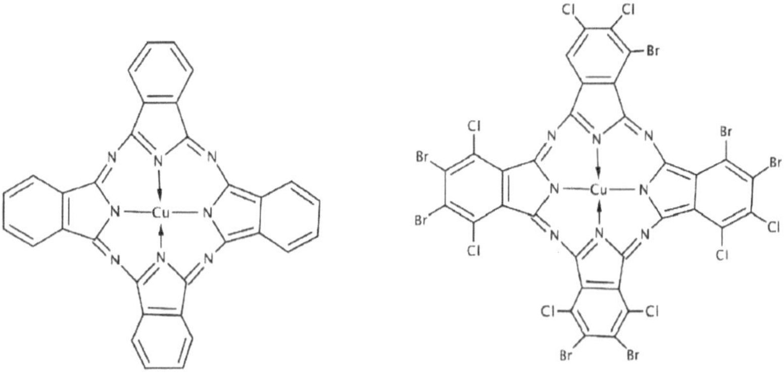

Phthalocyanine bluePB 15              Phthalocyanine green mixed isomers

**Figure 17.14:** Representative phthalocyanine pigments [5, 7].

### 17.2.9 Black pigments

Carbon black pigments are most commonly used in coatings and absorb in both the UV and visible regions of the electromagnetic spectrum [32]. Their manufacturing process (burning and/or cracking of petroleum derivatives) can be quite different and exerts an influence on the particle size distributions achieved and the chemical structure of the particle surfaces, and thus directly on the final intensity produced in the coating as well as the perceived depth of the black hue. Thus, in contrast to so-called furnace blacks, fine-particle high-color channel blacks are used to achieve intense black, glossy coatings. Formulation problems are usually caused by the fact that the particles have a very high surface area in relation to their volume, which can lead to the adsorption of resins and additives, causing undesirable effects, e. g., rheology or catalysis. Property-relevant parameters of these black pigments are particle size of the primary particles as well as their structure, porosity, and activity, and influence the processing (rheology and dispersibility) as well as final coating properties (e. g., surface conductivity).

A product overview of important grades for coatings can be found here [155].

So-called acetylene black is capable of increasing conductivity in films and is thus further used in primers for plastic parts to be coated, which are applied by electrostatic application.

Inorganic, functional so-called IR-reflecting pigments that reflect rather than absorb infrared radiation are now available in a wide range of colors. These pigments are known as complex inorganic colored pigments (CICPs) or ceramic colorants [115]. Many of them have excellent outdoor durability, and when used on building roofs and cladding, their IR reflectance significantly reduces solar heating, saving energy in hot climates. Combinations of colored pigments that absorb all visible wavelengths produce black; therefore, these pigments can be formulated to produce gray colors with much less IR absorption than those made with carbon black pigments.

### 17.2.10 Effect pigments

Effect pigments, in the two main classes special effect pigments and metal effect pigments, have been widely used for decorative and functional applications in systems such as paints, plastics, printing inks, and cosmetics for several decades. Figure 17.9 in section 17.1.1.3 provides an overview of the different classes. Their unique ability to achieve eye-catching optical effects, angle-dependent interference colors, pearl luster, or multiple reflection, has made them irreplaceable. Effect pigments show a number of advantages when compared to extended films, e. g., the wide variety of achievable optical effects, the ease of incorporation in all relevant application systems, the possibilities of blending them with other colorants, and the impression of "vivid" color and gloss effects [92, 99].

Effect pigments are made up of either substrate-free pigments or substrate-based structures. Many developments have taken place in the last 20 years, particularly in the field of substrate-based materials, for example, multilayers on mica or pigments based on alumina, silica, borosilicate glass or fluorophlogopite in addition to muscovite mica [99]. In the last five years, a notable innovation in the field of effect pigments has been achieved by the use of aluminum based VMPs (vacuum metallized pigments) as substrate material as shown in Figure 17.15. This new pigment technology results in significantly thinner effect pigment particles with a different property profile, which opens up completely new optical and application possibilities [156–159].

**Figure 17.15:** Left: Scanning electron micrograph of effect pigment particles based on ultra-thin particle technology (UTP) Right: Transmission electron micrograph (cross section) of an UTP-based effect pigment particle in an epoxy resin (inner layer: aluminum flake produced in a PVD (Physical Vapor Deposition) process used as a substrate material for effect pigment synthesis) [157].

One recent advance is the availability of effect pigments not only as free-flowing powders, but also as preparations (granulates, chips, pastes, color concentrates), which contain the highest possible concentration of pigment [92]. In addition to the pigment, the preparations are made up of binders or mixtures of binders based on solvent or waterborne systems. The advantages of such pigment-binder combinations are, for example, better pigment dispersibility, no dust formation during the introduction of the pigment into the application system, optimized wetting behavior, and improved color effects in the final products. An important, parallel research field is the improvement of optical and non-optical performance.

A property not possessed by conventional organic and inorganic pigments is what is frequently referred to as "flop." This is the change in color and/or gloss with the viewing angle. The origin of the effect lies in the almost two-dimensional, anisotropic nature of effect pigments [160, 161]. The anisotropic morphology of the particles explains why their use affects the appearance, particularly, because a change in processing technology results in a modified standard deviation of flake orientation [162–166].

For outdoor use, for example in automotive and architectural coatings, the weathering stability of many effect pigments needs to be improved by an additional surface treatment [92].

Pigments with diffractive properties have been known for quite a long time but have recently been used to demonstrate a new way of measuring and understanding the appearance for their use in industrial coatings [2, 167]. The diffractive pigments have been used to analyze the relative orientation of the particles in the coating layers by evaluating their behavior in liquid (solventborne) and powder coatings. The interference color seen in Figure 17.16 was measured using a high-resolution goniore-flectometer and the results analyzed using psychophysical and computational methods [167].

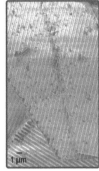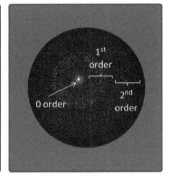

**Figure 17.16:** Left: Example of diffractive pigments used in coatings on a speed shape. Middle: A close-up view of diffractive pigment. Right: refraction orders in diffraction coatings exhibited on a spherical surface illuminated by a point-light as rainbow effects of decreasing intensity [167].

It should be briefly mentioned here, that gloss is a relatively little studied visual property of objects' surfaces and that the study of gloss re-emerged from research into other surface properties such as color and texture. Driven by the technological progress of state-of-the-art experimental techniques and measurements [168], it is part of the current vision and perception research [169–172].

Below can be found an overview of the many applications of effect pigments in coatings.

Automotive coatings and automotive refinishes [173–175]: Effect pigments are used in automotive coatings since more than 50 years. In fact, the first automotive effect paints based on nitrocellulose date back to the 1920s. Originally, the effect pigments were used to give the paints a metallic appearance. At that time, the paint systems offered relatively low weather resistance. This was before today's basecoat/clearcoat system was introduced on such legendary cars as the NSU RO 80 in the 1960s [176].

It was also at this time that pearlescent pigments, previously used only in the cosmetics industry, began to be tested for other applications. After the initial appearance problems were resolved, the first pearlescent automotive coatings appeared on the market. Today, original equipment manufacturers (OEMs) formulate the vast majority of their automotive coatings (80%) with effect pigments, depending on the vehicle model. Of these, 40–50% are mica-based coatings (examples are shown in Figure 17.17). The application method now most commonly used is the base-/clear-coat system, with so-called tri-coats [24, 28] being popular because of their strong, brilliant effect.

**Figure 17.17:** Pearl white stylings used both with a high gloss as well as with a matt clearcoat [160].

Since effect pigments are now commonplace in original equipment automotive paints, they are inevitably also used in repair or so-called refinish paints. For both OEM and refinish coatings, the effect pigments must have good weathering resistance properties and be surface treated [24].

Industrial coatings [177]: Effect pigments are also frequently used in industrial coatings. For examples in the architecture application please refer to Figure 17.18. Important areas of application are:

**Figure 17.18:** Effect pigments used in coatings, shown here in architectural applicants (left to right): powder coating, dispersion and coil coating application [160].

Can coatings: In can production, coating is often part of the initial production process. An entire sheet is treated in a pressure-like process before forming, a fact that must be taken into account for the pigment/binder ratio in the coating and the formulation of effect pigmentation in can coatings.

Furniture surfaces: With the advance of new trends in furniture design, effect pigments are increasingly being used in interior design, as well as for kitchen cabinets and furniture. The popular silk effect, for example, cannot be achieved with conventional pigments. For this reason, architects and designers are increasingly turning to coatings with effect pigments. This market therefore represents a development potential for manufacturers of effect pigments and is likely to lead to the emergence of new effect concepts.

Bicycle coatings: Since effect pigments can be used to create not only a wide range of colors but also effects, they are also very popular for bicycles, motorcycles, and other two-wheelers. Coatings for this market that contain effect pigments are formulated similarly to automotive coatings and require a similar level of weather resistance, durability, and high stability.

Other coating systems: In house paints, effect pigments are mainly found in interior and dispersion wall paints [178–182]. Special application processes can be used to produce various aesthetic effects up to and including glitter. In masonry paints and other paints for exterior surfaces, effect pigments must offer weather-resistant properties [183].

Light, antistatic coatings: While the following section will focus on functional pigments, mention should already be made here of the conductive pigments mentioned here, which are based on a platelet-shaped substrate similar to that used in the manufacture of effect pigments. These conductive pigments consist of thin platelets, usually muscovite mica, coated with Sn/Sb oxide. Their use enables the formulation of bright, conductive coating layers. Such coatings are applied electrostatically to plastics and are used in antistatic floor coverings [184, 185].

### 17.2.11 High temperature pigments

This pigment class consists of inorganic materials with a broad spectrum of chemical classes, with the CICPs or mixed metal oxides playing a central role [4, 115]. They are used in coating processes that require high (baking) temperatures, as well as in coatings with the highest requirements for weather, light and chemical stability. These include coatings for exterior applications, such as powder coating and the coil coating process, which are used for the manufacture of facades and roofing. It should be noted at this point that the use of many pigments remains severely restricted due to, for example, incompatibility reactions, as the high pH value of silicate paints, for example, precludes the use of ultramarine pigments.

Heat-resistant coatings indicate the use of temperature stable pigments such as spinel black based on copper chromite (PB 28) [4] as well as metallic pigments [186], which give furnaces, exhaust systems, and other objects an attractive, metallic glossy appearance.

### 17.2.12 Functional pigments

These pigments are used in coatings to influence both the appearance and the properties of the resulting films of the coatings [187–189].

This class includes corrosion protection pigments based on zinc or strontium chromate, as well as various metal silicates and phosphates used in primers. Active and passive corrosion protection pigments are used, with activity referring to their electrochemical effect. Zinc-based pigments, for example, provide corrosion protection through cathodic action and are used in so-called zinc-containing primers.

Other functionalities relate to properties such as matting to influence the gloss level of the final coating, with fine-particle silica, sometimes powdered polypropylene, often used as a matting agent for this purpose. The particles accumulate on the surface of the film due to convection and reduce the gloss level, keeping the overall pigmentation at a very low level [190].

The effect as a biocide or fungicide is realized, for example, by ZnO in coatings. Antifouling coatings are provided with various pigments to inhibit the growth of organisms on ship hulls, and in the past copper oxide and tin organics were used. Their use is subject to strict environmental regulations, and these pigments are also subject to a lengthy regulatory approval process. The use of finely dispersed silver in antimicrobial coatings should be mentioned for the sake of completeness.

In fire-retardant coatings, antimony(III)oxide ($Sb_2O_3$) is used in halogenated polymers, among other things, to minimize to the spread of flames through a chemical process.

Functional pigments are also used to control the rheological properties of coatings and influence sedimentation and storage stability by influencing viscosity. Representative examples are phyllosilicates such as bentonites and their use in solvent-based systems [191]. Furthermore, fumed silicas based on $SiO_2$ are available to achieve thixotropic properties in solvent- and waterborne coatings [192].

Infrared-reflecting pigments serve to reflect solar heat radiation and are used in various applications. The portfolio of these pigments has broadened considerably over the past few years, which is certainly due to the increased demand for roof and facade coatings with "cool roof" properties to save energy costs in hot climates, as well as the desire for additional color shades [115].

Combinations of colored pigments that absorb all visible wavelengths produce black; therefore, these pigments can be formulated to produce gray colors with much less IR absorption [193] than those made with carbon black pigments.

### 17.2.13 Nano-pigments

A review paper on nano-material technology applications in coatings is available here [194].

It is worthwhile to make a comment of the use of the prefix "nano," since with the emergence and success of nanotechnology about two decades ago, the use of this prefix became quite "en vogue," leading to misinterpretations.

According to the definition of the EU Commission, the particle size (1–100 nm) is the decisive criterion for a nano material where a nano form is present if ≥50% of the particles show a diameter in the range of 1–100 nm. This definition of "nano" is only based on the particle size of the material [195] and does not consider whether the material poses any risk or hazard. The assumption, that nano material generally represents a hazard is vague [196–200]. The inclusion of aggregates and agglomerates turns many pigments and fillers into nano materials per this definition – while concrete requirements are lacking for how to verify this in practice. Since the publication of this definition in the year 2011, public authorities and the stakeholder industries have been searching intensively for a straightforward and workable solution in order to enable a decision on whether there is a nano material or not. In a project it could be shown, that no universally accepted method exists for making this decision [195].

Comparing the different properties of "nano" with "standard" pigments of comparable chemistry, the special properties of the "nano" types result from the higher surface-to-weight ratio of the particles. They also provide a greater surface area for adsorption of resin, resulting in a greater amount of immobilized resin. Assuming an adsorbed layer of 10 nm, the fraction of resin adsorbed on 300 nm pigment particles was calculated to be 3% and 22% on 50 nm particles [195]. In some cases, two glass transition temperatures can be measured. When the particle size of the pigments is smaller than the wavelength of light, coatings pigmented with nanopigments are transparent [194].

Nanopigments, especially nanofillers, are said to improve resistance to scratching, scuffing, abrasion, heat, radiation, and swelling; reduce water permeability; and increase hardness, weatherability, modulus, and elongation at break while maintaining toughness.

Their use has gained importance in recent years [22], with various papers reporting their applications and benefits. For example, it has been shown that common pigments such as $ZnO$, $SiO_2$, $TiO_2$, $Al_2O_3$, etc., when added as nanopowders, can improve paint performance [201]. It was found that only 0.03% ZnO as nanopowder in an aqueous paint improved the mechanical, corrosion and UV properties many times [202]. Similarly, the addition of 0.3% nano-alumina improved the abrasion and scratch resistance of an alkyd-based aqueous coating by nearly 10% [203].

Dispersing nanopigments can be complex. At small particle diameters, especially if less than 10 nm, the physical properties of the materials change and can lead to a number of complications. Conventional pigment dispersion methods are often unable to reduce particle diameters to the nanoscale, but some success has been achieved [194].

Nano-color pigments used in metallic automotive coatings are prepared by grinding suspensions of the conventional colored pigments in butyl acetate with a commercially available dispersant containing very small, high-density spheres in a mill. The mill base is then used to make tinting pastes for matching metallic paints [204]. Ball mills are also used to disperse nano-aluminosilicate in an acrylic resin using a commercial dispersant. The dispersion is used in the production of scratch-resistant clear coatings [205].

The preparation of stable dispersions of nanopigments can also be difficult. The high surface energy of nanopigments makes them prone to agglomeration. If this occurs to a significant extent, the benefits of nanopigments cannot be realized. A variety of methods have been used to address this problem, including surface treatment of the pigments [206], in situ polymerization in the presence of the pigments, special pigment dispersants, and ultrasonic dispersion.

Submicron and especially nanodispersions with very narrow particle size distributions are much easier to disperse and can remain stable over longer periods of time if dispersants with low molecular weights and high equivalent weights of the functional groups are used. The type, number and physical arrangement of the polymers and/or oligomers are crucial influencing variables to obtain stable as well as cost-effective dispersions (see also the section on nanopigments).

Progress in the field of nanopigments is enormous and interesting developments of pigments with different chemistries and properties were reported over the past years. A few examples can be found below.

A green colored nano-pigment $Y_2BaCuO_5$ with impressive near infrared (NIR) reflectance (61% at 1100 nm) was synthesized by a nano-emulsion method. The UV-VIS spectrum of the nano-pigment exhibits an intense d-d transition associated with $CuO_5$ chromophore between 2.1 and 2.5 eV in the visible domain. Therefore, a green color has been displayed by the developed nano-pigment. The potential utility of the nano-pigments as "Cool Pigments" was demonstrated by coating on to a building roofing material like cement slab and PVC coatings [150].

ZnO nanoparticles (ZnO NPs) were obtained by a direct precipitation method. Transmission electron micrographs demonstrated that the synthesized ZnO NPs were of a pseudo-spherical shape and the average diameter of the particles is 20 nm. ZnO polyurethane nano composite (ZPN) coating films were fabricated by uniformly dispersing ZnO NPs in varying loading levels 0.1 to 2.0 wt.% in commercial two component polyurethane by ultrasonic treatment. The antimicrobial activity of ZPN-coating films was screened against Gram-negative and Gram-positive bacteria. Corrosion performance, physical and mechanical properties of ZPN-coating films was evaluated. The resulting perfect dispersion of ZnO NPs in polyurethane coating was revealed by scanning electron microscopy (SEM). The results showed a slowdown in growth of organisms on the ZPN coating surface, and also showed an improvement in the increased corrosion resistance, mechanical resistance at lower concentration, and this improvement increases with increased ZnO NPs wt.% [207].

The preparation of a series of orange Cr- and Sb-containing $TiO_2$ nano-pigments by the polyol process using triethylene glycol was reported. Field-emission SEM images proved the formation of discrete, non-aggregated Cr- and Sb- co-doped rutile nano-pigments with sizes below 100 nm after annealing precursors at 800°C. Zeta potential of aqueous dispersions of nano-pigments confirmed the feasibility to be used in the ink-jet printing decoration process [136].

Recent results on the preparation, characterization and color properties of the V-containing $ZrO_2$ inorganic yellow nano-pigment are reported. A series of monoclinic $V$-$ZrO_2$ solid solution nanocrystals were prepared by hydrothermal processing of mixtures of zirconyl nitrate and ammonium metavanadate in water. The high stability in aqueous dispersion and the structural and microstructural characteristics of the prepared and annealed nanocrystals up to 1200°C made them good candidates as nano-pigments for digital decoration. The advantages of the hydrothermal approach against other non-conventional methods for the preparation of $V$-$ZrO_2$ nano-pigments, such as sol-gel and polyol, are discussed [138].

### 17.2.14 Structural color and photonic crystals

Structural colors are closely related to the recent fast-growing fields of photonics [208, 209] and have attracted much attention in a wide variety of research areas because they arise from complex interactions between light and sophisticated microstructures. As an example, Figure 17.19 depicts color and flop created by polymer crystalline colloidal array pigments [210].

Their mechanisms are in principle of purely physical origin and differ considerably from the ordinary coloring mechanisms as in pigments, dyes and metals, where the colors are produced by the energy consumption of light [211–213].

A summary of scientific activities over different research fields from biology, chemistry, physics to engineering can be found here [208–216].

As reported in [216], photonic crystals have attracted great interest owing to its pigment-free, highly brilliant and fadeless characteristics. However, their poor mechanical strength as well as low adhesive force between these crystals and the substrate still limits its applications.

Another challenge is the reproducible reproduction of the periodically recurring units in the exactness physically required for this and to manufacture them on a large industrial scale [217]. Commercial applications already exist, e. g., in fabrics [218].

**Figure 17.19:** Color effects created by prototype polymer crystalline colloidal array pigments in coatings and scanning electron microscopy (SEM) cross-sectional images of the structures [210] (reproduced with permission from PPG).

## 17.2.15 Colorants for special coating applications

### 17.2.15.1 Thermochromic pigments
Thermochromic pigments are finding more and more functions/applications, also in coatings [2, 5].

Chromogenic polymers change their visible optical properties in response to an external stimulus. Related to the specific stimulus, they are classified as thermochromic (temperature), photochromic (light), electrochromic (electric field), piezochromic (pressure), ionochromic (ion concentration), or biochromic (biochemical reaction). The chromogenic phenomena enable the integration of sensor and actuator functionality or any kind of information into a material itself [2].

As reported in [55], thermochromism can occur in all different classes of polymers: Thermoplastics, thermosets, gels, inks, paints, or any kind of coatings. The polymer itself, an embedded thermochromic additive, or a supermolecular system formed by the interaction of the polymer with an incorporated non-thermochromic additive can cause the thermochromic effect. From a physical perspective, the origin of the thermochromic effect can also be diverse. It can originate from changes in light reflection, absorption and/or scattering properties with temperature [219].

There are limitations to the use of this technology, not the least of which is the generally poor UV resistance of thermochromic pigments in sunlight, which generally limits their use to items exposed to daylight for extended periods of time. Of course, they are inherently temperature sensitive and therefore items containing these materials must be protected from any undesirable heat source.

Fiber materials are a special case of thermoplastic polymers and thermochromic cellulose fibers were prepared by incorporating the commercial thermochromic pigment described in [220]. This thermochromic pigment is a microencapsulated reversible switching leuco dye developer-solvent system with diameters of the microcapsules ranging from 0.5 to 5 μm and a thermochromic color intensity change from magenta to colorless at a temperature of 32.7 to 32.9°C.

As far as thermoplastics are concerned, microencapsulated leuco dye developer-solvent systems and polyethylene and polypropylene masterbatches containing them are already commercially available from several manufacturers. Although these masterbatches exhibit excellent thermochromic switching behavior, the use of microcapsules results in a number of limitations, such as poor thermal and shear stability, which lead to processing difficulties. Incorporation of these microencapsulated leuco dye developer-solvent systems into high temperature polymers, such as polycarbonate, is not possible. The switching temperature of commercially available microencapsulated leuco dye developer-solvent systems ranges from −25 to 65° C. Efforts to extend this temperature range, especially toward higher switching temperatures, have been ongoing for many years [221]. The development of reversible and irreversible thermochromic organic pigments with higher switching temperature has been in focus in recent years [55].

In contrast to the thermochromic effect of organic composites, the thermochromism of inorganic materials, such as metal salts and metal oxides, has been known for a long time [220–224], where the occurrence of thermochromism in inorganic pigments originates from changes in the crystal structure. For example, the red to brown color change of copper-mercury iodide, $Cu_2(HgI_4)$, is caused by a change from an ordered to a disordered structure. In both modifications, the iodide ions form a face-centered cubic cell with tetrahedral holes partially occupied by the copper and mercury ions. In the low-temperature modification, all tetrahedral holes are definitely occupied, while in the high-temperature modification, the cations are randomly distributed over all tetrahedral holes. Another example is mercury(II) iodide. The transition from the α- to the β-modification at 127°C is associated with a color change from red to yellow. For most of the inorganic pigments considered, the thermochromic effect occurs at temperatures above 100°C and is irreversible. Examples of inorganic compounds with reversible thermochromic properties can be found here [55].

An unprecedented one-finger-push induced phase transition with drastic color change in an inorganic material is discussed in [225]. It is found that the original color reverses during annealing. Therefore, such metal complexes combine thermochromism and piezochromism in one material and could be called chromogens.

The use of poly(3-alkylthiophene)s as thermochromic pigments to introduce thermochromism into host polymers has been reported [226, 227]. This group of conjugated polymers exhibits thermochromism due to temperature-dependent conformational changes of the conjugated $\pi$-electron system. Modifications of their chemical structure is applied to vary the switching temperature of these pigments.

A comparison between the classes of thermochromic additives has been published in [55]. As explained by the authors for organic thermochromic materials the outstanding feature is, that the switching temperature and color can be customized and that the narrow absorption band allows the generation of multiple switching effects by the combined use of several thermochromic additives, properties important for applications as they allow flexible adjustment of the thermochromic properties for the specific application. The advantage of inorganic pigments is clearly their thermostability above 200°C and light stability, which is even suitable for outdoor applications. However, most inorganic pigments are toxic for these applications.

Future developments in the field of thermochromic polymers will cover subjects like improving the organic thermochromic systems with regard to thermal and mechanical stability of thermochromic pigments for doping of polymer matrices, increasing the UV and visible light stability of leuco dye developer solvent systems as well as the extension of the switching temperature range of thermochromic pigments, particularly beyond the current 69°C limit for commercial products [55].

### 17.2.15.2 Photochromic microcapsules

Photochromic dyes were introduced in the early 1990s. Unfortunately, photochromic dyes are inherently unstable and change their chemical structure when exposed to UV light. Because the dye is so sensitive in its excited state, stabilization is the biggest challenge for dye manufacturers because without stabilization, shelf life is very limited. However, microencapsulation can overcome this shortcoming. When photochromic crystals are irradiated with UV light, the dye undergoes a temporary chemical change, but when the UV source is removed, the molecules re-form their original bonding structure (Figure 17.20). In this example, the reversal reaction is predominantly thermally driven (the assistance of heat can be considered an example of thermochromism). In other photochromic dyes, the reverse reaction is photochemical. Because of their ability to exhibit resistance to thermal fading, fulgides are the class that has been most commonly used for this application [3].

Incorporated into coatings, the photochromic microcapsules can create interesting, optical effects. Indoors, away from UV light, the design on the fabric is white, but changes color when exposed to UV radiation in daylight. The palette ranges from yellow to purple to green. Despite the volatile nature of color formers to light, microencapsulated formulations are marketed in attention-grabbing product labeling and anti-counterfeiting. Other applications include textiles that develop new patterns when exposed to sunlight, or security devices that can be observed spectroscopically or with special instruments [3].

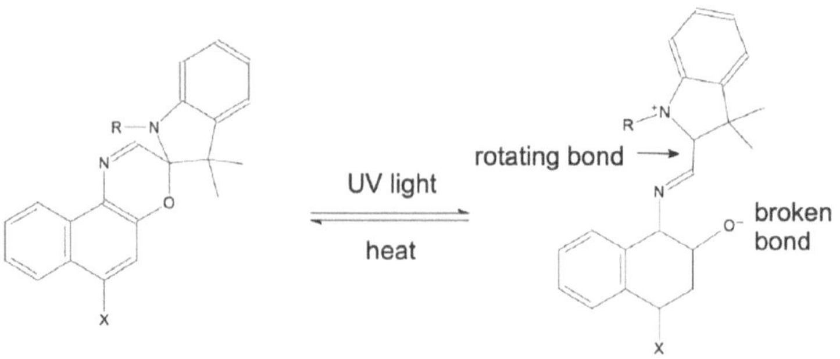

**Figure 17.20:** Mechanism of photosensitive dyes after [3] (Copyright Wiley-VCH GmbH. Reproduced with permission).

### 17.2.15.3 Electrochromic pigments

A large number of inorganic/organic materials show electrochromic properties, whereby these appear by the fact that color changes of these materials can be observed during an electron transition as well as during redox processes [228–230]. With regard to their use in coatings, Prussian blue or Fe(III)–hexacyanoferrate(II) is historically relevant, especially since, in addition to its use in varnishes and inks, it is responsible for the origin of the term "blue print" through its use in photographic prints. It produces the colorless Prussian white when reduced and is thus an anodic coloring electrochromic material [2, 231, 232].

### 17.2.15.4 Phosphorescent pigments

These are materials based on activated ZnS which, for example, produce a green luminescence when activated by copper. Luminous coatings for instruments and watches are typical applications for these pigments. When alkaline earth aluminates, mostly based on strontium doping, began to compete with conventional sulfide luminescent materials 30 years ago, and this also improved the availability of these raw materials, this stimulated the applications and led to further possible uses [2].

### 17.2.15.5 Solvent-soluble dyes for coatings

#### 17.2.15.5.1 Solvent dyes

A brief mention should be made of solvent dyes, for whose successful application in paints, stains and varnishes sufficient solubility in organic solvents is a prerequisite [6, 7, 233]. These raw materials are mostly red or yellow azo as well as complex compounds, which are produced by co-precipitation of acidic and basic dyes. As an example, Figure 17.21 shows the structure of the yellow, carcinogenic 4-aminoazobenzene (Solvent Yellow 2) and, in addition, that of Solvent Orange 1.

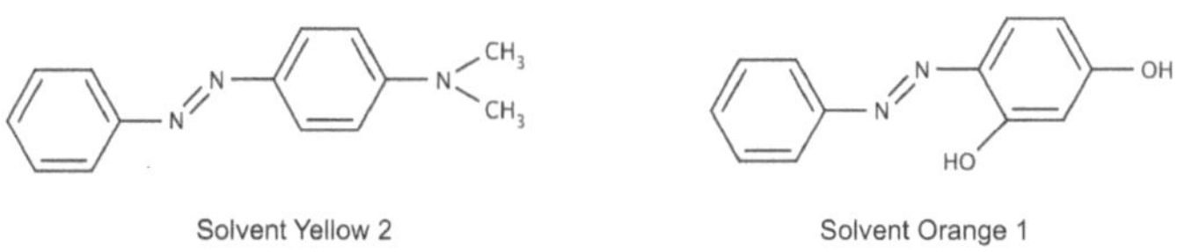

Figure 17.21: Structure of Solvent Yellow 2 and Solvent Orange 1 [7, 12].

Azo dyes that are soluble in polar solvents such as alcohols, glycols, esters, glycol ethers and ketones fall into this category, with the exception of the blue copper phthalocyanine derivatives. They are used in transparent protective coatings for metal and wood. From a chemical point of view, the most important alcohol- and ester-soluble azo dyes can be divided into three groups, which will not be discussed in further detail here. As a representative example of complex formation by an azo dye with a metal salt in organic solvent, Figure 17.22 shows, in this case, Solvent Red 8.

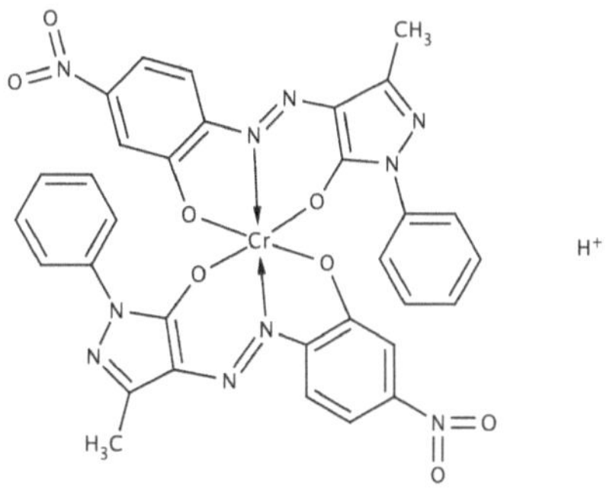

Figure 17.22: Structure of Solvent Red 8 [12].

### 17.2.15.5.2 Azo metal-complex dyes

It is known that solvent dyes based on a 1:2 chromium and cobalt complex without hydrophilic substituents are readily soluble in organic solvents such as alcohols, ketones, and esters [7]. However, improved solubility can be achieved by converting the metal complex sodium salts to salts of organic cations such as long-bed aliphatic ammonium ions [234,235,236]. Figure 17.23 shows representatively the structure of bluish red solvent dye [237]. These derivatives find a wide range of applications such as transparent coatings and wood stains [238].

### 17.2.15.5.3 Fat and oil soluble dyes

Fat- and oil soluble dyes are used in a wide range of coatings, which includes in particular the coloring of transparent coatings on aluminum foil. These dyes are soluble in a variety of waxes, resins and solvents except for water, but no clear distinction can be made between them and alcohol- and ester-soluble dyes. They are

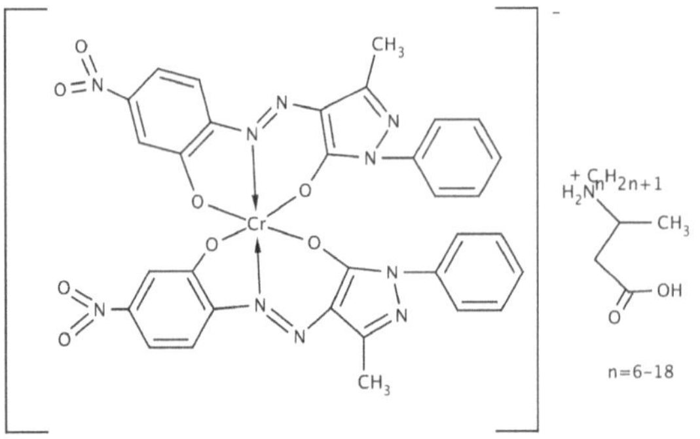

**Figure 17.23:** Structure of a bluish-red solvent dye [12].

azo dyes based on simple components, with the blue anthraquinone derivatives being an exception. As an example, the structures of Solvent Red 23 (7; R = H) and Solvent Red 24 (7; R = CH$_3$), are shown in Figure 17.24. These raw materials are typically distributed as powders or granules. In the case of liquids, they are highly concentrated solutions of fat soluble dyes in aromatic hydrocarbons, and in some cases solvent-free 100% liquid products [7].

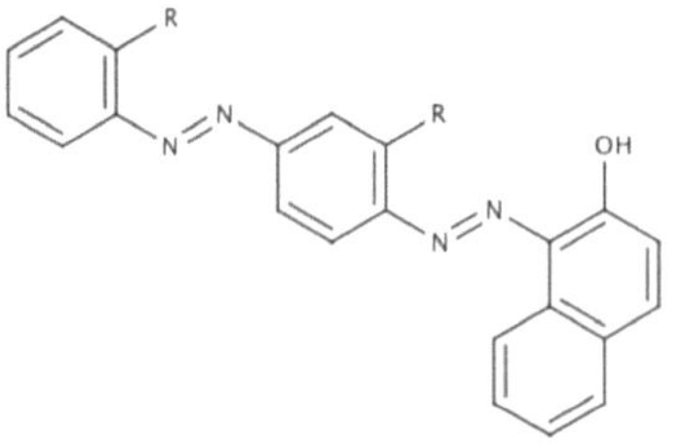

**Figure 17.24:** Structure of Solvent Red 23 (R = H) and Solvent Red 24 (R = CH$_3$) [7].

## 17.3 Colorants in different coating systems

### 17.3.1 Waterborne (emulsion) paints

Besides the polymeric binder and water as diluent, colored pigments, and inorganic fillers are the main constituents of emulsion paints besides solvents, dispersants, thickeners, preservatives, and defoamers. Different requirements are placed on interior paints than on exterior paints, like odor neutrality for interior and weather fastness for exterior paints. Interior paints are divided essentially according to the degree of gloss, into matt, semi-matt, satin, semi-gloss and gloss paints. Exterior paints are differentiated as masonry paints, elastomeric actings, wood coatings, plasters or textured finishes, silicate paints, silicone resin paints, and universal or house paints. Standard formulations of matt interior and exterior emulsion paints typically use $TiO_2$ and colored inorganic pigments along with fillers like chalk, heavy spar, talc, kaolin, mica [239].

In general, different pigments are used in emulsion paints in order to fulfil the application profile, as pigments may enhance the weather fastness of the coating, preferably by shielding the binder against UV exposure [1, 240, 241].

Without a doubt, the most important white pigments used in waterborne emulsion paints today, due to its high refractive index, is $TiO_2$, which is responsible for their opacity and whiteness. Depending on the type of $TiO_2$ used (rutile or anatase), it also provides chalking resistance. For chalking-resistant coatings, surface-treated rutile should be used, since anatase reacts photocatalytically with oxygen and moisture, leading to radical reactions, decomposition of the binder at the $TiO_2$ interface, and thus to chalking. Surface treatment can be either based on $ZrO_2$, $Al_2O_3$, $SiO_2$, or sometimes ZnO, as mentioned earlier in this chapter.

The alternative white pigments ZnO, ZnS, or lithopone have lower refractive indices. They are only used to a limited extent because of the associated poorer whiteness, lower hiding power and greater chalking tendency of the coatings. At most, they are used for the specific anti-fungal formulation of masonry paints.

For the sake of completeness, organic white pigments known as opaque particles should also be mentioned here. The polymer particles contain air-filled cavities that remain even when the paint dries. The difference in refractive index between the polymer and the air causes light to be scattered, which is how the paint provides opacity. These organic white pigments are marketed as a partial substitute for $TiO_2$.

In addition to white pigments, a wide variety of colored pigments are used. These colored pigments can contain both organic and inorganic materials. However, due to their higher lightfastness, better chemical stability, and easier dispersibility, aqueous coatings in practice predominantly contain inorganic pigments such as iron oxides or chromium oxide or, in the case of the latter, the non-toxic alternatives.

Organic pigments are normally used in the form of ready prepared pigment pastes usually only for sharing the colors.

The less expensive inorganic pigments (e. g., iron, cadmium, chromium or PbOs or lead sulfides, $PbMoO_4$, cobalt blue, carbon black) have much better UV stability (with the exception of Carbon Black) than the more expensive organic pigments (e. g., phthalocyanine, Azo pigments, quinacridones, perylenes, carbazoles). However, they do not usually result in the same color brightness. For outdoor applications, therefore, only the metal oxides are suitable, which in many cases also lead to good alkali resistance of the coating. For environmental reasons, iron oxides are the most important colored inorganic pigments used today. The lead and cadmium compounds, some of which are toxic, are replaced by bismuth vanadate [239].

### 17.3.2 Powder coatings

Powder coatings chemistry and its technology was described in detail by De Lange [39].

A detailed overview on the research progress in powder coatings which also highlights the activities in the field of pigments was published here [242].

A powder coating is a 100% solid coating applied as a dry powder and then formed into a film by heat. The typical components of a powder coating are binders (resins with, if necessary, hardeners and accelerators), colorants (mainly pigments), additives and, in some cases, fillers [1, 42, 243]. Colorants for powder coatings are almost exclusively pigments and must meet requirements such as thermal stability at the curing temperature, suppressed reaction with other components of the formulation and stability to shear forces during extrusion and grinding. The solid binder melts when heated, binds the pigment, and results in a pigment layer when cooled [242].

Inorganic pigments (e. g., $TiO_2$, iron oxides, and chromium oxides) largely meet these criteria. In order to achieve maximum resistance in powder coatings for exterior use, lead pigments must still be used in isolated cases. In Europe, the use of cadmium pigments is declining. Substitution by organic pigments generally limits the color spectrum in the red, orange, and yellow range. Metallic and pearlescent effects are achieved by using aluminum platelets or coated mica.

Organic pigments can be used in powder coatings, but since the crosslinking of the polymers takes place at high temperatures (>150°C), the heat stability of the organic pigments is very important.

The evaluation of the condition of a pigment dispersion in powder coatings is complicated by the lack of a simple and effective evaluation method, especially during the production process. Kunaver et. al. developed a method for evaluating dispersion quality in cured powder coatings or in samples taken immediately after the extrusion process [244]. They investigated differently pigmented powder coatings of yellow and orange–yellow color incl. $TiO_2$, red iron oxide, monoazo (benzimidazolone), tetrachloro-isoindolinone and yellow iron oxide as well as dis-azo yellow and diketopyrrolopyrrole pigments [145]. Quantitative information on pigment dispersion was obtained by plasma etch scanning electron microscopy image analysis. In order to correlate the differences in extrusion equipment and thus the energy input of the dispersion process, two different formulations were studied, each produced with three different extrusion systems and the results are compared with those obtained by color difference measurements. It was shown, that the plots of pigment size distribution correlate well with the measured color differences and with the energy input of the different extrusion units used.

Pigments impart color and functionality in powder coatings, whereas the desired effect is achieved by choosing the right pigment, the most important properties of which are good dispersion and thermal stability. $TiO_2$ is an inorganic pigment commonly used in powder coatings. There are studies that the surface of nano-$TiO_2$ was grafted with hydrophobic groups, so that the hydrophobicity was increased, which favored its dispersion in the coating matrix. Shi etal [245]. investigated the polyester powder coating modified with 2 mass percent nano-$TiO_2$. The results showed that the addition of nano-$TiO_2$ resulted in a delay of the melting point, and the starting and peak temperatures of the curing reaction of the system modified with nano-$TiO_2$ were decreased by more than 5°C. The addition of 2 mass percent nano-$TiO_2$ to the polyester powder coatings accelerated the curing process. Hadavand et al [246]. studied powder coating nano-ZnO with antibacterial properties. The surface was successfully modified with VTMS. The results showed that the thermal stability of modified nano-ZnO in the mixing process had excellent compatibility and good dispersion in the polymer matrix. The antibacterial experiments showed that the efficiency of coatings with modified nano-ZnO was higher than that of an unmodified one. In order to increase the ohmic conductivity of the resin system and modify the mobile charge carriers, Trottier et al [247]. introduced the ZnO and $SiO_2$ as pigments into the epoxy polyester resin system. The results were that the resin matrix was opened; the activation energy for dipole rotation was lowered; the frequency of relaxation peaks was shifted down; the $T_g$ of the powder coating was increased by the addition of the pigment. Puig et al. [248]. studied powder coating formulations with different contents of ZMP applied on galvanized steel. The corrosion protection properties were investigated by electrochemical methods. The various results showed that for this type of powder coating, the corrosion protection properties were improved when 10% or 15% ZMP was added, which was attributed to the improvement of the barrier properties and the inhibitory effect of the pigment. In the salt spray test, no differences in corrosion protection properties were observed between the samples with 10% ZMP and 15% ZMP. Phosphate pigments are considered highly effective and safe corrosion inhibitors in various coating applications. In the study of El-Ghaffar et al. [249]., anticorrosive hybrid epoxy/polyester and polyester powder coatings composites based on phosphate pigments have been formulated. The prepared powder coatings were applied to cold-rolled steel sheets and investigated for their physical–mechanical properties and evaluated for their anticorrosive properties using a salt spray chamber for 1000 h. The powder coatings were then tested for their corrosion resistance. The obtained results showed high-performance anticorrosive powder coating formulations for steel protection. There is no characteristic change in the physical and mechanical properties of the films.

When looking into the future trends in powder coatings, the following subjects can be identified [250, 251]:

Compared to liquid coatings, the excellent properties and the economic and ecological advantages contribute to the success of powder coatings. Although there are still disadvantages associated with the handling and use of the powder form, these shortcomings have now been further improved or minimized thanks to the further development of formulation and equipment. These developments are contributing to market growth in the powder coatings sector. Furthermore, new resin systems enable powder coatings to be cured at low temperatures around 120° C already. These systems will enable the application of powder coating to heat-sensitive substrates such as plastics, wood, etc., further stimulating market growth supported by new drying technologies such as radiation curing. High-temperature powder coating will be further optimized for its applications, too. Representative applications include gas and charcoal grills, exhaust components, fireplace, and lighting fixtures. Thin films of 20–70 microns can be produced in a wide range of gloss levels, color tones and effects, and surface textures. Ultra-thin powder films in the thickness range of about 20–30 micrometers are currently under development. These are of interest because they allow better film thickness control as well as wetting of recesses. Finally, developments in the field of equipment and machinery for processing powder coatings are also contributing to the progress of this environmentally friendly technology. The application efficiency of powder coatings is significantly higher compared to liquid coatings, which are often applied with air assistance, reaching values of up to 96%. To achieve these values, it would be necessary to apply liquid coatings by means of electrostatic application, as is common practice in automotive painting, for example, using high-rotation bells. The coating speed of powder coatings is being steadily increased by advances in radiation curing. Dwell times of 30 seconds are common in infrared curing, and 4–5 seconds in UV curing. Finally, challenges for further development of powder coatings are the well-known issues of color consistency and weathering stability over >15 years, especially with regard to architectural and facade applications. This will certainly lead to new pigment and surface treatment developments. In the case of effect powders, the focus will be on controlling the appearance and achieving, for example, metallic effects at the level of liquid coatings. As described in the chapter on automotive coatings, this represents a major challenge in connection with the control over the particle orientation of the effect pigment platelets in the dried paint film. This is part of current research [167, 252, 253].

Combined with stricter health, safety and environmental regulations, the use of raw materials and technologies in the paint and coatings industry is changing. One driver is the REACH regulation (Registration, Evaluation, Authorization & Restriction of Chemicals). The reduction or elimination of volatile organic compounds (VOCs) is the main focus here, which is why conventional organic coatings are in the process of being replaced because they contain a large percentage of VOCs. As a result, powder coatings are enjoying high market growth. Due to the commercial success and growing popularity of powder coatings, there is a great need for research in this field [254–257].

### 17.3.3 Radiation-curing systems

Radiation curing is a technology that uses electromagnetic radiation, mainly in the UV range, or ionizing radiation, mostly accelerated electrons, to trigger a chain reaction that converts mixtures of polyfunctional compounds into a crosslinked polymer network. One challenge in UV curing, especially for pigmented films, is to find the ideal parameters for the drying process and subsequent performance of the resulting coating film [258].

The main challenge in formulating coatings for UV-curable coating systems is that the properties of pigments and photoinitiators in a coating compete when exposed to UV light. It is understandable that an ideal pigment should offer UV transparency so as not to interfere with the reaction of the photoinitiators, but also good absorption in the 400 to 700 nm visible range to ensure good hiding power in the dried coating layer. With the right combination of photoinitiators that absorb UV light between 380 nm and 410 nm, and the right combination of lamps and curing conditions, most pigmented coatings can be fully cured. By using a dual curing strategy, the limitations can be almost eliminated [259].

A UV-curable coating can be designed to have satisfactory performance characteristics with a given equipment configuration, and identifying the UV-curing parameters with the greatest influence on the performance of the final coating is critical.

It is possible to identify these process design parameters: Total energy, irradiance profile, spectral distribution, and temperature reached by the coating from the infrared energy. Once these key parameters are identified, it is possible to define the "process window" and cure control and monitoring can then be done by measuring them.

New photoinitiators that provide for strong absorption at longer wavelengths are allowing formulators to incorporate more pigmentation into their coatings. Curing white pigmented coatings, for example, which had been a difficult task due to the strong pigment absorption in the longer UV region, is now accomplished with advanced photoinitiator blends of substituted phosphine oxides and phenyl ketones [260].

In [258], the absorption spectra of compounds contained in a pigmented paint are presented. This is a complicated situation with respect to the absorption behavior of the photoinitiators as usually at least two photoinitiators are used. One (acyl phosphine oxide, e. g., TPO) that absorbs at long wavelengths outside the absorption spectrum of the pigment and is responsible for through-curing, and another that absorbs in the short wavelength range and is responsible for surface curing. A more detailed description of radiometric methods for designing and monitoring UV processes can be found here [261].

The raw materials for the formulation of UV-curable ink and coating systems consist of medium molecular weight resins. The main types of resins are radical polymerizable unsaturated polyesters, acrylate-terminated molecules such as polyepoxides, polyesters, polyethers, and polyurethanes, as well as epoxides and vinyl ethers. Since most of these resins often have too high viscosity, they are diluted with reactive diluents to adjust the application viscosity. The reactive diluents used are mainly mono- or multifunctional acrylates, less frequently methacrylates or non-acrylate monomers such as styrene, vinyl pyrrolidone, divinyl ether, and a few others. In cationic curing coatings, vinyl ethers and monoepoxides are used as reactive diluents. In addition to these essential raw materials of a UV-curable formulation, additives such as surfactants, defoamers, leveling agents, flow regulators, flexibilizers, UV stabilizers, or fillers (clay, $CaCO_3$, and silica) as well as nanoparticles, which are transparent and can provide higher scratch resistance, are used depending on the requirements.

In North America, consumption of UV and EB formulated product usage increased from 77.3 to 165.1 (000´ metric tons) by volume in the period between 2001 and 2019, with current applications now focused on graphic arts (OPV, inks), wood, plastic coatings, printing plates, adhesives, optical fibers, and metal decoration. 3D printing, or additives for it, is mentioned as the application with the highest compound annual growth rate of 14.6% from 2021 to 2021. The growth of plastic coatings is forecasted at 6.3% [262].

Many new applications for radiation-curable coatings are emerging [262], 263]. Current and planned UV applications from the perspective of equipment suppliers have been frequently reported at the RadTech conferences [262]. In the automotive industry, for example, applications are coatings on headlight lenses, coatings on reflectors and coatings on instrument panels.

Pigments are mainly used in radiation-curing printing inks, but in the meantime, they are applied in areas such as coil coatings, although technical challenges like adhesion, flexibility and through-curing for these pigmented coating systems had to be solved [263]. Further evaluations of UV-curable systems for direct metal application [264–267] and through-curing of pigmented UV-curable coatings [267] have been reported.

Recent work investigated the crosslinking of pigmented coating material by UV light-emitting diodes (LEDs) enabling depth curing and preventing oxygen inhibition. Initiator systems were found that were able to cure thick layers of several millimeters with a conversion of 80–100% [268].

The use of metallic effect pigments in UV printing inks has been known for years, and in recent years this has also been successfully transferred to plastic coating and other applications [269, 270].

Driven by the benefits of UV curing and the market growth associated with the sustainability of this technology, pigment manufacturers are developing specific product portfolios for this application area [269, 271].

### 17.3.4 Coil coatings

Coil coating is the globally established term for the industrial process of continuously coating rolled steel and aluminum strip organically and can be summarized in the short formula: "Finish first, fabricate later." In this finishing process, the coils are coated with a paint or plastic film, rewound and then cut, punched, roll-formed, etc. at the processor according to their use as roofing elements, refrigerator walls, window profiles etc. [272].

The standard EN-10169:2012–06 [273] defines the typical process steps: cleaning and chemical pretreatment of the metal surface, single or double-sided, single or multiple application of liquid- or powder-coating materials with subsequent film formation under heat, or lamination of plastic films. The process step embossing is understood to mean the hot embossing of thick (plastisol) layers to produce a decorative surface structure. Post-treatment here involves the application of peelable protective films which additionally protect the coated surfaces during storage and transport and also facilitate difficult forming operations and assembly.

The success of coil-coated sheet is essentially based on the fact that the processor of this sheet does not have to carry out the coating process and all the associated costs, such as pretreatment, wastewater and exhaust air treatment, disposal of residual materials, and investment and maintenance costs for coating equipment and drying, because the coating is applied directly in the aluminum or steel rolling mill or in the refining store. The use of coil-coated sheet means that the finishing steps – in this case the coating of the individual pieces – are transferred to the upstream supplier. The processor can concentrate on his core competencies, mechanical processing and assembly. He no longer needs to worry about coating application. The coil-coated materials can be further processed with appropriately adjusted parameters without damaging the coating and thus impairing the service behavior.

The possible applications of pre-painted sheet are very diverse, because it can be processed using common techniques. Cutting, piercing, forming (bending, roll forming, deep drawing) are the main mechanical operations to which the sheet is subjected. Joining of sheet metal is done by stapling, flanging, folding, screwing, riveting, clinching, gluing, and by combinations of these techniques. Welding can only be used to a limited extent, because the coating is generally electrically non-conductive and decomposes at the high welding temperatures. The main applications of pre-painted steel and pre-painted aluminum are in the construction and architectural sectors.

An important segment is "white goods." This includes electrical household and commercial appliances of all kinds. Today, around 25% of the surfaces of household appliances are already precoated in the rolling mill. These include the bodies and doors of refrigerators and freezers, as well as housings for washing machines, dryers, dishwashers, ovens, microwave ovens, and stove hoods. Panels for information technology equipment, heaters, small appliances, and parts of coffee machines are examples of the general processing of coil-coated sheet metal.

There is also a wide range of applications in the automotive sector. For automotive applications, weldable galvanized steel sheets are pre-painted with anti-corrosion primer, which makes a significant contribution to corrosion protection of the car body. But coil-coated sheets are also used to replace traditional painting steps in automotive production, for example for bulkheads, oil filters or similar parts, some of which are also silenced. Further possibilities, for example sunroofs, result from the modular construction method now widely used in automotive engineering. Truck and bus bodies are also made from coil-coated materials. Caravans and bodies for motor homes are almost exclusively made from pre-painted aluminum. Other examples are parts of rail vehicles and ship interiors.

Coil coating is an efficient process, because it offers the advantages of a continuous process. The efficiency lies in the high throughput rates of large-capacity lines or the flexibility of comparatively slow lines. On the large-capacity lines, areas of more than 200 $m^2$ per minute can be coated on both sides. With this order of magnitude, coil coating occupies a top position in metal painting. The roller application also ensures almost 100% transfer of the paint. It allows thinner paint coat thicknesses to be applied with the same quality compared with piece coating.

The principle of coil coating has also been used for decades for the continuous coating of packaging sheet, stainless steels or electrical sheet, which is provided with an insulating lamination. This is partly carried out on the same coil lines.

Coil coating is accompanied by the worldwide standardization of flat products, their test methods and the corresponding component development. In the course of globalization and the intensive exchange of information, the cooperation of the regional/national trade associations and the two globally active umbrella organizations, the European Coil Coating Association in Brussels and the National Coil Coating Association in Cleveland, Ohio, should be emphasized.

Four major market segments are served by these materials, each market segment primarily uses one or two specific resin types:
- Construction (roofing, panels, garage doors),
- Transportation (truck and bus body parts, automobiles, exterior trim parts, gas tanks, engine components).
- Consumer goods ("white goods" such as refrigerators or washing machines, office furniture, shelving,
- Computer components, signs, industrial equipment)
- Packaging (cans, containers, crowns, drums, barrels) [272].

An organic coating is applied after the inorganic pretreatment of the sheet. Usually, a primer is applied first to the visible side of the sheet and then a top coat. This coating must be matched to the intended use of the sheet. For example, the coated surface of sheets that are processed into refrigerator housings must be flexible to withstand deformation. It must be hard so as not to suffer damage during processing by tools, but also resistant to foods such as ketchup, mustard, mayonnaise or orange juice. Parallel to the coating of the visible side, the reverse side is also coated. Here, too, a primer is applied first, depending on the intended use, and a functional back coat is applied in the second step. Frequently, however, a reverse side coating is applied in a single layer. To protect the surface during further processing, a film is often laminated at the end, which is removed again after processing. Films are also laminated in small quantities as an alternative to the topcoat.

Solventborne coatings contain film formers, additives, pigments, fillers, and solvents as basic building blocks. The film former gives the coating its system-determining basic properties such as adhesion, elasticity, hardness, durability, and corrosion resistance. Additives, fillers and pigments provide special properties such as coloration, hiding power, weldability, and the solvents guarantee the processability of the coating material. Solventborne coatings typically have a solids content of 40% to 70% and are usually thermosetting systems. During the baking process, the solvent evaporates and, in a chemical reaction, the polymers crosslink to form a film on the surface. The reaction is usually accelerated by catalysts.

Plastisol coatings have a solids content of almost one hundred percent. Plastisol is a paste-like mixture of polymers and plasticizers. Thus, for the PVC plastisols, PVC is dispersed in the plasticizer and mixed with the necessary additives, pigments and fillers. The plastisols gel under the influence of heat and form a thick surface film. An organosol is a mixture of polymer and plasticizer that also contains organic solvents. For example, the polyvinylidene fluoride polymer used in coil coating is processed as an organosol. Organosols belong to the group of physically drying film formers, which solidify by releasing the solvents without any chemical reaction.

Powder coatings and radiation-curing coatings are solvent-free. Here, crosslinking of the basic building blocks is induced thermally or by radiation.

Foils are laminated as finished products directly onto the pre-treated or onto the already lacquered metal strip. The bond to the metal is produced with the aid of an adhesive or by hot laminating the film.

Decoration and protection are provided by the topcoat layer. When designing the coating for coil coating applications, four aspects must be taken into account, which must be optimally coordinated with each other:
- It must be possible to manufacture the coating consistently.
- When processed on the coil coating line, it must withstand the high mechanical stresses in the coating process itself. For example, it must not foam excessively. The shear conditions on the rollers must not cause the pigments to become even more wetted, because heavier wetting would result in undesirable color shifts. The coating must be designed in such a way that, during curing, the solvent can first escape from the surface within about 10 seconds before the surface closes. Curing too quickly will result in unevenness on the surface.

The paint must be both elastic and hard so that the paint surface is not damaged during further processing by cutting, bending, edging, deep drawing, assembling, and fixing.

The paint must also be optimized for the final application. Roof and facade elements, for example, must be resistant to weathering and at the same time meet aesthetic requirements in terms of gloss and color tone for years to come. The surfaces of refrigerators and washing machines should also look good for many years in the household and must not be permanently discolored by vinegar, coffee, or ketchup.

The coating material is selected on the basis of the requirement profile specified by the processing and use of the coil-coated sheets.

It is even possible to apply powder coatings to a flat metal strip transported at high speed, although this is technologically difficult. For this reason, powder coating is only applied to a very small extent at low strip speeds in coil coating processes. The powder-coating layer is applied with the aid of guns, as in industrial coating. New application methods such as the use of rotating or electromagnetic brushes have not yet become established. Powder coatings are based on polyester or polyurethane. They do not contain any solvents and are thermally cured in a conventional or infra-red dryer. During curing, the powder particles first melt and then combine to form a smooth coating film. In the second step, the cross-linking reaction, the physical properties of the paint film are formed. The use of powder coatings for coil coating is interesting in terms of coating thickness, because this technology allows coatings of up to 100 µm to be achieved, which can only be achieved with liquid coatings and roll application in coil coating with PVC plastisol. The higher layer thicknesses are interesting for applications in highly corrosive environments.

Almost all colors can be reproduced in the coil coating process, therefore a wide range of pigments are being used in coil coating formulations [272] in order to provide the desired color and effect. But as coil-coated coatings are typically cured by an acid-catalyzed process that can react with some types of pigments and coil-coating parts are bound by long manufacturer warranties of up to 25 years, pigments with the highest performance are being used.

Only a few high-quality organic pigments are suitable for use in coil-coating paints, as they must be stable up to 250°C PMT (peak metal temperature [272]). Therefore, coil coating application can be considered as a domain for inorganic "high temperature" pigments. These CICPs are preferred where colored objects are exposed to high temperatures, UV light, or harsh chemical environments either during the manufacturing process or in use.

Oxide mixed-phase pigments which are lightfast, weather-, acid-, alkali-, and temperature-resistant, and also resistant to most chemicals with use in coil coatings can be found here [15].

An important group of pigments used in coil coatings are ceramic pigments with high reflectivity in the NIR region [115]. These colored inorganic pigments are produced synthetically and are specially designed for use in the building sector, for the "cool roof". In particular, they reflect thermal radiation and thus prevent the rapid whitening of buildings. They also have the advantage that they retain their high reflectivity even in dark shades and that their color tone changes over time when exposed to sunlight over years [274].

Inorganic pigments such as $PbCrO_4s$ are no longer widely used due to their heavy metal content, whilst organic pigments have gained in importance, as described earlier. They are hydrophobic and more finely divided than inorganic ones. They have a high, selective absorption capacity, which manifests itself in very pure shades in a wide range of color nuances. However, their scattering and hiding power are lower than those of inorganic pigments. Therefore, they often have to be combined with inorganic pigments to cover the substrate. In addition to lightfastness and weather resistance, the organic pigments considered for coil coating must also be able to withstand the drying process at temperatures of up to 250°C. This also severely restricts the choice of pigments.

The interplay of functionality, aesthetics and harmony is in great demand, especially in architecture. Therefore, inorganic luster pigments, as explained in an earlier section of this chapter, have gained in importance in the coil coating industry as they can achieve special, appealing and eye-catching effects setting new trends in coil coatings [272, 275]. Metallic pigments fulfill the requirements of the coil coating industry and are frequently used by paint manufacturers to generate color shades like, e. g., RAL 9006. Also, interference or pearlescent pigments are used in coil coatings and produce a hue shift dependent on the viewing angle [99].

Anti-corrosion pigments are not used because of their coloring effect, but because – as the name implies – of their special properties with regard to corrosion protection. They achieve their effect in two ways, physically and chemically: They extend the diffusion path of water and aggressive substances from the coating surface to the metal substrate. They can ensure good adhesion of the coating material to the substrate or also prevent the film former from being destroyed by reflecting or absorbing UV radiation. This passive corrosion protection can be achieved with the aid of aluminum silicates or mica, among others. Some anti-corrosion pigments cause a chemical effect through reactions at the interface between substrate and coating. They create an alkaline environment on the substrate surface in which acidic compounds are neutralized and others are converted to poorly soluble compounds [276, 277]. As a result, a protective layer is built up on the metal surface. Pigments with a high oxidation potential retard corrosion [190, 278]. Common anti-corrosion pigments are zinc phosphate, zinc dust, zinc aluminum phosphate, tungstates, zirconates, and vanadates.

Conductivity is achieved by carbon black pigments or, in the case of light-colored coatings, by mica pigments coated with antimony or tin oxide. Pigment carbon blacks with specific surface areas of up to 1,000 $m^2/g$ are chemically resistant, light, and weather stable and exhibit high color depth and color strength [185, 188, 279].

### 17.3.5 Plastic coatings

Pigments are introduced into automotive plastic coating basecoat layers to impart color, gloss control, and/or UV protection. The amount of pigment utilized depends upon the color and the hiding power required of the coating.

Solvent-based basecoats are applied in dry film thicknesses from about 13 µm up to 45 µm, depending on the hiding power of the coating. The hiding power is defined as the lowest film thickness that covers a black and white or gray and white coverage diagram. The hiding power depends on the paint due to the composition of the pigments that make up the paint. For example, if the paint consists mainly of transparent iron oxides (e. g., red basecoat with a straight tint), the required film thickness is higher than for a black, carbon black basecoat with very good hiding power. Poorly covering colors, such as red, yellow and white solid colors, require a higher film thickness to achieve opacity than metallics and darker shades [260].

Special effect pigments, as described earlier in this chapter, have allowed color stylists to develop more innovative colors that are richer, have high color intensity and show sharper hues.

Solvent-based basecoats are typically applied and allowed to flash off (equivalent to solvent evaporation) at room temperature or slightly elevated temperature for three to five minutes before the clearcoat is applied. This process of applying one coating over another, with the first coating not in a crosslinked or cured state, is known as the "wet-on-wet" process.

Waterborne basecoats are used to achieve better orientation of the anisotropic effect pigment particles used in the effect coating. As already mentioned in the effect pigments part of this chapter, this is still part industrial and academic research [252, 253]. The basecoats consist of urethane or acrylate dispersions, which may or may not be crosslinked. The basecoat is often sprayed on and baked in a so-called "heating flash" (with infrared, microwave or convection ovens) to remove most of the water (>90%) before clearcoat application [260].

Standard blacks in combination with $TiO_2$s of the rutile type are used for the widely used contract basecoats. For the formulation of high-gloss coatings, the entire range of inorganic and organic colored pigments is used. They are selected depending on the intended application and the associated light and weather fastness with optimum economy. In this respect, there is no difference between plastic coatings and their equivalents on other substrates. The formulation of waterborne or solventborne pearlescent and metallic effect paints follows the same principles as for basecoats. Special attention must be paid to the protection of the effect pigments by the resin matrix [280].

Fillers are able to improve important properties such as adhesion, chemical, or stone chip resistance. Commonly used fillers include talc (hydrated magnesium silicate), barium sulfate, or chalk ($CaCO_3$). Due to its platelet shape, talc is said to have the property of distributing any impact forces acting on it in the layer (energy dissipation) and to protect the film from damage by this mechanism [280].

Pigments and extenders or fillers, in combination with the resin systems, have a significant influence on the thermal expansion, the glass transition temperature and the mechanical properties of the paint film. Further technical criteria for the selection of pigments and fillers are their influence on adhesion and sanding behavior. In view of the alkali sensitivity of some plastics, pH-neutral pigments and fillers are usually selected for the coatings that are in direct contact with the substrate surface. Electrically conductive pigments enable the electrostatic coating of primers. The first choice of such pigments are carbon blacks with a high dibutyl phthalate adsorption value. The color of these primers can range from medium to dark gray. To achieve better color matching at low basecoat thickness, light-colored primers containing special electrically conductive mica pigments have been developed [280].

For the overall film performance of all automotive and industrial coatings, the pigment/binder ratio is of importance. The selection of colored pigments is oriented to the OEM master color panel and the feasibility of color matching with the car body.

At this point it should be mentioned that powder coating technology is also used for plastic coatings [280] and that their application is mainly limited to the powder IMC technology. As described in the part 4.2. above, film forming and curing temperatures are about to reach level of 120°C opening new possibilities in the coating of plastics [250, 251].

### 17.3.6 Automotive coatings

The performance of an automotive coating can be evaluated from various perspectives, such as the durability of the exterior and interior finishes or the aesthetic characteristics [5]. However, there are limits to the paint properties, process capabilities, and most importantly, the cost that can be spent to improve the paint finish [22]. Consequently, each automotive company defines its color and appearance standards to meet or exceed competitor levels and customer expectations. Uniformity – or harmony – between all components is particularly important. Exterior painting requires that add-on parts such as bumpers, spoilers, mirror housings and other trim match adjacent body panels. Differences in color and appearance are especially apparent in car models where sheet metal matches very closely [139].

The quality of coatings can be judged by three main criteria: Protection against harsh environments, durability and quality of appearance. A harsh environment refers to damage caused by falling objects (falling out), UV rays, heat (above 80°C) or cold (below 20°C), scratches, stone chips (chipping), and the rust from salt or road deicer. A measure of longevity is a car with rust protection for more than 20 years and good paint/gloss appearance for more than a decade. The three parameters used to determine appearance quality are color, paint smoothness, and gloss [281].

As products enhanced with effect paints become more attractive to potential buyers, effect paints can be described as a tool used to increase sales and create brand value. Therefore, it is understandable, that their properties should be maintained over the entire product life cycle in order to preserve brand reputation and avoid customer complaints [282].

In order to meet these requirements, only raw materials that fall into the high-performance category must be used in automotive coatings (please refer to Figure 17.5). The components of paints that form the paint layers on a car include the pigments, resins, and additives, as well as solvents that impart flowability. Pigments primarily impart color and gloss; they also help produce the thickness of coatings. Resins, including synthetic and cross-linking materials as well as hardeners, form the paint film. Finally, additives, which act as anti-settling agents, stabilize the paint and make it easy to process. The factors responsible for the quality of the appearance are the visual quality (aesthetics), which is determined by the sprayed layer and includes color and brightness, and is influenced by the presence or absence of metallic flakes; appearance (smoothness), which refers to the film's unevenness with wavelengths in the range of 1–5 mm; gloss, which is relevant to the unevenness with a wavelength of 0.01–0.1 mm; and color and floating behavior, or the film's ability to exhibit color-changing properties depending on the viewing angle [139].

As described earlier, it is the effect pigments that provide these color-changing, unique visual effects, resulting from their chemical composition and physical properties, with the particle morphology/size and orientation of the effect pigment flakes playing a crucial role in the expression of color and effect in the dried paint film [99].

Industrial (liquid) paint application processes always make allowance for parameters like air pressure, pistol type, nozzle size, spraying pistol to object distance as they control flake orientation within the dried paint layer as well. They account for the geometry of the paint droplets, the impact speed and the spraying pattern generated on the substrate. The development of the paint film layer or coalescence of these polymer droplets loaded inter alia with flaky pigments is biased, another factor which affects the standard deviation of particle orientation within the polymer matrix. Finally, also the evaporation of (co-) solvents in solvent-/waterborne paints and the stratification of the polymer film should be mentioned for the sake of completeness. All these factors explain, why for example car body parts are being repaired by experienced body painters, as the repair of effect pigmented car surfaces is a real challenge and reproducibility of all application steps is the key [283].

A lot of effort is put into better understanding the interaction of all influencing parameters in order to achieve maximum control over the developing color and the effect in the dried paint layer in advance.

An insight into these efforts provides [139]. It describes main influencing factors on color tone which are pigment alignment and concentration, where pigment alignment is affected by atomization, the viscosity of the paint after coating a surface, disturbance by dust, and the impact velocity of the spray on a surface. The concentration of a pigment on a surface is also affected by the transfer efficiency of the spray and the pigments on the substrate. An acceptable surface with a good metallic sheen is achieved when the metallic or mica effect pigments are aligned parallel to the surface; if this is not the case, the surface will appear dull and have a lower value (i. e., it will appear darker, as already reported in [99]). In practice, the colored pigments behave like fine flakes moving in a wet film. Therefore, it is difficult to ensure that the pigments are aligned with their largest particle area parallel to the surface. For this reason, various additives are used to ensure that the aluminum flakes are oriented parallel to the surface and are evenly distributed; generally, these additives increase the paint viscosity immediately after spraying a surface, thereby preventing pigment movement and disorientation. To ensure surface-parallel alignment of the aluminum flakes, it is also necessary that the impact velocity of the spray droplets on a surface is high enough; the velocity is influenced by the viscosity and diameter of the paint droplets. If the droplets are too small, the concentration of pigments in them will also be small, resulting in a low transfer efficiency. When the transfer efficiency is high, the concentration of aluminum flakes in the coating is high, as is the gloss level.

Research in the field of effect coatings is ongoing, highlighting the interest of the (automotive) industry and academia to the same extent. The focus of the work is on the analysis, control and prediction of the orientation of effect pigment particles and their influence on the final color and appearance using state-of-the-art analytics [167, 252, 253].

The automotive industry has for many decades been the driving force behind innovative painting concepts must still be regarded as such today. As important trends in automotive coatings, the use of powder coating as well as the application of alternative coating layer sequences such as 3-wet painting must be mentioned. Both technologies offer sustainability or environmental advantages, such as reduced VOC and $CO_2$ emissions in particular [139].

Powder coatings have been explained in a section above. Their compositions contain very low concentrations of volatile solvents, on the order of 2%, much less than any other coating system. The automotive industry uses powder coatings for wheels, bumpers, hubcaps, door handles, trim and accent parts, truck beds, radiators, filters, and numerous engine parts [284]. A clear powder topcoat has also been developed; BMW and Volvo are using it on their new car models, and GM, Ford and Chrysler have formed a consortium to test it on their production lines. Powder coatings are an advanced method of applying a decorative and protective finish to almost any type of metal and can be used by both industry and consumers. The result is a uniform, durable, high-quality, and attractive surface. Powder coating can be used to produce much thicker coatings than traditional liquid paints without running or falling off. Items that are powder coated have less difference in appearance between horizontally and vertically coated surfaces than liquid coated items. In addition, powder coatings can be used to achieve a variety of special effects that are not possible with other coating processes. While powder coatings have many advantages over other coating methods, there are also some disadvantages to the technology, one of which occurs when powder coatings are used in conjunction with particulate metal particles such as aluminum flakes [244]. The reason for this is that the process of film formation in powder coatings is fundamentally different from liquid coatings, as the formed coating layer has a significantly higher viscosity which has a tremendous impact on flake orientation of the effect pigment particles. Therefore, color matching of parts sprayed with the identical color shade using different paint technologies like liquid and powder is a huge challenge, as described in detail already.

The 3-Wet Paint process is striving to eliminate or minimize the amount of spray application processes, space and curing ovens, as this can result in significant material and energy savings, like the successfully established wet-on-wet (3-Wet) system that largely eliminates the primer oven in the painting process [139]. Waterborne 3-wet paint systems have been developed to reduce VOCs and $CO_2$ emissions by two-thirds. Recently, an aqueous 3-wet paint system was developed with an appearance similar to that of a conventional 3-layer two-bake (3C2B) paint system by using base resins with a low glass transition point to promote leveling and then reducing the melamine content of the paint to minimize shrinkage during curing [285].

To illustrate the innovative power of automotive manufacturers with regard to new painting processes that enable completely new color shades and effects, three current developments should be mentioned here.

In [286], a successful transfer of color shades from individual to series production was reported, which now enables this OEM to offer innovative red and orange colors with increased color depth and purity using "tinted clears" within a special paint layer sequence.

In [287], the development of an economical and ecologically beneficial single-paint process for the realization of bi-tone designs for automobiles was realized, enabling this car manufacturer to meet the bi-tone market trend thanks to this smart and efficient paint process.

In [288], a sustainable, lean and flexible eco-paint process for truck cabins has been developed. As challenges which had to be tackled during the development were: to develop a new pigment and resin strategy to enable high solid contents, to control the curing behavior and wetness of the scrubbed material and to realize a wet-on-wet film thickness with more than 70 microns. As a result, an efficient and flexible, sustainable and environment-friendly paint concept with premium quality could be realized.

Pigments are used to add color and effects to car exteriors and interiors. In automotive topcoats, mixtures of inorganic/organic pigments, metallics, pearlescent pigments, $TiO_2$, and carbon black are used to create color and effects, plus functional pigments (for corrosion protection), supplemented by extenders, nanoparticles and matting agents. In terms of organic pigments, 5,000 t of organic pigments were used in the automotive sector, coloring approximately 60 million cars produced in 2006, as reported in [24].

Besides coloristic attributes, important technical properties of pigments used in automotive coatings are dispersibility, rheology, flocculation, sedimentation and storage behavior in the liquid phase followed by light and weather fastness, solvent-, chemical-, heat-, and bleed resistance in the dried film [289].

Among the inorganic pigments, $TiO_2$ white is by far the most important. The photocatalytic activity of $TiO_2$ leads to rapid degradation of the organic binder matrix, so the surface of the $TiO_2$ particles must be covered with an inorganic coating to prevent photooxidation of the polymer matrix. This is done by applying layers of $SiO_2$ and $Al_2O_3$ [186].

The metallic effect is caused by the reflection of light from the surface of the aluminum particles. Larger particles are better reflectors, resulting in higher flop and brightness, while smaller particles show less flop because the amount of light scattered off the edges as an undirected reflection is increasing. With coarser aluminum pigments, the individual particles become more visible, resulting in graininess or texture. Since aluminum reacts with water to form aluminum hydroxide and hydrogen gas, the aluminum surface must be passivated for use in waterborne basecoats. Chromium treatment of the aluminum pigments results in very good gassing stability, but chromium has become a substance of concern for some automotive manufacturers. Silica treatment results in stable protection of aluminum particles from hydrolysis even under conditions of shear stress in a circulation line. Organophosphates can be applied to the aluminum paste or added during the paint manufacturing process.

Newer effect pigments based on aluminum are so called VMPs manufactured in a physical vapor deposition process [99]. Although the use of these ultra-thin pigments of <50 nm thickness can be challenging in coatings, they allow to realize highly reflective coatings with an almost chrome-like appearance [290]. A method of producing a polished metal effect finish on a vehicle using VMPs was reported here [291].

More recent effect pigments are based on light interference with layers of materials with different refractive indices [99, 292], where the layer thickness is of the order of the light wavelength of about 500 nm [293].

An overview on organic pigment classes used in automotive coatings can be found in Figure 17.25.

## 17.4 Conclusion and future outlook

The paint or coating industry is changing and thus exerts a direct influence on research and development in the pigment industry [294–300].

In addition to quality, technology and commercial factors, the issue of sustainability dominates research, and development activities, leading to increased demand for high-solids, waterborne, powder coatings, and radiation-curing coating systems.

As described in this chapter, the diverse requirements for the successful use of the different pigment classes in these systems may well be regarded as a challenge. For this reason, development work on colorants used in the above-mentioned coating systems is aimed at synthesizing novel products that open up new color spectra. At the same time, they must strictly comply with all technical and safety requirements for their handling during paint production, subsequent use and the subsequent recycling process.

Ergo, product quality may well be mentioned as a constant in this equation, which is why coatings will continue to be subject to a continuous improvement process in the future. The focus here is on further increasing the service life of coated surfaces in order to reduce environmental impact, waste and raw material and energy consumption.

The advancing automation and increased use of intelligent technology and artificial intelligence in the pigment and coatings industry should also be seen in this context, as these support further process improvements and further improve the stability, quality consistency and reproducibility of the products produced [301–304].

If, at the end of this chapter, we again consider Figure 17.5, which summarizes the requirements profile for high-performance pigments for automotive coatings, the criterion "Autonomous Driving" may be listed here as an example. This has only recently been integrated into this overview and is directly related to the sensor technology that is finding its way into the automotive industry. This leads to special requirements for the coatings and colorants used, which can only be mastered with expert knowledge and high formulation expertise and thus represents a prime example of the constant change under the influence of technology, such as 3D printing [305], on the coating and pigment industry.

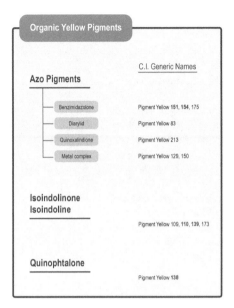

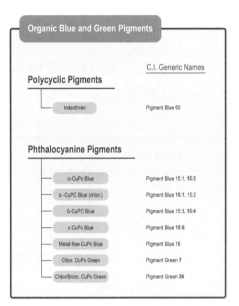

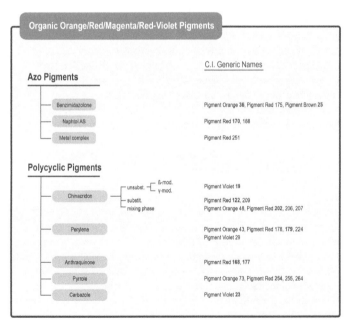

━━━━ = important representatives of this group

**Figure 17.25:** Organic pigments used in automotive coatings incl. C.I. generic names [140].

# References

1. Ullmann's Encyclopedia of Industrial Chemistry. Paints & coatings, section 4.3. 6th ed. Weinheim: VCH Verlagsgesellschaft, 2006.
2. Bamfield P. Chromic phenomena - technological applications of colour chemistry. The Royal Society of Chemistry, 2001.
3. Ghosh SK, editor. Functional coatings. Weinheim: WILEY-VCH Verlag, 2006.
4. Tracton AA, editor. Coating materials and surface coatings. Taylor & Francis Group; 2007.
5. Wicks ZW Jr, Jones FN, Pappas SP, Wicks DA Organic coatings science and technology. 3rd ed. John Wiley & Sons, 2007.
6. Shore J, editor. Colorants and auxiliaries, vol. 1 colorants. 2nd ed. Bradford: Society of Dyers and Colourists, 2002.
7. Herbst W, Hunger K. Industrial organic pigments. 3rd ed . Weinheim: VCH, 2004.
8. Zollinger H. Color chemistry. 2nd ed . Weinheim: VCH, 1991:347.
9. Gordon PF, Gregory P. Organic chemistry in colour. Berlin: Springer Verlag, 1987.
10. Nassau K. The physics and chemistry of color. New York: Wiley-Interscience, 1983.
11. Tilley RJ. Colour and the optical properties of materials. Chichester: Wiley, 2000.
12. Hunger K. Industrial dyes. Weinheim: VCH, 2003.
13. Nassau K, editor. Color for science, art and technology. Amsterdam: North-Holland, Elsevier Science, 1998.
14. Zollinger H. Color - a multidisciplinary approach. Weinheim: Wiley-VCH, 1999.
15. Goldschmidt A, Streitberger H-J. BASF-handbook basics of coatings technology. 2nd ed. Hannover: Vincentz Network, 2007.
16. Pfaff G. Inorganic pigments. Berlin/Boston: Walter de Gruyter GmbH, 2017.
17. Klöckl I. Chemie der Farbmittel, Bd. 1, Grundlagen, Pigmente und Farbmittel. 2nd Aufl. Berlin/Boston: Walter de Gruyter GmbH, 2020.
18. Winkler. Titanium dioxide. 2nd ed. Vincentz Network, 2013.
19. Müller B, Poth U. Coatings formulation. Hannover: Vincentz Network, 2006.
20. Rössler A. Lackentwicklung mit statistischer Versuchsplanung. Vincentz Network Hannover, 2011.
21. Gueller R. The workflow management software Arksuite, FARBE + LACK WebKonferenz Industrie 4.0. Vincentz Verlag Hannover, 2020.
22. Khanna AS. High-performance organic coatings. Cambridge, UK: Woodhead Publishing Limited, 2008.
23. Lin L. Book review - organic coatings science and technology. Dyes Pigm. 2000;45:85–86.
24. Streitberger H-J, Dössel K-F, editor. Automotive paints and coatings, 2nd completely revised and extended edition. Wiley-VCH, 2008.
25. Flick EW. Water-based paint formulations. Vol. 4. Westwood/USA: Noyes Publications, 1997.
26. Benzing G, et al. Pigmente für Anstrichmittel. Grafenau: expert verlag, 1988.
27. Poth U. Autolacke formulieren. Hannover: Vincentz Network, 2007.
28. Richter G. Trends bei der Farb-Applikation und Effekt-Lackierung. JOT. 2005;3:38.
29. Finkenzeller M. PPCJ. 2005;11:22.
30. Gangloff C. PPCJ. 2009;25.
31. World dyes & organic pigments. The Freedonia Group Inc., 2009.
32. World dyes & organic pigments. The Freedonia Group Inc., Study #3264, 2015.
33. Die Köpfe hinter dem Design. Automobil-Produktion. 2005;11:58.
34. Dworschak M. Der Spiegel. 2006;25:126.
35. Lindstrom M. Buyology: Truth and lies about why we buy and the new science of desire. Bantam Dell Pub Group, 2008.

36. Häusel H-G. Brain view - Warum Kunden kaufen. Rudolf Haufe Verlag, 2009.
37. Adams R. Global overview of the $TiO_2$ & coloured pigment industries' (page 15), Smithers $TiO_2$ and colour science symposium. Berlin, 2019.
38. Koleske JV. Meyers RA, editor, In encyclopedia of analytical chemistry. Chichester: John Wiley & Sons Ltd, 2000.
39. De Lange PG. Powder coatings chemistry and technology. Vincentz Network: Coatings Compendia, 2004.
40. Verkholantsev VV. Functional variety. Effects and properties in surface-functional coating systems. Eur Coat J. 2003;9:18.
41. Wulf M, Wehling A, Reis O. Coatings with self-cleaning properties. Macromol Symp. 2002;187:459.
42. Hegedus CR. A holistic perspective of coating technology. JCT Res. 2004;1:5.
43. Parkin IP, Palgrave RG. Self-cleaning coatings. J Mater Chem. 2005;15:1689.
44. Nun E, Oles M, Schleich B. Lotus-effect-surfaces. Macromol Symp. 2002;187:677.
45. Kuhr M, Bauer S, Rothhaar U, Wolff D. Coatings on plastics with the PICVD technology. Thin Solid Films. 2003;442:107.
46. Perez M, Garcia M, Del Amo B, Blustein G, Stupak M. Core-shell pigments in antifouling paints. Surf Coat Int Part B Coat Trans. 2003;86:259.
47. Zhou LC, Koltisko B. Jct Coatings Tech. 2005;2:54.
48. Tiller JC, Liao CJ, Lewis K, Klibanov AM. Designing surfaces that kill bacteria on contact. Proc Natl Acad Sci USA. 2001;98:5981.
49. Johns K. Surf Coat Int Part B Coat Trans. 2003;86:91.
50. Provder T, Baghdachi J, editors. Smart coatings II; ACS symposium series, vol. 1002, Washington, D.C: American Chemical Society, 2009.
51. Provder T, Baghdachi J, editors. Smart coatings; ACS symposium series, vol. 957. Washington, D.C: American Chemical Society, 2007.
52. Zhang H, Zeng W, Du H, Ma Y, Ji Z, Deng Z, Zhou Q. Comparison for color change between benzodifuranone and benzodipyrrolidone based epoxy coating. Dyes Pigm. 2020;175:108171.
53. Rossi S, Simeoni M, Quaranta A. Behavior of chromogenic pigments and influence of binder in organic smart coatings. Dyes Pigm. 2021;184:108879.
54. Nanguo L, Brinker CJ. Smart light responsive materials - azobenzene containing polymers and liquid crystals. In Smart Light-Responsive Materials, Zhao Y, Ikeda T, editors, Hoboken: John Wiley & Sons, 2009.
55. Seeboth A, Lötzsch D. Thermochromic phenomena in polymers. Shrewsbury: Smithers Rapra Technology Limited, 2008.
56. Kaluza U. Physical/chemical fundamentals of pigment processing for paints and printing inks, edition lack und chemie. Elvira Moeller GmbH, 1981.
57. Mollet H, Grubenmann A. Formulierungstechnik. Wiley-VCH, 1999.
58. http://www.borchers.com.
59. https://solutions.covestro.com/en/materials/coatings.
60. http://www.basf.com/global/en/products/segments/industrial_solutions/dispersions_and_pigments.html.
61. https://www.byk.com/en.
62. Funke W. Problems and progress in organic coatings science an technology. Progr Org Coat. 1997;31:5.
63. Schulz U. Accelerated Weathering. Vincentz Network; 2008.
64. Taft Jr WS, Mayer JW. The science of paintings. New York: Springer-Verlag, 2000.
65. Welsch N, Liebmann CC. Farben – Natur, Technik, Kunst. Berlin: Spektrum Akademischer Verlag, 2003.

66.  Reichel A, Hochberg A, Köpke C. Plaster, render, paint and coatings. Munich: Institut für internationale Architektur-Dokumentation GmbH & Co KG., an edition DETAIL book, 2004.
67.  Gordon PF, Gregory P. In developments in the chemistry and technology of organic dyes, Griffiths J, editor. Oxford: Blackwell Scientific Publications 1984:66.
68.  Gray GW. Chimia. 1980;34:47.
69.  Okawara M, Kitao T, Hirashima T, Matsuoka M. Organic colorants: A handbook of data of selected dyes for electro-optical applications. Oxford: Elsevier, 1988.
70.  Matsuoka M, editor. Infrared absorbing dyes. New York: Plenum, 1990.
71.  Dürr H, Bauas-Laurent H, editors. Photochromism molecules and systems. Elsevier B.V., 2003.
72.  Fox R, Flory PJ. Second-order transition temperatures related properties of polystyrene. I. Influence of molecular weight. J Appl Phys. 1950;21:581.
73.  Gysau D. Fillers for paints. Hannover: Vincentz Netzwerk, 2019.
74.  Calbo LJ, editor. Handbook of coatings additives. New York: Marcel Dekker, 1987.
75.  Karsa DR, editor. Additives for waterbased coatings. Cambridge: The Royal Society of Chemistry, 1991.
76.  Shore J, editor. Colorants and auxiliaries, vol. 2 auxiliarie. 2nd ed. Bradford: Society of Dyers and Colourists, 2002.
77.  Bieleman J, editor. Additives for coatings. Weinheim: Wiley-VCH, 2002.
78.  Davison G, Lane B, editors. Additives in water-borne coatings. The Royal Society of Chemistry, 2003.
79.  Kittel H. Lehrbuch der Lacke und Beschichtungen. vol. Band 4. Stuttgart: Hirzel-Verlag, 2006.
80.  Boxall J, von Fraunhofer JA. Paint formulation. New York: Industrial Press Inc., 1981.
81.  Patton TC. Paint flow and pigment dispersion. New York: J. Wiley & Sons, 1978.
82.  Stoye D, Freitag W, editors. Paint, coatings and solvents. 2nd ed. Weinheim: VCH Verlagsgesellschaft, 1998.
83.  Heilen W et al. Additives for waterborne coatings. In European coatings TECH FILES, Vincentz Network|, 2014, ISBN: 9783748602187. DOI: 10.1515/9783748602187.
84.  Vash R. Pigment wetting and dispersinig additives for water-based coatings and inks. In Surface phenomena and additives in water-based coatings in printing technology, Sharma MK, editor. Boston, MA: Springer, 1991. DOI: 10.1007/978-1-4899-2361-5_10.
85.  Bugnon P. Surface treatment of pigments - treatment with inorganic materials. Prog Org Coat. 1996;29:39.
86.  Xiao B, Wu H, Guo S. Regulating the properties of C.I. Pigment Red 170 by surface modification via hydrous alumina. Dyes Pigm. 2016;127:87.
87.  Wu H, Gao G, Zhang Y, Guo S. Coating organic pigment particles with hydrous alumina through direct precipitation. Dyes Pigm. 2011;92:548.
88.  Kumar KR, Ghosh SK, Khanna AS, Waghoo G, Ansari F, Yadav K. Silicone functionalized pigment to enhance coating performance. Dyes Pigm. 2012;95:706.
89.  Tadros TF. Solid/Liquid dispersions. London: Academic Press, 1987.
90.  Lyklema J. Fundamentals of interface and colloid science. vol. 1 – 3. London: Academic Press, 1991–2000.
91.  Parfitt GD, editor. Dispersion of powders in liquids (with special reference to pigments). London: Applied Science Publ., 1981.
92.  Pfaff G, Maile FJ, Kieser M, Maisch R, Weitzel. In special effect pigments. Zorll U, editor. 2nd revised ed, Hannover: Vincentz Verlag, 2008.
93.  Napper DH. Polymeric stabilisation of colloidal dispersions. London: Academic Press, 1983.
94.  Hays BG. Am Inkmaker. 28 Jun 1984, 13 Oct 1986, 28 Nov 1990 .
95.  Legrand P. Stabilizing properties of polymers. ECJ. 2005;3:90.

96. Schaller C, Schauer T, Eisenbach CD, Dirnberger K. Synthesis and stabilizing properties of amphipolar polyelectrolytes. Eur Phys J. 2001;E 6:365.
97. Reynders P, Maile FJ. Sedimentation-behavior of substrate-based effect pigments. In: XXVII Fatipec Congress, vol. 2, Aix-en-Provence, France, Apr 2004.
98. Mischke P. Filmbildung. Vincentz Network, 2007.
99. Maile FJ, Pfaff G, Reynders P. Effect pigments – past, present and future. Progr Org Coat. 2005;54:150.
100. Estlander T, Jolanki R. Paints, lacquers, and varnishes. In Rustemeyer T, Elsner P, John SM, Maibach HI, editors. Kanerva's occupational dermatology. Berlin, Heidelberg: Springer, 2012. DOI: 10.1007/978-3-642-02035-3_61.
101. Clarke EA, Steinle D. Health and environmental safety aspects of organic colorants. Rev Prog Color. 1995;25:1.
102. https://colour-index.com/definitions-of-a-dye-and-a-pigment.
103. Kindly reproduced by permission of ETAD – The Ecological and Toxicological Association of Dyes and Organic Pigment Manufacturers. http://www.etad.com.
104. Kindly reproduced by permission of CPMA – Color Pigment Manufacturers Association, Inc. http://www.pigments.org.
105. Lewis PA. In color for science, art and technology, Nassau K, editor. Amsterdam: Elsevier, 1998:285.
106. Büchner W, Schiebs R, Winter G, Büchel KH. Industrielle anorganische Chemie. Weinheim: Verlag Chemie, 1984.
107. Pfaff G. In industrial inorganic pigments. Buxbaum G, Pfaff G, editors. 3rd ed, Weinheim: Wiley-VCH Verlag, 2005:230.
108. Winnacker-Küchler, Chemische Technologie. 4th ed, vol. 3, München: Carl Hanser Verlag, 1983.
109. Lewis PA. Pigment handbook. New York: J. Wiley & Sons, 1988.
110. Smith AE, Comstock MC. Spectral properties of UV absorbing and Near IR reflecting blue pigment, Yin10xMnxO$_3$. Dyes Pigm. 2016;133:214.
111. Li J, Smith AE, Jiang P, Stalick JK, Sleight AW, Subramanian MA. True composition and structure of hexagonal "YAlO$_3$", actually Y$_3$Al$_3$O$_8$CO$_3$. Inorg Chem Article ASAP. 2015;54:837.
112. Smith AE, Sleight AW, Subramanian MA. Synthesis and properties of solid solutions of hexagonal YCu$_{0.5}$Ti$_{0.5}$O$_3$ with YMO$_3$ (M = Mn, Cr, Fe, Al, Ga, and In). Mater Res Bull. 2011;46:1.
113. Smith AE, Mizoguchi H, Delaney K, Spaldin NA, Sleight AW, Subramanian MA. Mn$^{3+}$ in trigonal bipyramidal coordination: a new blue chromophore. J Am Chem Soc. 2009;191:17084.
114. Subramanian M. Benign by design: advanced inorganic color pigment design through materials chemistry. Prague: The Pigment & Colour Science Forum, 2015.
115. Comstock MC, Complex inorganic colored pigments: comparison of options and relative properties when faced with elemental restrictions. In : 56th SCAA Conference, Melbourne, VIC, Sep 2016.
116. Baughman GL, Banerjee S, Perenich TA. Dye solubility. In Physico-chemical principles of color chemistry, advances in color chemistry series, vol. 4, Peter AT, Freeman HS editors. Dordrecht: Springer, 1996. DOI: 10.1007/978-94-009-0091-2_5.
117. https://colour-index.com/introduction-to-the-colour-index.
118. Giles CH. The surface properties of dyes. Rev Prog Color. 1981;11:89.
119. Patton TC, editor. Pigment handbook, vol. 3 . New York: Wiley-Interscience, 1973.
120. Lewis PA. editor. Pigment handbook. 2nd ed, vol. I, New York: Wiley-Interscience, 1988.
121. Solomon DH, Hawthorne DG. Chemistry of pigments and fillers. New York: Wiley-Interscience, 1983.

122. Braun JH. White pigments, federation of societies for coatings technology. PA: Blue Bell, 1995.

123. Challener C. Jct Coatings Tech. 2005;2:44.

124. Faulkner EB, Schwartz RJ, editors. High performance pigments. 2nd ed. Weinheim: Wiley-VCH Verlag GmbH & Co. KGaA; 2009.

125. Büschel KH, Moretto -H-H, Woditsch P. Industrial inorganic chemistry. Weinheim: Wiley-VCH, 2000.

126. U.S. Department of Commerce, Bureau of Census, Current Industrial Reports: Paint and Coating Manufacturing. http://www.census.gov.

127. Lambourne R. In paints and surface coatings, theory and practice, Lambourne R, Strivens TA, editors. 2nd ed, Chem Tech Pub. Inc., 1999.

128. Pfaff G, Reynders P. Angle-dependent optical effects deriving from submicron structures of films and pigments. Chem Rev. 1999;99:1963.

129. Fasano DM. Use of small polymeric microvoids in formulating high PVC paints. J Coat Technol. 1987;59:109.

130. Vanderhoff JW, Park JM, El-Aasser MS. Preparation of soft hydrophilic polymer core/hard hydrophobic polymer shell particles for microvoid coatings by seeded emulsion polymerization. Polym Mater Sci Eng. 1991;64:345.

131. Lewis PA. Organic pigments, federation of societies for coatings technology. PA: Blue Bell, 1995.

132. van den Haak HJW, Krutzer LLM. Design of pigments dispersants of high-solids paints system. Progr Org Coat. 2001; 43:56.

133. Patent US 9,062,216 (The Shepherd Color Company). 2014.

134. Ryan M. Expanding the durable color envelope, RTZ Orange & NTP Yellow. Rome: The Pigment & Colour Science Forum; 2013.

135. Cao L, Fei X, Zhao H. Environmental substitution for $PbCrO_4$ pigment with inorganic-organic hybrid pigment. Dyes Pigm. 2017;142:100.

136. Calatayud JM, Pardo P, Alarcón J. Cr- and Sb-containing $TiO_2$ inorganic orange nano-pigments prepared by a relative long hydrocarbon chain polyol. Dyes Pigm. 2016;134:1.

137. Tsukimori T, Oka R, Masui T. Synthesis and characterization of $Bi_4Zr_3O_{12}$ as an environment-friendly inorganic yellow pigment. Dyes Pigm. 2017;139:808.

138. Calatayud JM, Alarcón J. V-containing $ZrO_2$ inorganic yellow nano-pigments prepared by hydrothermal approach. Dyes Pigm. 2017;146:178.

139. Akafuah NK, Poozesk S, Salaimeh A, Patrick G, Lawler K, Saito K. Evolution of the automotive body coating process—A review. MDPI Coat. 2016;6:24.

140. Wilker G. Praktische Anwendung der Farbrezeptierung von Automobilfarbtönen, GDCh Seminar "Pigmente". Darmstadt, 2008.

141. Rosenhahn C. Highly targeted color spaces with made-to-measure iron oxide pigments. Boston: The Pigment & Colour Science Forum, 2018.

142. Patent US 10,570,288 (The Shepherd Color Company). 2018.

143. Gramm G, Fuhrmann G, Wieser M, Schottenberger H, Huppertz H. Environmentally benign inorganic red pigments based on tetragonal $\beta$-$Bi_2O_3$. Dyes Pigm. 2019;160:9.

144. Jaffe EE et al. J Coat Technol. 1994;66:47.

145. Iqbal A, Jost M, Kirchmayr R, Pfenninger J, Rochatand AC, Wallquist O. The synthesis and properties of 1, 4-diketo-pyrrolo[3, 4-c]pyrroles. Bull Soc Chim Belg. 1988;97:615.

146. Zeng W, Zhou Q, Zhang H, Qi X. One-coat epoxy coating development for the improvement of UV stability by DPP pigments. Dyes Pigm. 2018;151:157.

147. Jose S, Reddy ML. Lanthanumestrontium copper silicates as intense blue inorganic pigments with high near-infrared reflectance. Dyes Pigm. 2013;98:540.

148. Kim SW, Sim GE, Ock JY, Son JH, Hesegawa T, Toda K, Bae DS. Discovery of novel inorganic $Mn^{5+}$-doped sky-blue pigments based on $Ca_6BaP_4O_{17}$: crystal structure, optical and color properties, and color durability. Dyes Pigm. 2017;139:344.

149. Kim SW, Saito Y, Hesegawa T, Toda K, Uematsu K, Sato M. Development of a novel nontoxic vivid violet inorganic pigment e $Mn^{3+}$-doped $LaAlGe_2O_7$. Dyes Pigm. 2017;136:243.

150. Sheethu J, Aiswaria P, Laha S, Natarajan S, Reddy MLP. Green colored nano-pigments derived from $Y_2BaCuO_5$: NIR reflective coatings. Dyes Pigm. 2014;107:118.

151. Moser FH, Thomas AL. The phthalocyanines, vol. 1: properties. Boca Raton, Florida: CRC Press; 1983.

152. Moser FH, Thomas AL. The phthalocyanines, vol. 2: manufacture and applications. Boca Raton, Florida: CRC Press, 1983.

153. Booth G. In the chemistry of synthetic dyes. Venkataraman K, editor. Academic Press, 1971:241.

154. Leznoff CC, Lever AB. Phthalocyanines: properties and applications. vol. 1/2. New York: VCH, 1989.

155. Musche N. Ruß - Das kleine Schwarze. FARBE LACK. 2019;01:40.

156. Patent EP 3283573 (Schlenk Metallic Pigments GmbH) 2016.

157. Huber A, Maile FJ, Binder Y. New ultra-thin effect pigments. In: European Coatings Conference, Nuremberg, 2019.

158. Huber A, Maile FJ. Alles Gold, was glänzt? FARBE + LACK. 2020;12:40.

159. Bies T, Hoffmann RC, Stöter M, Huber A, Schneider JS. Environmentally benign solution based procedure for the fabrication of metal oxide coatings on metallic pigments. Chem Open. 2020;9:1251.

160. Maile FJ, Reynders P. A colourful menagerie of platelets for transparent effect pigments. APCJ. 2010;14.

161. Maile FJ, Reynders P. Substrates for Pearlescent Pigments. ECJ. 2003;4:124.

162. Maile FJ, Rösler M, Huber A. The macroscopic appearance of effect coatings and its relationship to the local spatial and angular distribution of reflected light. In: Proceedings of the American Coatings Conference. Charlotte, NC, USA, Jun 2008.

163. Maile FJ, Rösler M. Local gloss and sparkle caused by effect pigments: psychophysics, measurements and simulations. In: XXVI. FATIPEC-Congress, Dresden, 2002.

164. Kirchner E, Njo L, de Haas K, Rösler M, Gabel P. More than colour ECJ 2006;(11):46.

165. Siemen A. Untersuchung der Bronceverteilung bei der Hochrotationszerstäubung von Metalleffektlacken im Hinblick auf die Effektausbildung, Diplomarbeit. Universität Paderborn, 1990.

166. Geilen S. Korrelation zwischen den optischen Eigenschaften von Effektbeschichtungen und der Orientierung der enthaltenen Effektstoffe, Diplomarbeit. Universität Paderborn, 2002.

167. Filip J, Vávra R, Maile FJ. Optical analysis of coatings including diffractive pigments using a high-resolution gonioreflectometer. JCT Res. 2018.

168. Filip J, Vavra R, Maile FJ. BRDF measurement of Highly-Specular materials using a goniometer. In proceedings of 33th Spring Conference on Computer Graphics (SCCG 2017), Mikulov, 2017:131, (Honorable mention).

169. Chadwick AC, Kentridge RW. The perception of gloss: A review. Vision Res. 2015;109:221.

170. Filip J, Vávra R, Maile FJ, Kolafová M. Framework for capturing and editing of anisotropic effect coatings. Computer Graphics Forum. 2020. DOI:10.1111/cgf.14119, https://onlineli brary.wiley.com/doi/10.1111/cgf.14119.

171. Fleming RW. Visual perception of materials and their properties. Vision Res. 2014;94:62.

172. Cheeseman J, Ferwerda J, Maile FJ, Fleming RW. Scaling and discriminability of perceived gloss. J Opt Soc Am. 2021;A 38:203.

173. Rink H-P, Mayer B. Water based coatings for automotive refinishing. Progr Org Coat 1997;34: 175.
174. Betten P. Kolloidchemische Untersuchung zur Optimierung von Autoserienlacken, Dissertation. Universität Kiel, 2003.
175. Slinckx M, Henry N, Krebs A, Uytterhoeven G. High-solids automotive coatings. Progr Org Coat 2000, 38: 163.
176. Hoffmann P, Duschek W. Berichtsband DFG. 1999;41:123.
177. Huber A. Use of pearl lustre pigments in industrial coatings. The Coatings Agenda Europe, 2003.
178. Maisch R, Stahlecker O, Kieser M. Mica pigments in solvent free coatings systems. ECJ. 1994;9:582.
179. Bendel A. Der Maler Lackierermeister. 1993;10:894.
180. Bendel A. Der Maler Lackierermeister. 1997;3.
181. Bendel A. Der Maler Lackierermeister. 1999;12 .
182. Bendel A. Der Maler Lackierermeister. 2000;7.
183. Thometzek P, Ludwig A, Karbach A, Köhler K. Effects of morphology and surface treamtent of inorganic pigments on waterborne coating properties. Progr Org Coat 1999; 36:201.
184. Brückner H-D, Glausch R, Maisch R. Neuartige helle, leitfähige Pigmente auf Glimmer/ Metalloxid-Basis. Farbe + Lack. 1990;96:411.
185. Glausch R, Pfaff G, Maisch R. New results with light colored conductive pigments. In: XXIst FATIPEC Congr., Amsterdam, Bd. 2, 1992:33.
186. Wheeler IR. Metallic pigments in polymers. Rapra Technology Ltd, 1999.
187. Anliker R, Moser P. The limits of bioaccumulation of organic pigments in fish: Their relation to the partition coefficient and the solubility in water and octanol. Ecotox Environ Saf. 1987;13:43.
188. Pfaff G, Kuntz M, Rüger R. Electro-conductive pigments for coating applications. In: 8th Int. Conf. Advances in Coatings Technology, Warsaw, Poland, Nov 2008.
189. Bohem ME, Pook N-P, Adam A, Tran TT, Halasyamani PS, Entenmann M, Schleid T. Luminescence and scintillation properties of $La_2[Si_2O_7]:Ce^{3+}$ functional pigment: a concept for UV-protection of coatings. Dyes Pigm. 2015;123:331.
190. Ruf J. Organischer Metallschutz. Hannover: Vincentz Verlag, 1993.
191. Memnetz SI et al. Jct Coatings Tech. 1989;61:47.
192. Christian H-D. Proc. Waterborne High-Solids Powder Coat. Symp., New Orleans, LA, 2004, Paper 17.
193. Matsuoka M, editor. Infrared absorbing dyes. New York: Plenum, 1990.
194. Fernando RH, Sung L-P, editor. Nanotechnology applications in coatings, ACS Symposium Series 1008. American Chemical Society, 2009.
195. https://www.vdmi.de/en/. (Position Paper of the Pigment and Filler Industry in the Nano Discussion 9 2019). Accessed: 28 Mar .2021, 8:27.
196. Brown DM Johnston HJ, Gaiser BK, Pinna N, Caputo G, Culha M, Kelestemur M, Altunbek M, Stone V, Roy JC, Kinross JH, Fernandes TF. A cross-species and model comparison of the acute toxicity of nanoparticles used in the pigment and ink industries. Nano Impact. 2018; 11:20.
197. Delaval M, Wohlleben W, Landsiedal R, Baeza-Squiban A, Boland S. Assessment of the oxidative potential of nanoparticles by the cytochrome c assay: assay improvement and development of a high-throughtput method to predict toxicity of nanoparticles. Arch Toxicol. 2017;91:163.
198. Brzicova T et al. Toxicol Vitro. 2019;54:178.

199. Joonas E, Aruoja V, Olli K, Kahru A. Environmental safety data on Cuo and $TiO_2$ nanoparticles for multiple agal species in natural water: Filling the data gaps for the risk assessment. Sci Total Environ. 2019;647:973.

200. Spengler A, Wanninger L, Pflugmacher. Oxidative stress mediated toxicity of $TiO_2$ nanoparticles after a concentration and time dependent exposure of the aquatic macrophyte Hydrilla verticillata. Aquatic Toxicology. 2017;190:32.

201. Baer DR, Burrows PE, El-Azab A. Enhancing coating functionality using nanoscience and nanotechnology. Prog Org Coat. 2003;47:342.

202. Dhoke SK, Mangalsinha TJM, Dutta P, Khanna AS. Formulation and performance study of molecular weight, alkyd-based waterborne anticorrosive coating on mild steel. Prog Org Coat. 2008;62:183.

203. Dhoke SK, Sinha TJM, Khanna AS. Effect of nano-$Al_2O_3$ particles on the corrosion behavior of alkyd based waterborne coatings. J Coat Technol Res. 2009;6:353.

204. Perera DY. Effect of pigmentation on organic coating characteristics. Prog Org Coat. 2004;50:247.

205. Patent US 6,875,800 (PPG Industries Ohio, Inc.). 2005.

206. Patent US 6,916,368 (PPG Industries Ohio, Inc.). 2005.

207. Elsaeed AM, El-Fattah MA, Azzam AM. Synthesis of ZnO nanoparticles and studying its influence on the antimicrobial, anticorrosion and mechanical behavior of polyurethane composite for surface coating. Dyes Pigm. 2015;121:282.

208. Sibilia C, Benson TM, Marciniak M, Szoplik T, editors. Photonic crystals physics and technology. Italia: Springer-Verlag, 2008. DOI : 10.1007/978-88-470-0844-1 .

209. Lourtioz J-M, Benisty H, Berger V, Jean-Michel Gerard Daniel Maystre J-MGD, Tchelnokov A, editors. Photonic crystals - towards nanoscale photonic devices. Berlin Heidelberg: Springer-Verlag, 2008. DOI: 10.1007/978-3-540-78347-3.

210. Decker E, Purdy S, Xu X, Munro C. Polymer crystalline colloidal array color technology. Boston: The Pigment & Colour Science Forum, 2018.

211. Ohnuki R, Sakai M, Takeoka Y, Yoshioka S. Optical characterization of the photonic ball as a structurally colored pigment. Langmuir. 2020; 36:5579.

212. Kinoshita S, Yoshioka S, Miyazaki J. Physics of structural colors. Rep Prog Phys. 2008;71:076401 30. DOI: 10.1088/0034-4885/71/7/076401.

213. Kinoshita S. Structural colors in the realm of nature. World Scientific Publishing, 2008.

214. Srinivasarao M. Nano-optics in the biological world: beetles, butterflies, birds, and moths. Chem Rev. 1999;99:1935.

215. Zhao T, Parker R, Vignolini S. Cellulose photonics: from nature to pigment applications. Berlin: The Pigment & Colour Science Forum, 2019.

216. Shi X, He J, Xie X, Dou R, Lu X. Photonic crystals with vivid structure color and robust mechanical strength. Dyes Pigm. 2019;165:137.

217. Yu X, Ma W, Zhang S. Hydrophobic polymer-incorporated hybrid 1D photonic crystals with brilliant structural colors via aqueous-based layer-by-layer dip-coating. Dyes Pigm. DOI 10.1016/j.diepig.2020.108961.

218. Dushkinaa N, Lakhtakiab A. Structural colors, cosmetics and fabrics. In Proceedings of SPIE - The International Society for Optical Engineering · Aug 2009.

219. Seeboth A, Lötzsch D. In encyclopedia of polymer science and technology. Kroschwitz JI, editor. 3rd ed. vol. 12, New York, NY, USA: John Wiley & Sons, 2004:143.

220. Rubacha M. Thermochromic cellulose fibers. Polym Adv Technol. 2007;18:323.

221. Rosenzweig M. Modern Plastics Int. 2003;338.

222. Day JH. Thermochromism of inorganic compounds. Chem Rev. 1968;68:649.

223. Kirk-Othmer encyclopedia of chemical technology. 5th ed, vol. 6. New York, NY, USA: John Wiley & Sons, 2004:130.
224. Hughes GJ. Thermochromic solids. J Chem Educ. 1998;75:57.
225. Gaudon M, Deniard P, Demourges A, Thiry AE, Carbonera C, Nestour A, Largetea A, Létard JF, Jobic S. Unprecedented "one-finger-push"-induced phase transition with a drastic color change in an inorganic material. Adv Mater. 2007;19:3517.
226. Lucht BL, Euler WB, Oj. Gregory in Proceedings of the 22 3rd American Chemical Society National Meeting, Orlando, FL, USA, 2002, oral contribution POLY 307.
227. Beildeck C, Lucht BL, Euler WB. Polym Prepr. 2001;42:211.
228. Green M. Chem Ind Sept. 1996;2:611.
229. Monk PM, Mortimer RJ, Rosseinsky DR. Electrochromism: fundamentals and applications. Monk PMS. The viologens: synthesis, physicochemical properties and applications of the salts of 4,4′-bipyridine, Wiley, Chichester, 1998. Weinheim: VCH, 1995.
230. Aegerter M, Mennig M. Sol-gel technologies for glass producers and users. 2004. DOI: 10.1007/978-0-387-88953-5.
231. Itaya K, Uchida I, Neff VD. Electrochemistry of poly nuclear transition metal cyanides: prussian blue and its analogues. Acc Chem Res. 1986;19:162.
232. Neff VD, Electrochemical oxidation and reduction of thin films of prusian blue. Electrochem J. Soc. 125:886.
233. Dien C, Evans NA, Stapleton IW. In the chemistry of synthetic dyes. Venkataraman K, editor. vol. 8. New York: Academic Press,1978 :81.
234. Patent DE 743 848 (IG Farbenindustrie AG). 1944.
235. Patent DE 746 839 (IG Farbenindustrie AG) . 1944.
236. Patent DE 12 60 652 (BASF). 1968.
237. Patent US 4,204,879 (Williams Hounslow Ltd). 1980.
238. Patent EP 0 177 138 (ICI). 1985.
239. Schwartz M, Baumstark R. Waterbased acrylates for decorative coatings, European coatings literature. Vincentz Network Hannover, 2001.
240. Dörr H, Holzinger F. Kronos Titandioxid in Dispersionsfarben. Leverkusen: Kronos Titan; 1989.
241. Nägele E. Dispersionsbaustoffe. Rudolf Müller Verlag, 1989.
242. Du Z, Wen S, Wang J, Yin C, Yu D, Luo J. The review of powder coatings. J Mater Sci Chem Eng. 2016;4:54.
243. Pietschmann J. Industrielle Pulverbeschichtung, Vieweg und Teubner. Wiesbaden: GWV Fachverlage GmbH, 2010.
244. Kunaver M, Klanjsek Gunde M, Mozetic M, Hrovat A. The degree of dispersion of pigments in powder coatings. Dyes Pigm. 2003;57:235.
245. Shi Q, Huang W, Zhang Y, Zhang Y, Xu Y, Guo G. Curing of polyester powder coating modified with rutile nano-sized Ti-tanium dioxide studied by DSC and real-time FT-IR. J Therm Anal Calorim. 2011;108:1243. DOI:10.1007/s10973-011-1855-4.
246. Hadavand BS, Ataeefard M, Bafghi HF. Preparation of modified Nano ZnO/Polyester/TGIC powder coating nanocomposite and evaluation of its antibacterial activity. Compos Part B Eng. 2015;82:190. DOI: 10.1016/j.compositesb.2015.08.024.
247. Trottier EC, Affrossman S, Pethrick RA. Dielectric studies of Epoxy/Polyester powder coatings containing Titanium Dioxide, Silica, and Zinc Oxide pigments. JCT Res. 2012;9:525. DOI:10.1007/s11998-012-9405-y.
248. Puig M, Gimeno MJ, Gracenea JJ, Suay JJ. Anticorrosive properties enhancement in powder coating duplex systems by means of ZMP anticorrosive pigment. assessment by

electrochemical techniques. Prog Org Coat. 2014;77:1993. DOI: 10.1016/j.porgcoat.2014.04.031.
249. Abd El-Ghaffer MA, Abdel-Wahab NA, Sanad MA, Sabaa MW. High performance anti-corrosive powder coatings based on phosphate pigments containing poly(o-aminophenol). Prog Org Coat. 2015;78:42. DOI: 10.1016/j.porgcoat.2014.09.021.
250. https://www.chinapowdercoating.com/future-development-powder-coating/.
251. https://www.chinapowdercoating.com/technicals/powder-coating-trends/.
252. Filip J, Vávra R, Maile FJ, Eibon B: Image-based discrimination and spatial non-uniformity analysis of effect coatings.In proceedings of the 8th conference on pattern recognition applications and methods. February, Prague 2019.
253. Feng H, Xu H, Zhang F, Wang Z. Color prediction of metallic coatings from measurements at common geometries in portable multiangle spectrophotometers. JCT Res. 2018;15(22). DOI: 10.1007/s11998-017-0026-3.
254. Dumain ED, Agawa T, Goel S, Toman A, Tse AS. Cure behavior of polyester-acrylate hybrid powder coatings. Jct Coating Tech. 1999;71:69. DOI: 10.1007/BF02697908.
255. Iwamura G, Agawa T, Maruyama K, Tekeda H. A novel acrylic polyester system for powder coatings. Surf Coat Int. 2000;83:285. DOI:10.1007/BF02692728.
256. Takeshita Y, Sawada T, Handa T, Watanuki Y, Kudo T. Influence of air-cooling time on physical properties of thermoplastic polyester powder coatings. Prog Org Coat. 2012;75:584. DOI:10.1016/j.porgcoat.2012.07.003.
257. Lafabrier A, Fahs A, Louarn G, Aragon E, Chailan J-F. Experimental evidence of the interface/interface formation between powder coating and composite material. Prog Org Coat. 2014;77:1137. DOI: 10.1016/j.porgcoat.2014.03.021.
258. Schwalm R. UV Coatings. Amsterdam: Elsevier B.V., 2007.
259. Menzel K, et al. UV & EB technology, RadTech North America, el5, Charlotte, NC, May 2004. In: Technical Conference Proceedings.
260. Ryntz RA, Yaneff PV, editors. Coatings of polymers and plastics. Marcel Dekker Inc, 2003.
261. Stowe RW. Jct Coatings Tech. March 2003. www.radtech-europe.com.
262. Weber E. North American UV+EB market overview, RadTech. Orlando FL: USA, 2020.
263. Dvorchak MJ, Clouser ML. Aerospace UV cured coatings; yesterday, today & tomorrow, RadTech. Orlando FL: USA, 2020.
264. Heylen M. New developments in UV resin for metal coatings, RadTech Europe 2005. In: Conference Proceedings, Vol. I 2005, 181.
265. Weikard J. Urethane acrylates on metal substrates, RadTech Europe 2005. In: Conference Proceedings, Vol. I 2005, 187.
266. Amigo J. Innovative developments in UV pigmented low viscosity (100 % solids) for the automotive and metal coatings industry, RadTech Europe 2005. In: Conference Proceedings, Vol. I 2005, 195.
267. Pietschmann N. UV curable metal coatings - Special possibilities and problems, RadTech Europe 2005. In: Conference Proceedings, Vol. I, 2005, 523.
268. Schmitz C, Poplata T, Feilen A, Strehmel B. Radiation crosslinking of pigmented coating material by UV LEDs enabling depth curing and preventing oxygen inhibition. Prog Org Coat. 2020;144:105663.
269. Müller M. Metallpigmentierte UV-Systeme, Seminar Strahlenhärtung. Würzburg: Farbenlabor, 2019.
270. Beck E. Strahlenhärtung - Innovative Anwendungen, Farbenlabor: Seminar Strahlenhärtung, Würzburg, 2019.
271. Dvorchak MJ, Clouser ML. Aerospace UV cured coatings; Yesterday, Today & Tomorrow, RadTech 2020. Orlando FL, USA.

272. Meuthen B, Jandel A-S. Coil Coating: Bandbeschichtung: Verfahren, Produkte und Märkte. 2 Aufl ed. Wiesbaden: Friedrich Vieweg & Sohn Verlag, GWV Fachverlage GmbH, 2008.
273. Bianchi S, Broggi F. Coil coating: the advanced finishing technology. Key Eng Mater. 2016;710:181.
274. ASTM G173–03. Standard tables for reference solar spectral irradiance: direct normal and hemispherical on 37° tilted surface. West Conshohocken, PA: ASTM International; 2021. ASTM Home Page, http://www.astm.org. Accessed: 18 Jan 2021.
275. European Coil Coating Association (event): 38th Annual General Meeting (Salzburg 2004). Brussels: ECCA, 2004, N. Brown, Putting colour into coil coating.
276. Maile FJ, Schauer T, Eisenbach CD. Evaluation of corrosion and protection of coated metals with local ion concentration technique (LICT). Progr Org Coat. 2000;38:111.
277. Maile FJ, Schauer T, Eisenbach CD. Evaluation of the delamination of coatings with scanning reference electrode technique. Progr Org Coat. 2000;38:117.
278. Laible R. Umweltfreundliche Lackiersysteme für die industrielle Lackierung. Expert-Verlag, 1989.
279. Smith A. Inorganic primer pigments. Blue Bell, PA: Federation of Societies for Coatings Technology, 1989.
280. Wilke G, Ortmeier J. Coatings for plastics, european coatings tech files. Vincentz Network Hannover; 2012.
281. Gómez O, Perales E, Chorro E, Burgos FJ, Viqueira V, Vilaseca M, Martinez-verdú FM, Pujol J. Visual and instrumental assessments of color differences in automotive coatings. Color Res Appl. 2016;41:384.
282. Lindstrom M. Buyology: Truth and lies about why we buy and the new science of desire. Bantam Dell Pub Group, 2008. ISBN 0-385-52388–2.
283. Maile FJ, Riva L, Casagrande R. Testing the repairability of Xirallic® OEM Paint finishes. Pitture e Vernici, 19/2004.
284. Patent US 4,346,144 (EI Du Pont de Nemours and Co). 1980.
285. Weiss KD. Paint and coatings: A mature industry in transition. Prog Polym Sci. 1997;22:203.
286. Farion F, Lavisse K. Tinted clear coats for mass production: challenges and success. In: Automotive Circle Conference, Vincentz Network, Nov 2020.
287. Heylen A., Smart and efficient single process paint method for new Yaris bi-tone design. In: Automotive Circle Conference, Vincentz Network, Nov 2020.
288. Steigleder T, Groenewolt M. Eco-paint process: a sustainable, lean & flexible paint process for truck cabs. In: Automotive Circle Conference, Vincentz Network, Nov 2020.
289. Gee P, Gilligan S. Eur Coat J. 2006;5:44.
290. Maile FJ, Martins A. Chrome-like appearance without using chromium. Picture E Vernici - Eur Coat. 2015;1:16.
291. Patent US 8,512,802 (Axalta Coating Systems). 2013.
292. Maile FJ, Reynders P, Sharrock S. Breathing life into cars. PPCJ. 2004; 194:15.
293. Liu W, Caroll JB. Jct Coatings Tech. 2006;3:82.
294. Seubert CM, Nichols ME. The Future of transportation mobility and its effect on coatings, color, and pigments. Boston: The Pigment & Colour Science Forum; 2018.
295. Plüg C. Recent developments in the field of pearlescent pigments. Berlin: Pigment & Color Science Forum; 2019.
296. High performance pigments, paints and coatings, market insights to 2023. Boston: Pigment & Color Science Forum, 2018.
297. Schulte S. The best technical papers on pigments for high-performance coatings published in the European coatings journal within the past three years. In: EUROPEAN COATINGS dossier. Vincentz, 2018.

298. Gagro D. Gute Aussichten trotz Herausforderungen. Farbe Lack. 2019;125:1.
299. Kumar V, Bhattacharya A. The smart future of the coatings industry, paint & coatings industry PCI online. 8 Feb 2019.
300. Fiuza TER, Borges FM, da Cunha JBM, Antunes SRM, de Andrade AVC, Antunes AC, de Souza ECF. Iron-based inorganic pigments from residue: Preparation and application in ceramic, Polymer, and Paint. Dyes Pigm. 2018;148:319.
301. Labaziewicz P. Cars are becoming rolling sensor platforms. In: TI E2E Community, 25 Sept 2014, Texas Instruments Inc.
302. The future of coatings in a world of autonomous vehicles. American Coatings Association. 2017. https://www.paint.org/article/future-coatings-world-autonomous-vehicles/.
303. Decker E. High-Tech coatings to enable autonomous vehicles. SAE Conference Detroit, 2018.
304. Liese M. Lacke für radarbasierte Fahrerassistenz-Systeme. JOT. 2019;9:36.
305. Koerner M, Radek S. Druckfrische Autos? Farbe Lack. 2019;125:1.

Ghita Lanzendörfer-Yu

# 18 Colorants in cosmetic applications

**Abstract:** The color cosmetic market, even though highly dependent on color, is driven by texture. Pigment and color innovation are predominantly taking place in the field of inorganic effect pigments, as colorants are regulated within the different cosmetic directives and toxicological profiles have to be established prior to use. Therefore, the formulation and the packaging are the relevant innovative factors. Nevertheless, color cosmetics are driven by fashion and trends. One of the main drivers for sales is social networks. They are image based and therefore the ideal platform to spread these trends, boosting sales of color cosmetics recently.

Even though, color cosmetics seem to be very simple in composition, the development, production and quality control are far from that. In color cosmetics, all cosmetic disciplines cumulate: dispersion strategies, emulsion technologies, molding and extruding of sticks and pencil leads, mixing and compressing of powders, liquid inks and so on. And almost every discipline requires separate production vessels, resulting in a rather complex manufacturing process.

**Keywords:** color cosmetics, pigments, dyes, fashion, formulas, application, makeup

## 18.1 Introduction

Cosmetics globally are a highly profitable market with many players that strive to attain market shares and profits. A continuous growth of the global color cosmetic market is forecasted until at least 2023 [1].

The cosmetic market is not only split globally into different cultural surroundings, but also due to distribution channels such as mass market, prestige market, door to door selling or social media marketing and influencers worldwide. Additionally, cosmetics are also sold by beauty therapists and dermatologists; these practices are different for every country.

Consumer demands create further diversification of the markets, start-ups are common and local brands also have loyal consumers. Regulation, even though many efforts have been made for global harmonization, is still different in the main markets such as EU, USA and Asia-Pacific. Most prominent examples therefore are ultraviolet (UV) filters and sun protection factor (SPF) measurements as well as the ban of animal testing.

This article has previously been published in the journal Physical Sciences Reviews. Please cite as: G. Lanzendörfer-Yu, Colorants in Cosmetic Applications *Physical Sciences Reviews* [Online] 2021, 5. DOI: 10.1515/psr-2020-0161

https://doi.org/10.1515/9783110588071-018

### 18.1.1 Market

Due to the diversity of sales channels, the complete turnover of (color) cosmetics is difficult to access. This has to be kept in mind when analyzing the data.

The global cosmetics and toiletries market has experienced a steady growth over the past two decades. It always was positive, even in the years 2008–2009 with the financial crisis in place. Usually growth rates range from 3.4% to 5.5% [2, 3]. Color cosmetics are a vital part of the cosmetic market worldwide. Its turnover volume ranges from 20,6 bn USD in 2002 [4] to a predicted 75 bn USD in 2021 [5].

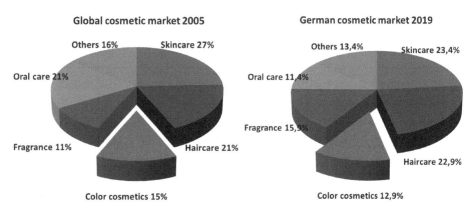

**Figure 18.1:** Comparison of the segmentation of the global and German cosmetic market in 2005 and 2019, respectively. Numbers are taken from [4] and IKW publishing [6] and show a relatively stable segmentation, where color cosmetics have a share of about 13–15%.

Color cosmetics generally accounts for approximately 15% of the cosmetic turnover (Figure 18.1). By comparing the numbers of 2005 with 2019, therefore using financial reports and other publications [6, 7], the fraction of that category of color cosmetics remains the same more or less. Notes should be taken on the fact that sun care products are grouped separately in different markets [8, 9, 12].

With the growth of the cosmetics and toiletries market, color cosmetics has a stable share of about 15%. Nevertheless, on a national basis these numbers can vary noticeably. The segment of color cosmetics has grown significantly in Germany from 2013 to 2017 with 6.1% [10]. In Europe in 2017, color cosmetics ranged on the fifth rank with a turnover of 11,17 bn euros [11].

The color cosmetics market usually is further segmented into four categories defined by application areas: face, lips, eye and nail. The size of these segments differs considerably in the regions considered (Table 18.1).

Growth rates and market shares of the four segments further differ according to global and fashion trends. Most sensitively react eye, lip and nail category, because

**Table 18.1:** Comparison of the segmentation of color cosmetics in different markets in %. Globally the biggest share is taken by products for the application on the face, followed by lip products. The different market shares of these categories vary according to cultural preferences and fashion trends.

| Market / Year | N.A. 2004 (4) | Europe 2004 (4) | Asia 2004 (4) | Global 2007 (3) | Global 2012 (10) | UK 2012 (12) |
|---|---|---|---|---|---|---|
| Face | 25 | 19 | 32 | 36 | 37 | 38 |
| Lips | 26 | 24 | 25 | 27 | 28 | 29 |
| Eyes | 32 | 26 | 21 | 27 | 24 | 16 |
| Nails | 18 | 31 | 18 | 10 | 11 | 16 |

of their dependence on color. Over the last decade, social media and the role of influencers gained importance to market makeup and other products as well.

However, product categories do not tell much about the variety of products sold and product formats; preferences vary intercultural. Small lipstick formats, for instance, are favored in Asia.

As a color cosmetic product consists of two essential parts: the formula and the packaging/applicator, one has to consider both when discussing trends and likeability of the products.

Some products, such as pencils, would be impossible to realize without the appropriate packaging/casing as well as suitable production lines.

As the packaging/casing is an important part of the product, it not only affects the applicability of the formula but also shows off a certain status or lifestyle. For instance, nobody would use a body lotion in public, but the use of a lipstick is a common picture in all locations.

One can discern two groups of color cosmetic products, those which are not affected by fashion styles so strongly (job products and products with a general beautifying action) and those which are highly affected by fashion and trends (fashion products and products that underline individuality). The first group of products serves to maintain even skin or enhance expression of eyes; the others convey the trendiness of the user and give color and optical effects.

Lifespan of those products generally differs accordingly. The so-called job products stay in the market up to 3–5 years. Trend products can change on a rate of 3–18 months.

## 18.1.2 Regulation

All cosmetic products are regulated either in Europe by the EEC cosmetic directive, in North America by the Food and Drug Administration (FDA) and in Japan by the Japanese Cosmetic Ingredient Codex (JICIC).

A "cosmetic product" shall mean any substance or preparations intended to be placed in contact with the various external parts of the human body (epidermis, hair system, nails, lips and external genital organs) or with the teeth and mucous membranes of the oral cavity with a view exclusively or mainly to cleaning them, perfuming them, changing their appearance and/or correcting body odors and/or protecting them or keeping them in good condition [13].

Additionally, the use of colorants is regulated separately with a so-called positive list. All colorants for cosmetic use are indicated, assessed and have a color index number (Table 18.2). The numbers which should be indicated on the final packaging can vary from the regions Europe, North America and Japan [14].

**Table 18.2:** Color Index (CI): All colorants are listed on a positive list, where also a detailed description of application to various body sites is available. As the CI unfortunately is not used globally, also FD&C numbers are used as are trivial names. Color Index usually gives no information about the manufacturing process, coating materials or the substrate on which a dye has been precipitated on. CI strictly refers to the chemistry of the colorant [15].

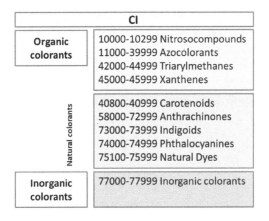

Pigments of the same chemical group do not have to behave identical when it comes to dispersion or pay off characteristics. Depending on the production and/or coating process, characteristics can vary significantly.

As the market for natural cosmetics is growing, the demand for natural colorants also rises. As natural pigments are regarded:

Mineral pigments

Natural colorants

As natural colorants are difficult to transfer into insoluble matter, only have reduced light fastness and generally less color intensity, their main application is in food industry [16, 17].

### 18.1.3 Tattoo and permanent makeup

If there had been a trend which not had been properly foreseen, this will be tattooing and permanent makeup (PMU). Within the last decade, millions of people received multiple tattoos and probably as many used permanent makeups. Both are based on the same technology, namely using a device that transports pigments UNDER the epidermis of the skin. Therefore, this technique is considered to be invasive and to be regulated separately from cosmetics. Regulation actually took place in 2009 [18, 19].

Even though there are many publications suggesting differently, tattoo and PMU colors are based on the same pigments used in color cosmetics. Some adverse reactions have been reported. Most of them are related to (red) pigments. But it is not clear if they could be caused by the dispersing agent as well [20].

With the tattooing technique in place, pigments generally are viewed differently. Tattoo colors are shot into the dermis by injection techniques, they can come into contact with blood, lymph and immune competent cells, accumulate in lymph nodes and therefore can cause different reactions than those known from makeup. A general overview is given by Henrik Petersen and Klaus Roth [21].

Nevertheless, tattooing and PMU should be kept in mind, the following parts will concentrate on the products conventionally applied on the surface of the skin.

## 18.2 Makeup in general

### 18.2.1 The changing demands for a perfect makeup

Even though makeup in a general sense is one of the cultural achievements of mankind, I do not want to focus on prehistoric, antique and medieval practices and directly start from the beginning of the industrial revolution.

First lipstick prototypes were exhibited on the world exhibition in Amsterdam 1883. They were named "wand of Eros" or "stylo d'amour," thus clearly defining their use as a seduction agent. Lipstick cases did not exist at that time and therefore the "packaging" consisted of paper wrapped around the bullet. It is said that the first lipstick case was invented in the 1920s in the USA when redesigning a bullet [22, 23].

A milestone in the use of color cosmetics has been World War I. At that time more women joined the workforce and had their own money at hand. The use of lipsticks in that context can be seen as significant as well as for the emancipation of women as well as for reflecting the cultural changes and attitudes. In the 1920s in Europe coloring of the lips in bright red was connected with loose women and therefore not regarded as appropriate for a faithful wife. The women's lib used those colors in protest and to fight for their rights of equality.

Red still is a color that evokes many emotions. The color red has been something special since the beginning of time and, in addition to white and black, was the first of the colors to be given a name [24]. Red means blood, emotion, beauty, rule, but also anger and danger. Red is definitely not a color for wimps [25, 26].

Wearing makeup today is neither connected with protest nor social well behavior but with the expression of moods or personality of a self-confident modern woman. Makeup has also to do more than deliver color. Women expect to be able to choose from a huge variety of products that satisfy different requirements ranging from luxury to convenience, from caring to long lasting, from shining to matte.

The variety of products on the market reflects also those requirements. They can be grouped in needs. Internally directed needs are well-being, self-confidence, mood orientation and protection. Those characteristics are mainly delivered by the formula (i.e. contains vitamins, protective oils, UV filters), the texture (feels comfortable, does not dry out the skin) and last but not least a beautiful casing and presentation of the product.

Externally directed needs are beauty, attractiveness, femininity, fashion and perfection. These characteristics are transported by a fashion-oriented brand, huge color assortment, product claims like "stay on" or "long lasting" and last but not least a packaging that conveys value and allows accurate application.

Modern consumers use makeup for beautifying aspects of the face. Competition especially in the social media channels is high. Consumers want to present their most beautiful self, therefore using special camera settings, unique makeup or even surgery.

Additionally, modern makeup has to cope with the demands of new visualization technologies. As like makeup did in the past when movies were made and then changed colors from black and white to technicolor: Makeup did not only have to withstand the lightening conditions it also had to convey a natural look. This nowadays is even more important as HD television makes every flaw visible and users are broadcasting their own videos in the internet.

The Internet is flooded with makeup tutorials and whole industries live from explaining and promoting makeup [27].

### 18.2.2 The human face and attractiveness

The human face is stated to be "the most interesting continent of the world" (Georg Christoph Lichtenberg, 1742–1799). Attractiveness is influencing choices such as dating, hiring or in judicial conviction. A lot of positive attributes are connected with physical (facial) attractiveness [28]. Additionally, we humans have an extraordinary well-developed ability to process, recognize and extract information from other's faces. Therefore, making up the face and keeping it attractive is more than fashion it has many social implications.

### 18.2.2.1 Attractive features

Systematic investigations of the features that contribute to attractiveness have gained interest in the past years. Modern tools for the processing and modifying of images helped to promote results, too. This research is located in the field of social psychology and related disciplines [29,30,31]. One major aspect of attractiveness is symmetry which is preferred over apparent asymmetry. But assessing attractiveness, a slight (natural) deviation from symmetry is most attractive [31]. Attractiveness is also influenced by personal, individual preferences that might be rooted in cultural surroundings [32].

The measuring of the face, determining the correct parameters and correlating them with attractiveness, is a tedious task. As most measurements showed no specific size or proportion in the face to be responsible for determining attractiveness, the grouping of certain aspects was more successful [33]. There are three main attractive features.

*Neonate features*: They comprise all features that indicate youth like the evenness of the skin tone and the size of the eyes (large).

*Sexual maturity*: These features are defined eyebrows, high cheekbones, defined chin and mouth.

*Expressiveness*: Features such as health and emotional stability.

These features are the same in all cultural groups [34]. We assessed effects of makeup, asking groups of people to judge the attractiveness (and other characteristics that they connote with a person's face) according to different makeup stages using a standardized questionnaire.

Some interesting results were found: Lipsticks have a strong individual effect on the beauty of the person wearing it. Only when the color matches the type of person a positive effect regarding the attractiveness could be observed. But color independently people (means women) wearing lipstick were judged as emotionally more stable and more self-confident [35]. This effect of lipsticks was also found by Cash et al [36].

Foundation and Mascara were found to be products that worked independently of the type and always had a strong beautifying effect. The socio-psychological aspects of makeup are also very important, as wearing makeup can add self-confidence and clearly influences the way people look at the other [36].

### 18.2.2.2 Mobile and fixed areas of the face

Another grouping divides the face in mobile (communication) and fixed (structuring) areas (Figure 18.2). To enhance the communicative area of the face leads to culturally related differences in perception of beauty. Also, making up the mobile areas of the face is enhancing the evocations linked to these areas [37].

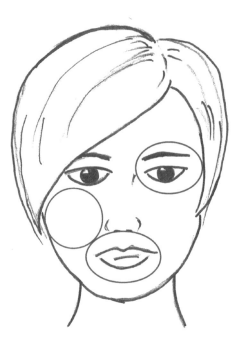

**Figure 18.2:** Facial areas can be divided in mobile (red circles) and fixed (blue circle). The mobile parts are important for communication (eye, lips) and the fixed parts are important for structuring the face.

### 18.2.2.3 The Lips

The lips surround the mouth, the most important opening in the face. They are essential for the intake of food one of the most important sense organs of the body. Feelings connected with the mouth/lips are highly pleasurable (nice food, kisses) and at the same time is essential for verbal communication. Cultural differences exist in respect of the attractiveness of lip size.

Considering the many functions of the mouth and the lips (i.e. symbolism of fertility, expressive organ, communication), it is obvious that the coloring of the lips can also symbolize many things.

Considering the symbolism of the color red, the coloring of the lips consequently becomes a highly erotic action. The use of a lipstick in public was despised only some years ago. Also striking red lips signal sexual openness.

For the modern women of today coloring of the lips is underlining individuality: It is reflecting life style as well as fashion (using the right color for the clothing/trend) or personal moods. The use of special (offending or provocative) colors and coloring of the lips in a special way stands for expression of protest, provocation or rebellion against existing (social) rules. Teenagers use to provoke their surroundings in that way.

### 18.2.2.4 The eye area

In contrary to the mouth, the eye area is rather "silent" when it comes to its expressiveness. It does not need bright colors to display its provocative potential on the other hand makeup seldom consists of enhancing both areas – lips and eyes – at the same time.

Nevertheless, the eyes also have a strong sensual meaning: Emphasizing the eye area increases perceived largeness and expressiveness. As the eye is anchored in innocence (neonate figure), it means that the makeup result is an innocent and intellectual seduction and emotionally appealing.

Eyes are regarded as a window to the soul and to the intellect, seduction is not perceived as "vulgar" as when making up the lips. Eye makeup enhances communicativeness and makes the face emotionally appealing. It is the "innocent" seduction with doe like eyes, slightly enticing.

Dark-circled eyes were the counterpart in the 1968 protests, the sexual revolution and Woodstock festival. In the 1980s heavily made-up eyes were in fashion again like cat eyes or smoky eyes often alongside a gothic look. Today, eye makeup is used to enhance the glamorous look and to answer the yearning of consumers for catwalk and dramatic looks.

It is interesting to note that eye makeup, especially eye liners, is one item of the color cosmetics products that also can be used by men, because it does not affect the perception of the male gender. Prominent examples are Buster Keaton and Johnny Depp.

Coloring the eyes is more a fashion statement than a sign of protest. As the eye makeup can be varied in more ways than a lip makeup, one can make the eyes look larger by using light colors or add intensity of expression by using darker colors. Eye makeup can be shaded and many colors can be used to achieve the desired effect – from playful to a mysteriously veiled appearance.

### 18.2.2.5 The face

As the skin of the face is regarded as the structuring part, no "color" is needed to enhance its "action" in attractiveness. Most important are even skin tone, the absence of wrinkles (indicates youth) and pimples (health) as the main attractive features. They can be influenced – to a certain amount – by cosmetics.

Nevertheless, for facial applications it is extremely important to use the right hue of the preparations, means that it should match the underlying skin tone. This can be cool, warm or neutral. It is identified by the small blood vessels that shine through the skin and either appear blue – cool skin undertone or green – warm skin undertone. If both colors are visible one speaks of a neutral skin tone.

The identification of the right skin undertone is vital for makeup artists who then choose the appropriate colors accordingly. This measure enables them to create natural looking makeups.

The skin undertone is fixed and does not vary like the skin tone which is determined by the amount of melanin in the skin and is influenced by ethnicity and UV radiation.

Blushers, highlighters and contouring products on the other hand are used to enhance sexual maturity as they are applied to contour the cheekbones, highlight them or modelling the face. Some experience in applying these products helps to obtain the desired effect.

### 18.2.3 The making of makeup

#### 18.2.3.1 Pigments

In color cosmetics the active ingredients are pigments or precipitated dyes. The choice of pigments should be made according to the intended application format. It should be also considered whether to use coated pigments or predispersed preparations that facilitate production and give more uniform finished products. The choice definitely will be made due to production capacity and milling equipment available. Figure 18.3 gives an overview on inorganic pigments and organic colorants used in makeup formulations

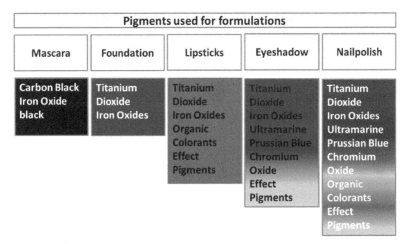

**Figure 18.3:** Inorganic pigments and organic colorants used in the diverse makeup formulations. Whereas mascara and foundations mainly depend on the use of inorganic pigments, in lipsticks, eyeshadows and nail polishes a wider variety is used. For lipsticks it is important also to incorporate food grade colorants, in eyeshadows the pigments should have a small and even particle size distribution to avoid scratching. In nail polishes the limitations are less as nails are considered impermeable.

### 18.2.3.2 Effect pigments

Effect pigments (often referred to as pearl or interference pigments) are synthetically derived inorganic pigments. They are based on flaky materials like mica or its synthetic variants with different layers of an optical dense oxide, mostly $TiO_2$. Depending on the thickness and multitude of the layers, different optical effects such as luster, shimmer and colors based on interference or goniochromaticity can be obtained. By using Iron Oxides for layering, different colors from golden to dark red can be achieved. Other pigments can be precipitated on the flakes as well giving an almost complete color palette [38].

The various optical characteristics and the different particle sizes are important criteria for the selection of an effect pigments (Figure 18.4).

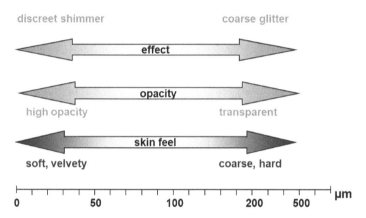

**Figure 18.4:** Effect pigments are a vital part of many color cosmetic formulations not only delivering interference colors and shimmer but also textural effects. Therefore, they should be chosen with deliberation.

For mica and mica-based pearls, the FDA does not allow the use of particles with a size larger than 150 µm in the eye area. This regulation meant as a precaution also has a practical implication. The larger the pigments and the better they adhere on the eye lid the more mechanical impact they can have due to the continuous movement of the eyelid.

In the manufacturing process of colored products, effect pigments should be added last as they are sensitive to shear forces. Color matching of the bulk should be completed before adding effect pigments.

### 18.2.3.3 Fillers

Whereas fillers are often regarded as cheap ingredients that are formulated into a product just for cost reasons, they are actually the hidden champions of all color cosmetic formulations. They enable homogenous application of colors, help formulation stability, dispersion of pigments and influence the texture of the finished product. Usually they do not have color properties but can have optical effects which for instance are used for wrinkle reduction. Their main benefit is for the formulator to help to adjust the portions of solids in color development. As formulations come in a variety of colors, fillers are used to balance the solid phase of the formulation when developing a color range.

Most commonly used fillers that are essential for powder formulations are talc, kaolin and mica. Other inorganic fillers are boron nitride, BiOCl, amorphous or spherical $SiO_2$, silica dimethyl silylate, and barium sulfate. Important organic fillers are natural and modified starch, nylon-12, polytetrafluoroethylene, polypropylene, vinyl dimethicone cross-polymer, polymethyl methacrylate, and lauroyl lysine.

In the course of the discussion of microplastic, almost all organic fillers have been taken out of the existing formulations.

### 18.2.3.4 Dispersion strategies and production methods

In order to produce a uniform and stable color cosmetic preparation the pigments have to be milled, dispersed and stabilized in the formulation. Often the same cosmetic ingredient is used to achieve this task. For instance, this can be achieved with castor oil for stick formulations or certain emulsifiers in foundations. An overview on different dispersion systems used in color cosmetics is shown in Figure 18.5.

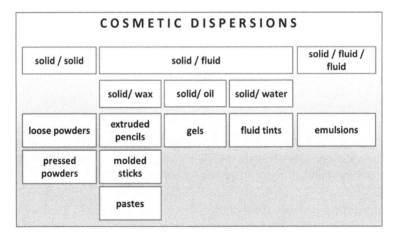

**Figure 18.5:** Overview on different dispersion systems used in color cosmetics.

Milling is used to break down the agglomerated pigment particles in order to obtain a uniform size distribution and reproducible color. If milling goes so far to break down primary particle size that will strongly affect the color of the pigment and most often is not desired. Therefore, when pigments are purchased in different hues such as iron oxide red with blue or yellow hue, milling should be done with caution.

Dispersion is important to sufficiently wet the pigment surface in order to prevent reagglomeration. Often the best way is to use a three-roll mill, which combines milling and dispersion in one step.

Stabilization is important to slow down the settling of the pigments, or creaming of the oils, thus leading to a uniform intermediate product that gives reproducible results and/or can be used for nuancing the colors.

Production methods definitely depend on the formula, batch size and the frequency of production. The production vessel only should be used for colored products as there are dead spots in the machines where tiny amounts of pigments can settle and contaminate "white" products.

In order to find the right production vessel, equipment manufacturers offer in-house trials where the formula developers and production people can try out the machines with their own formulations.

The different methods of production for the diverse product types will be discussed together with the respective product.

### 18.2.3.5 Color reproduction

The challenge that affects all products is the reproduction of the color originally selected for the assortment. Fortunately, color cosmetics are not applied to big areas (in contrast to car lacquers) and also no aged and new product applied at the same time. Therefore, color reproduction is somehow easier than for technical applications but nevertheless important.

The color of the final product is determined by the following:

(a) Pigments/colorants itself as they can have color deviations originating from their production process. Quality control and the assessment of tolerable deviations is essential [39].

(b) Production process: Huge efforts are made on the dispersion process of the pigments to make it reproducible and to get the final shading of the product. As the quality of dispersion is depending on the shear forces applied, it is common that the color varies from lab scale to production scale. During the development process, usually rules are established how to handle these (sometimes dramatic) color changes by adapting the pigment composition. It should be noted here that also – of course – the change of the production vessel (even if it is the same type and size) can have an effect on the final color.

(c) Formula: There is a significant difference whether the formula is a liquid dispersion of the pigments in water, a wax based or an emulsified one. The same pigment composition will look differently in the different systems, due to the different wetting properties of the dispersion medium. In the wax-based composition more gloss can be obtained due to the incorporation of oils with a high refractive index, whereas in the emulsion system opacity will be added to the optical aspects. Also changes in the composition can lead to undesired changes in color.

(d) Storage and aftertreatment: This is very important, as color cosmetics are not sold in bulk size but in small quantities like mg or ml. Therefore, the bulk has to be stored, quality checked and then filled into the appropriate casings. The filling process – for wax-based systems often connected with reheating – again poses shear forces on the product that in the worst case can lead to color changes. Thus, it is important to check the color during the whole production process.

(e) Color standards: Storage of the product also can lead to color changes as waxes tend to crystallize or re-crystallize and emulsion droplets increase in size over time. Pigments have a high tendency to re-agglomerate or to aggregate in an environment that suits them best. This is the reason why the perceived color of the product can change over time. This is extremely important to keep in mind, when comparing a freshly produced product with the color "standard" that has already aged.

(f) Color assessment: Comparing the colors of products usually is done by drawdowns where the sample and the standard are applied next to each other on a white/black paper and drawn down with a Japan spatula or a special device [40, 41]. This method is appropriate when comparing products that are also worn in a certain thickness and "drawdown," i.e. nail polishes or lip color. For all other products this method only gives a rough estimate whether production resulted in the right color direction or not. Nevertheless, for experts this method is sufficient to judge which pigments should be used for color shading.

Comparing the sample and the standard on the lower arm is a method that is closer to the final application conditions and also due to the application eliminates color changes to a certain extent that resulted from aging of the product. Sample and standard are applied on the lower arm in the following fashion (Figure 18.6).

This is extremely important when assessing colors containing high amounts of pearls or interference pigments that also can change color according to gazing angle.

It is important to consider that the skin of every person is different and also application methods are highly individual. Therefore, it is best to always have the same person(s) doing the color assessment or to train a group.

The method giving the most objective data is a colorimetric measurement [36]. It gives numbers and allows the definition of acceptable areas of color deviations. Unfortunately, in cosmetics we do not deal with pure pigment dispersions but with complex systems. Due to high gloss of interference and pearl pigments this method is not suitable for products containing high amounts of these materials. Furthermore, the colorimetric measurement has to be established for every formula, thus being a relatively laborious method that competes with the quick and readily available drawdown or application tests.

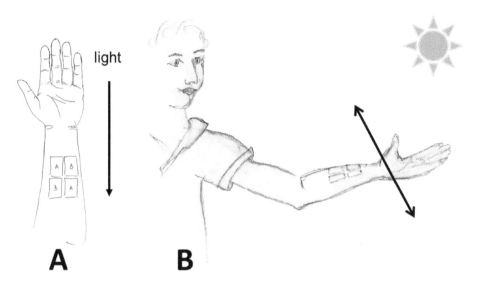

**Figure 18.6:** Color assessment on skin. (A) On the volar forearm two products, A and B are placed as depicted. The light source is coming from the front as shown in (B). With flexing the forearm and rotational movements color shades can be assessed (specular and aspecular angle). Usually products A and B are standard and newly produced product. They should be placed as closely together as possible and the same amount should be used.

## 18.3 The landscape of formulations

In this article, the formulations that are used in the different application areas or segments are covered. Dispersion technology usually does not correlate with application area. For consumers packaging and application characteristics are very important. I will cover those only when the formula/technology is unique for that application.

### 18.3.1 Facial products

The decorative products applied on the facial skin are foundations, powders, concealers, blushers and highlighters. In these groups, different technological approaches are used to obtain the desired effects. Many new products have been developed, which due to their enhanced skin feel or different application strategies are difficult to group into the "traditional" application formats. These are camouflage products or water-based makeup for artists.

Additionally, in the area of professional makeup, concealers and camouflage products are used. Due to their high load of pigments they differ considerably from formulas in mass or prestige market.

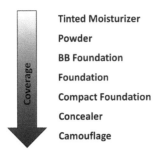

**Figure 18.7:** Skin coverage in different cosmetic products.

Coverage and pigment load are increased with the duty of the product (Figure 18.7). Tinted moisturizers deliver only little color and coverage, which is highest in camouflage products. They contain up to 45% of pigments. Besides pigment load application, uniform coverage and lastingness are important. Consequently, further ingredients are incorporated, such as waxes and film formers.

### 18.3.1.1 Foundations

Concerning the term foundation, there is some confusion. Often all makeup products are referred to as foundation. Makeup and foundation are used synonymously. Additionally, so-called primers, which are applied on the skin before any makeup are called foundation. In this article foundations are referred to as products that have a color that resembles skin complexion.

Foundations usually are liquid to creamy formulations. They are used to give a uniform layer on the skin. This either can be used as a foundation for further products or are used already as the product that gives the desired skin tone. Normally foundations are applied after the regular care products have been applied to the skin. Applicators, brushes or fingers are used for this purpose. Table 18.3 depicts a general composition of a foundation.

Foundations initially were developed in co-operation with the evolving movie industry where a product was needed that covered the skin evenly, withstood the hot lamps unimpaired and showed a natural skin tone when viewed on the film. Today such products are still found in the theatre or for special effects.

Consumers today expect more than just covering of the skin. They want their foundation to care (skin moisturization) and have a pleasant texture. The product should feel light on the skin and last all day.

A foundation composition, respectively, formulation is selected according consumers' expectations. As a pleasurable skin feel and rather long-lasting action is required today, much effort should be spent on the selection of the appropriate emulsification system and oils. In general, oils with a good spreadability and/or volatile oils are preferred. Less spreadable oils can give better stabilization of the pigment dispersion and add a cushioning skin feel. As the range of available raw materials is huge, only a few examples should be given to get the general idea (Tables 18.4 and 18.5).

**Table 18.3:** General composition of a foundation. Usually they are emulsified systems and can be manufactured as oil-in-water, water-in-oil emulsions or silicone emulsions. The latter are very popular for long lasting products and outstanding texture. Cosmetic foundations generally achieve their hue due to inorganic pigments such as titanium dioxide and iron oxides. In order to change the hue of the color also ultramarine or chromium oxide as well as sometimes also organic pigments are used.

| Substance | Concentration |
|---|---|
| Pigments (titanium dioxide, iron oxides) | 3–15% |
| Fillers | 0–10% |
| Waxes or wax-like substances | 0–10% |
| Emulsifier, Co-emulsifier | 2–10% |
| Oils with good spreadability | 0–10% |
| Dispersing aids | 1–5% |
| Moisturizer | 2–8% |
| Gel formers | 0,3–1,5% |
| Film formers | 0–2% |
| UV filters, preservatives, perfume, antioxidants | 0–5% |
| Water | 40–80% |

**Table 18.4:** Composition of an O/W foundation with medium coverage. Stearic acid emulsion systems are long proven systems for pigment dispersions.

| Function | Ingredient (INCI) | % |
|---|---|---|
| Emulsifier system | Stearic acid | 3,0 |
| | Glyceryl stearate and PEG-100 stearate | 2,0 |
| Oils | Mineral oil | 1,0 |
| | Dimethicone | 3,0 |
| Volatile oil | Cyclohexasiloxane | 3,0 |
| Oil with medium spreadability | Caprylic/capric triglyceride | 5,0 |
| Oil with high spreadability | Isostearyl isostearate | 2,0 |
| Dispersing aid | Acetylated lanolin alcohol | 0,8 |
| Fillers | Talc | 1,0 |
| Pigments | Titanium dioxide CI 77,891 | 4,5 |
| | Iron oxide yellow CI 77,492 | 1,2 |
| | Iron oxide red CI 77,491 | 0,3 |
| | Iron oxide black CI 77,499 | 0,2 |
| | Interference pearl | 2,5 |

**Table 18.4** (continued)

| Function | Ingredient (INCI) | % |
|---|---|---|
| Gel former | Xanthan gum | 0,8 |
| Stabilizer | Magnesium-aluminum Silicate (Veegum®) | 1,5 |
| Moisturizer | Glycerol | 5,0 |
| | Propylene glycol | 3,0 |
| Neutralizing agent | Triethanolamine | 0,9 |
| Others | Antioxidants, preservatives, fragrance | q.s. |
| Water | Water | Ad 100 |

**Table 18.5:** Composition of a silicone foundation with medium coverage from dow corning, similar formulation available from Happi [42].

| Function | Ingredient (INCI) | % |
|---|---|---|
| Emulsifier system | Cyclopentasiloxane and PEG/PPG 18/18 Dimethicone | 12,0 |
| | Polysorbate 20 | 0,5 |
| Silicone gum | Cyclopentasiloxane and dimethiconol | 12,0 |
| Volatile oil | Cyclohexasiloxane | 4,0 |
| Fillers | Nylon | 3,0 |
| | Cyclopentasiloxane and dimethicone crosspolymer | 5,0 |
| Pigments | Titanium dioxide CI 77,891 | 6,0 |
| | Iron oxide yellow CI 77,492 | 1,8 |
| | Iron oxide red CI 77,491 | 0,8 |
| | Iron oxide black CI 77,499 | 0,3 |
| Stabilizer | Sodium chloride | 1,2 |
| Moisturizer | Glycerol | 5,0 |
| Others | Antioxidants, preservatives, fragrance | q.s. |
| Water | Water | Ad 100 |

Coverage of the foundation can be modified by varying the pigment load. Titanium dioxide contributes most to coverage as it has the highest refractive index. In general, to reduce coverage, more fillers are used as their refractive index is rather low.

Also foundations with medical indication have emerged. Most widely used are foundations with anti-acne activity, which contain salicylic acid as active ingredient.

As foundations should achieve an even coverage, products usually have a lower viscosity. Also for the application with brushes, a fluid foundation is favored. Packaging usually is a tube or a bottle with or without dispenser.

For the production of foundations, the procedure is followed like for many emulsion products. In a vessel the water phase is heated and the pigments are dispersed. Then the other phases are added, homogenized and cooled.

### 18.3.1.2 Compact foundations

Compact foundations are offered in a cast or molded form. Cast compacts usually need a sponge applicator as they are solid. Normally, they are packed in a casing that also contains a mirror and the applicator. They give the product a more valuable appearance. This product type is preferred by consumers wanting a more sophisticated makeup.

An example of a compact foundation is shown in Table 6. Depending on the formulation (with or without water or volatile oils) also the packaging has to fulfill certain requirements. The material has to be sufficiently resistant to diffusion of the volatile compounds and the closing should be tight. The production process demands certain pre-requisites of the casing material, i.e. when the stick is to be molded directly into the casing a certain thermal stability is required to avoid deformation. As compacts usually are molded in a separate godet, the formula should be stable throughout the whole process of assembling the product. The casing of a compact foundation usually is a multicompartment one, consisting of the godet with the product, a mirror and a separate compartment for the sponge.

**Table 18.6:** Source: Composition of a compact foundation [43].

| Function | Ingredient (INCI) | % |
|---|---|---|
| Dispersion agent | Bis-Hydroxypropyl dimethicone | 16,8 |
| Silicone gum | Dimethicone/vinyl dimethicone crosspolymer and silica | 5,0 |
| Fillers | Mica | 2,05 |
| | Talc | 10,0 |
| | Polymethyl methacrylate | 4,0 |
| Coated pigments | Titanium dioxide CI 77,891 and triethoxycaprylsilane | 8,0 |
| | Iron oxide yellow CI 77,492 and triethoxycaprylsilane | 1,5 |
| | Iron oxide red CI 77,491 and triethoxycaprylsilane | 0,35 |
| | Iron oxide black CI 77,499 and triethoxycaprylsilane | 0,1 |
| Waxes | Beeswax | 3,0 |
| | Ozokerite | 6,0 |
| Oils | Squalane | 10,0 |
| | Petrolatum | 2,0 |
| | Dimethicone and trisiloxane | 25,0 |
| | Cyclopentasiloxane and dimethicone copolyol | 1,0 |
| Others | Preservatives | q.s. |
| Water | Water | 5,0 |

Molded sticks are products preferred by younger consumers or those who like to have their product at hand easily to refresh their makeup when needed.

Therefore, the bulk can be either molded directly into the packaging or into special forms and transferred to the casing afterwards like for lipstick bullets.

### 18.3.1.3 Concealers

Concealers are offered in a huge variety of formats. The range lasts from products for makeup artist who are working with palettes, to products in mass market, where mostly pencils are sold. Consumers usually want to cover single spots. Therefore, pencils or applicators are suitable. The main difference to foundations is, that concealers are used primarily to cover skin discoloration. Usually they are solid, have a higher pigmentation and come in different colors than other facial products. The main colors for concealers and their actions in covering skin discolorations are shown in Figure 18.8.

| | | |
|---|---|---|
| | Violet / Purple | Covers yellowish discolorations, gives frehness |
| | Green | Covers reddish discolorations (pimples, veins) |
| | Yellow | Covers blueish discolorations (veins) |
| | Peach/ Pink | Freshens up the skin tone |
| | Orange / Beige | Highlights the skin tone (very dark circles around the eyes) |
| | Dark Beige / Brown | Freshens up the skin tone, contours |

**Figure 18.8:** Main colors for concealers and their actions in covering skin discolorations.

Concealers and pencils (which are also used to conceal) are applied only to small areas of the face in order to conceal discoloration or pimples. Therefore, often active ingredients are incorporated which can be applied right on the spot where they are supposed to work.

These products depend very much on their casings which also serves as the application system at once. Whereas concealers mostly are semisolid formulations either emulsion or wax based, pencils only come as wax based varieties. They depend on a lead that is either extruded and placed in the wooden pencil or molded and cast in special twist up devices (for details see Eyeliner 3.2.3).

The advantage of injection molded plastic containers for concealers is that they can be much better designed to the customers' ideas than pencils [44].

### 18.3.1.4 Camouflage products

Camouflage products are highly pigmented and used to cover up dermatological (vitiligo) or cosmetic (tattoo) flaws. They definitely do not belong to the mass or prestige market but to dermatological or cosmetological applications. The application of a good camouflage cannot be done within minutes like a makeup, but needs at least 30 min or more. Camouflage products contain up to 45% Pigments and filmformers for the lastingness [45].

### 18.3.1.5 Ultra-liquid foundations

When color should be applied to large areas, ultra-liquid formulations are used. Usually they are formulated without emulsifier. Therefore, the dispersion separates over time and has to be redispersed by shaking the bottle. Special bottles with metal balls are doing the job [46].

Another example is the cushion foundation, which originally was developed in Asia as sun protection. The ultra-fluid composition is placed in a cushion like sponge and can be applied with another applicator (Figure 18.9). Here the packaging is the essential part of the product as it is airtight and therefore reduced evaporation of water [47].

**Figure 18.9:** Cushion foundation consisting of a packaging that resembles a compact foundation, but with a liquid formulation.

### 18.3.1.6 Loose and compact powders

Powders most likely belong to the oldest makeup products and also served medical purposes. As talc is the basis for most powders it is important to note, that talc powders are still the most widely sold powder products. Especially in humid and hot climate, talc absorbs access water from the skin and gives a cooling feeling. The absorption of fluids also can be disadvantageous as it leads to "caking" and resulting in an uneven finish over time.

These are aspects not wanted in facial makeup, where an even skin surface has to be maintained over a long period of time.

Loose powders are mainly used in order to highlight and even out the skin, whereas pressed powders often are used alternatively or additionally to foundations. Their main purpose is to mattify the skin, improve lastingness of the product. The general formulation of a powder is summarized in Table 18.7.

**Table 18.7:** General formulation of a powder.

| Substance | Concentration |
| --- | --- |
| Effect pigments | 0–25% |
| Iron oxides and titanium dioxide | 0–15% |
| Fillers | 0–20% |
| Perfume, preservatives, antioxidants, UV Filters, Oils | 0.5–4% |
| Talc | Add 100 |

Huge differences occur in the production process of powders that contain classical pigments and those which do not, as the classical pigments have to be grinded together with the talc to ensure de-agglomeration and as a consequence a uniform particle size. The sole use of pearlescent pigments only requires a mixing process and therefore is much less work intensive. Pearlescent pigments should not be milled or ground as they break down and lose their luster and color.

Pressed or compacted powders on the other hand need more oily substances and magnesium stearate as compressing aid to be compacted. When classical pigments should be applied, coated pigments are used as they adhere better to skin and – depending on the kind of coating – are easier to formulate.

### 18.3.1.7 Blushers, contouring and highlighting

Blushers like concealers are applied only to restricted areas of the face, but their main action is to give a healthy glow to the face and sculpture the features. They come to market as loose or compacted powders, emulsions or wax based molded formulations and liquids.

This is a special variety of blushers, which also can be applied all over the face or – according to color – on the lips. These products are no dispersions but solutions of water- soluble dyes. Application is a little difficult as they have no hiding capacity and cannot be blended very well. Therefore, they are most suitable for smaller areas. For the comfort of wearing of such products, humectants are added. A composition of a liquid blusher with soluble dyes is given in Table 18.8.

**Table 18.8:** Composition of a liquid blusher with soluble dyes.

| Function | Ingredient (INCI) | % |
|---|---|---|
| Dispersion agent | PEG-40 hydrogenated castor oil | 0,7 |
| Water soluble dyes | CI 14,700<br>CI 19,140<br>CI 28,440 | 0,4 |
| Thickener | Xanthan gum | 0,5 |
| Neutralizing agent | Sodium hydroxide 45% | 0,8 |
| Film former | Gum Arabicum | q.s. |
| Moisturizer | Glycerol | 5,0 |
| Others | Preservatives, water-soluble UV filters | q.s. |
| Water | Water | Ad 100 |

In order to enhance facial attractivity sexual maturity is an important feature (see chapter 2.2.5), therefore, not only blushers are used. Common are contouring materials that usually have a darker tone than the skin but no other color. They are placed on the areas that should be optically reduced. They are used before a foundation is applied and usually are blended in. Highlighters on the contrary are products that are applied at the end of the makeup process and are placed on areas that should be highlighted. They consist of larger amounts of effect or shimmer pigments to achieve that action. Figure 18.10 shows contouring and strobing as popular makeup trends.

### 18.3.2 Eye products

The market of eye products consists of mascara, eyeshadows, eye liners, and eyebrow preparations. More than 50% of the sales volume of these products are related to mascara makeup. Eye liners and shadows together carry about the same volume of market. Eyebrow preparations have become very trendy only recently. As only eye shadows are fashion driven products, it is very interesting to note that the eye

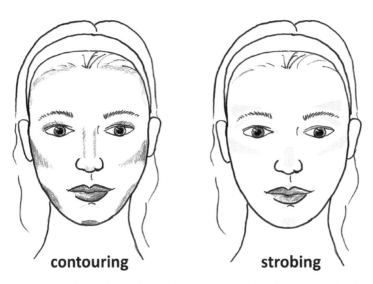

**contouring**     **strobing**

**Figure 18.10:** Contouring and strobing are very popular makeup trends that differ in the areas the product is applied. For contouring products with rather darker colors are used, i.e., blushers or concealers, whereas for strobing products high amounts of pearl or shimmering pigments are applied.

segment mainly is made up by so-called "job" products with general beautifying effects. The areas of the eye, which can be colored, are shown in Figure 18.11.

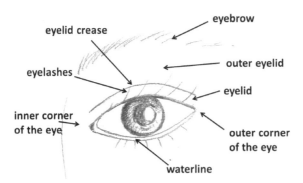

**Figure 18.11:** Areas of the eye, which can be colored.

As the "job" products (mascara, eyeliners) do not offer exciting colors, much more attention is paid to the application and the packaging format of the product as well as the technological aspects of the formula.

Nevertheless, the eye area is very sensitive. Allergies often manifest first on the eyelid and also adverse reaction to ingredients can occur here more easily. On the other hand, contamination of the product has to be prevented, so that no infection takes place. This is especially difficult as many products for eye makeup consist of a format where an applicator is introduced into the mass on a regular basis. This aspect also is taken care of by the note for shelf life after opening [48]. This is usually 6 months and therefore much lower than for other products.

Products, which are applied in the eye area usually are formulated without perfume.

### 18.3.2.1 Mascara

Mascara is a product to be applied on the eye lashes in order to darken and thicken them. The main color for mascara is black. Other colors such as brown, green or blue are of minor importance.

Nevertheless, consumer demands are high. Mascara should emphasize the eyes dramatically by enlarging the lashes to their extremes in order to give an open and wake look. Mascara should be tolerated well, adhere to the lashes and neither stick together nor smudge. Smudging means the transfer of the color to the lower lid which is not desirable. Mascara should dry in a relatively short time, stick to the lashes all day and come off with regular makeup removers at the end of the day. Mascaras come to market in a tube with a wand applicator. The applicator, also referred to as brush, is the point of difference and heavily patented.

Formulations usually vary from "regular" to "waterproofed." The first one is a stearate-wax-pigment dispersion with water and the latter one is a wax-dispersion with a volatile oil. Table 18.9 contains a comparison of regular and waterproofed mascara formulations.

The regular formulation generally is a TEA-saponified stearate-system with high amounts of waxes, pigments and film formers. The stearic acid salts also act as very efficient pigment dispersers. This formulation is in the market more than 50 years with only minor variations. Differentiation of the products takes place by incorporation of different waxes or film formers, in most cases, however, via packaging, and/or application, especially the brush.

Waterproof formulations generally are formulated under completely different rules. Here waxes, film formers and pigments are dispersed in volatile oil, such as cyclomethicone or isododecane.

As all formulas consist of a considerable amount of waxes, the production process requires pigment dispersion at high temperatures (above the melting point of the waxes). Also filling of the bulk into the containers requires reheating, as the finished product is a paste like product, which is difficult to fill.

Nowadays also some mascaras are formulated with carbon black. Carbon black itself is very difficult to disperse due to its small particle size and huge surface area. In general, much lower amounts of carbon black are necessary to achieve the same color intensity as with iron oxides. The color of carbon black mascaras however tends more to black and blue-black than that of black iron oxides.

**Table 18.9:** Comparison of regular and waterproofed mascara formulations clearly show that the main difference lies in the emulsifying system, film formers and volatile fluid.

| Function | Substances | Regular % | Waterproof % |
|---|---|---|---|
| Pigments | Iron oxide black | 10 | 10 |
| Fillers | Talc | q.s. | 5 |
| Waxes | Beeswax | 10 | 8 |
| | Carnauba Ozokerite | 3 | 4 |
| | | 7 | 6 |
| Emulsifier, Co-emulsifier | Stearic acid | 5 | |
| | Glyceryl Stearate | 5 | |
| Neutralizer | TEA | 1,5 | |
| Humectant | Propylene glycol | 5 | |
| Film formers | Gum arabic | 0,5 | 2 |
| | PVP/Eicosencopolymer | | |
| Stabilizer | Stearalkonium hectorite | | 2 |
| preservatives, Antioxidants | | q.s | q.s |
| Volatile fluid | Isododecane | | Ad 100 |
| | Water | Ad 100 | |

## 18.3.2.2 Eyebrow products

Eyebrows are the outer, hairy border of the eye sockets and – like the eyelashes – serve to protect the eye. They follow the shape of the skull and therefore cannot be shaped into any shape. But just as our faces are not "ideal," nor is the shape of the brow.

Eyebrows are regarded the frame of the face and a decisive factor in attractiveness. At the same time, they say a lot about character and personality and are important in the expression of emotions. As eyebrows consist of hairs, they therefore can be made up such as mascara, trimmed or colored or even tattooed.

Hair, especially a lot of hair is a sign of vitality that also is valid for eyebrows. The shape and thickness of eyebrows underlies fashion and cultural trends. Thick eyebrows are considered a sign of strength of character, fun and success. Women with thin eyebrows are considered flexible. Fused eyebrows (Frieda Kahlo) are considered beautiful only in some cultures.

In the past years, eyebrows were reinvented in makeup and a huge variety of products came to the market in order to shape, thicken and color them [49].

Most important are formats that resemble mascara and pencils. With both products not only color can be applied, but also individual hairs can be "made up."

These techniques are especially important for people who have undergone cancer treatments and lost their hairs. Here tattooing of the eyebrows serves as beautifying and mood lifting agent.

Figure 18.12 shows the ideal brow of a human face, which is defined with a line from the outer part of the nostril to the inner corner of the eye and ends at the imaginary line from the nostril to the outer corner of the eye (blue arrows). The brow should have a "kink" at the point where the line of the nostril to the middle of the pupil passes (red arrow). Deviations are common and often also quite attractive.

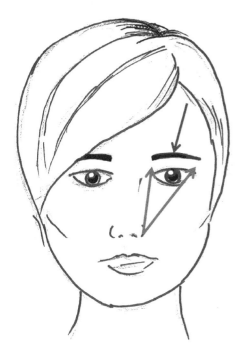

**Figure 18.12:** Ideal brow of a human face.

### 18.3.2.3 Eyeliner

The traditional version of an eyeliner was called "khol" or "kajal", which is the Hindu word for carbon. The use of carbon black dispersion in fact is very old. The product was applied around the eye and on the inner eyelid mainly for medical considerations i.e. to repel insects and in consequence avoid conjunctivitis. It can be assumed that the combination with antimony or lead was used in order to obtain a certain antibacterial activity and also give stimulus to enhance lacrimation, so that the constant flow eliminated particles from the eye. Lacrimation also gives gloss to the eye and as a result the eye looks bigger and healthier. At that time with no hygienic standards in force, blindness posed a big threat [50].

The classical format is a pencil liner in a wooden format that derived from the pencil industry and firstly was marketed in 1927 as the so-called dermatograph. Originally designed for surgeons to draw the line for exact incisions, it was used by women as a makeup product. This pioneering work was done by August Schwanhaeusser as he experimented in obtaining a softer lead and maintaining good sharpening characteristics of the final pencil.

The pencil today is still the most important format. The production process is a multistep process and cannot be completely automatized [51]. It involves the preparation of the lead: heating the waxes, addition of the fillers and pigments, dispersing the pigments and cooling down the bulk. After that the bulk is transferred into a cartouche which is stored for a certain period of time. Afterward the lead is extruded, placed in wooden slats that are glued, sawed to appropriate form, polished, lacquered decorated, sharpened and capped. The general composition of pencil leads is given in Table 18.10.

Table 18.10: General composition of pencil leads.

| Substance | Concentration |
| --- | ---: |
| Pigments and fillers | 30–60% |
| Hydrogenated palm- or cocglycerides | 30–50% |
| Oils | 5–15% |
| Hydrogenated vegetable oil | 2–12% |
| Ceresin | 1–5% |
| Beeswax | 1–5% |
| UV filters, preservatives, antioxidants | 0–5% |

In general, extruded leads have a higher pigment load than molded leads, which also contain more fluid components. Nevertheless, molded leads are of high interest as a plastic casing can come in different shapes. The lead usually needs no sharpening. For the production process of these products, the appropriate machines and tailor-made casings are essential and described in patents [52, 53].

Wax based pencil (leads in wooden or plastic casings) contain relatively high amounts of pigments and due to the production process differ in the hardness of the leads. Application generally gives fine or wide lines that can be blended easily. This kind of product also very often is applied in the inner lid of the eye (waterline, see Fig. 11), where they give different makeup effects.

In contrast, fluid inks give sharp lines that cannot be blended. Ink liners usually are marketed in a bottle with an applicator system. Those formats are favored by more experienced consumers who want an accurate makeup. The general composition of cosmetic inks is shown in Table 18.11.

**Table 18.11:** General composition of inks.

| Substance | Concentration |
|---|---|
| Pigments | 20–30% |
| Thickeners | 1–5% |
| Emulsifier, coemulsifier, solubilizer | 0.5–2% |
| Filmformers | 2–25% |
| UV filters, preservatives, antioxidants, neutralizing agents, humectants, chelating agents | 0–5% |
| Water | 60–80% |

For the production of those systems usually a base is prepared and the pigments are added accordingly (similar to nail polishes). The dispersion of the pigments has to be ensured by either applying high shear force or milling. The choice of the appropriate equipment is one of the challenges when formulating an ink.

Pencil formats for those inks are liked very much but due to the technical considerations of a pencil (which has to have a reservoir) only pigments with small particle size (i.e. carbon black) can be used.

Eyeliners – pencils and inks – are best tested on the ball of the thumb (Figure 18.13). Here not only pay off, but also accuracy of application and the hardness of the lead can be evaluated best.

### 18.3.2.4 Eyeshadow

Eyeshadows are fashion-oriented products. Their task is not only to improve the looks and widen the eye but also to give a color statement. Therefore, a huge color variety is very important. Formulations and application formats have to follow fashion trends.

As most eyeshadows are compacted powders, they contain a compaction aid, i.e. magnesium stearate. Eyeshadows often contain high amounts of pearls and interference pigments. Therefore, attention has to be paid to the mixing process to avoid the breakdown of the pearls. Thus, a formula which has neither to be milled nor pressed is useful. This can be obtained for instance with silicone-based gels. A composition of an eye shadow with a high amount of effect pigments is shown in Table 18.12.

**Figure 18.13:** Test of eyeliners on the ball of the thumb.

**Table 18.12:** Composition of an eye shadow with a high amount of effect pigments [54].

| Function | Ingredient (INCI) | % |
|---|---|---|
| Pigments | Effect pigments | 23,0 |
| Silicone gel | Cyclopentasiloxane and dimethicone copolymer | 49,5 |
| Filler | Trimethoxysilicate | 9,0 |
| Others | Preservatives, antioxidants | q.s. |
| Volatile fluid | Cyclopentasiloxane | Ad. 100 |

For eye makeup, it is usual to have palettes, especially for compacted products or applicators for paste like products. As the formula besides the color does not offer much space for innovation, the packaging is a relevant part of these products. This usually is designed individually and with 3-D printing easily accessible. Nevertheless, for the consumers the application charcteristics are of high relevance, this includes a suitable application format.

Eye shadow palette with molded or cast powders is a good example how packaging is affecting the perception of the product itself. Due to Internet tutorials the demand for palettes has risen considerably (Figure 18.14).

**Figure 18.14:** Eye shadow palette with various colors.

### 18.3.3 Lip products

Lip products are mainly wax or fat-based dispersions, sometimes water-based inks are used. As the skin of the lips is very delicate, attention should be taken when formulating with coarse pigments or fat components that have phase transitions points that could lead to unwanted particles. A main issue of lip products is the application that should be smooth and easy and should be possible with no pressure, summarized as pay-off. Consumers also do not want to have oily films on the lips that in worst case could make the pigments "feather" into the lines of the lips.

As the skin of the lips not only is thinner than skin on other body parts it also lacks sebaceous and sweat glands. In order to keep our lips moist, we are licking them unconsciously many times a day. Therefore, also parts of the formulation gradually are ingested and therefore all lip formulations should be food grade. Additionally, preservation is an important issue as the product can be contaminated with oral bacteria.

But lips are also very sensitive to intrinsic and extrinsic changes, i.e. herpes often manifests itself at the lip and also irritating substance can cause unwanted side effects i.e. preservatives. No melanocytes are present in the basal layer and so lip-products ideally are formulated to give certain UV protection.

Last not least the casing/packaging is an essential part of the product and should be kept in mind when developing the formulation.

#### 18.3.3.1 Pigment dispersion strategies

In lip formulations, organic pigments and lakes are used to obtain bright red colors. Those pigments are more difficult to disperse than inorganic pigments, especially iron oxides. In order to obtain reproducible colors, pigments have to be ground and dispersed very carefully. State of the art for this task is still the three-roll mill.

The normal procedure implies that the pigments firstly are wetted by an oil (generally castor oil, INCI: Ricinus Communis Seed Oil; Lanolin Oil or other polar oils), which has a good dispersion ability and good stabilization characteristics on the finished dispersion (rather high viscosity). Those pre-ground dispersions can be stored, which allows a rather easy preparation of the final product and also facilitates color shading. Alternatively, predispersions can be obtained commercially.

Nevertheless, lip products can be produced with roller ball mills or other milling equipment without pre-dispersion of pigments. Color shading is done at the end of the production process. Color shading, if necessary, is done with aliquots of the preparation in which the additional pigments are dispersed.

### 18.3.3.2 Lipsticks

The stick is the "classical" format of the lip color and is unsurpassed by any other application form due to its convenience. Additionally, the lipstick casing transports a lot of life style. So even if the formula is remaining the same, a different packaging can add value to the product and can make it highly desirable. Therefore, the lipstick is and will be the most used form of lip color.

The molding process is a crucial step in production of a lipstick and all other stick products as well. The bulk is (re)heated in a vessel and aliquots are poured into the casts. Cooling process takes from about only a few to several minutes depending on the product or capacity of the machine. Then the so-called bullet will be ejected from the cast using air or mechanical devices and transferred into the casing. The crystallization of the waxes has to be rather quick so that the bullet gets enough mechanical stability to be ejected. Final hardening of the stick takes about 24 hours or even longer.

A stick usually is molded, that means the mass is heated to a temperature where it is liquid and then poured into molds or the final packaging. In order to obtain high-quality products, the cooling of the poured mass is an essential production parameter. The solidity of the poured stick evolves even after the production has completed and last for days to weeks.

A general composition for lipsticks is depicted in Table 18.13, the proportion of oil/wax/pigment is usually around 4/2/1. The wax composition is of major importance as it defines stability, oil binding capacity and ease of production.

For the classical lipstick, the composition is of vital importance, as the wax composition mainly contributes to the structure and stability of the stick. The wax composition of molded sticks has to contract when cooled down to allow easy removal. The production process of lipsticks needs some consideration. Generally, the base is heated to the temperature where all solid waxes are molten. Pigments (generally pre-dispersed) and active ingredients are added successively. Then the

**Table 18.13:** General composition of a lipstick formulation. Important for the stability is the choice of the waxes and oils with good pigment dispersion properties. Micro-waxes belong to ingredients of petrochemical origin. These cannot be certified under natural cosmetics. Reformulation of sticks is difficult as micro-waxes add to thermal stability and also are used to replace natural waxes such as carnauba or candellila wax.

| Substance | Concentration |
|---|---|
| Natural waxes, wax-like substances | 10–20% |
| Microwaxes | 0–10% |
| Oils with good spreadability | 15–30% |
| Oils with low spreadability | 40–60% |
| Surface active substance, emulsifier | 0–5% |
| Pigments, fillers, film formers, UV filters, preservatives, aroma, antioxidants | 5–20% |

bulk is cooled and stored or processed directly afterward. An example for a classical lipstick formulation is shown in Table 18.14.

**Table 18.14:** Example for a classical lipstick formulation.

| Function | Ingredient (INCI) | % |
|---|---|---|
| Waxes | Candelilla wax | 2,75 |
| | Carnauba wax | 1,25 |
| | Beeswax | 1,0 |
| | Ceresin | 5,9 |
| | Ozokerite | 6,75 |
| | Microcrystalline wax | 1,4 |
| Dispersing aid | Acetylated lanolin alcohol | 2,5 |
| | Sorbitan triisostearate | 2,0 |
| | Oleyl alcohol | 3,0 |
| | Bis-Diglyceryl polyacyladipate-2 | 2,0 |
| Oil with medium spreadability | Caprylic/Capric triglyceride | 5,0 |
| Oil with high spreadability | Isostearyl isostearate | 5,0 |
| | Isostearyl palmitate | 7,5 |
| Fillers | Nylon-12 | 1,0 |
| Pigments | Inorganic, organic and effect pigments | 8,0 |
| Others | Antioxidants, preservatives, fragrance | q.s. |
| Main dispersing oil | Ricinus communis seed oil | Ad 100 |

### 18.3.3.3 Emulsion sticks

In a wax-based stick formulation usually a few percent of hydrophilic ingredients or even water can be easily incorporated. But higher amounts are difficult to obtain. Even though formulations containing up to 30% of water were marketed, the production process is difficult to handle and has to be closely followed. In contrast to water free formulations the bulk cannot be cooled down, stored and reheated whenever needed. The mass has to be prepared directly before molding. As the capacity of the molding machine generally is the limiting factor; logistics have to be adapted and the amount of bulk prepared should be processed within a working shift.

Additionally, color stability is the main issue with sticks containing water. Firstly, organic pigments are extremely difficult to incorporate, ideally colors are developed using iron oxides and titanium dioxide. Secondly as the mass has to be stirred and heated over several hours and water is most likely to evaporate, consistency of the final sticks is difficult to achieve if at all. Water content will vary from the first to last sticks as will color, thus making it extremely difficult to produce a uniform batch [55].

### 18.3.3.4 Gloss and pomade

Lip gloss became very popular in the last years. It is not only used to highlight lip makeup but became a product of its own. It is offered as colorless or colored product and can be used alone or as finish for the lip color. It is marketed either in a tube with a wand applicator or in a squeezable tube or in a combination of both. It usually comes with a fruity flavor.

The glossy effect is obtained by incorporation of substances with high refractive index. The base normally consists of a polymer such as polybutene or polyisobutene. For the stabilization only minimum amounts of waxes should be used, more common is the stabilization with inorganic gel formers like hectorites [56]. Gloss formulations tend to have a poor lasting effect and therefore have to be applied several times.

### 18.3.3.5 Matte lipsticks

Matte lipsticks and long-lasting colors are very popular since many years. Usually they are formulated with volatile oils, thus making it difficult to obtain molded sticks (see emulsions sticks). Therefore, a new generation of formats has evolved, which is marketed as fluid lipsticks, but technically are pastes.

When developing such formulations some aspects are important: the use of fillers and the absence of pearl or effect pigments, the use of oils with low refractive index, the use of silicone polymers and volatile oils and the use of film-forming materials, which can be resins or waxes.

As the application and wear of such products sometimes is not as comfortable as regular lipsticks, a colorless stick can come with the product and is applied on top of the color [57].

### 18.3.3.6 Lip tint
Fluid compositions are more often used for light color and a natural look on the lips than for care properties. They do not give a perceivable film on the lips and therefore are not liked by all lipstick users.

### 18.3.3.7 Lip liner
Lip liners are offered in form of pencils and the lead consists of a blend of waxes, oils and pigments. They are used to define the borders of the lips more accurately and due to their harder consistency, they also help to diminish feathering of lipstick colors. They are an essential tool, as they help to sculpture the lips. Figure 18.15 shows the effect of different lip liners on the perceived size of lips. The comparison of different lip colors is depicted in Figure 18.16.

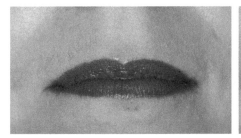

**Figure 18.15:** Comparison of the effect of different lip liners on the perceived size of lips. Use of dark (left) and light (right) lip liner with a lipstick in the natural color of the lips. With a dark lip liner, the lips look bigger, with the light ones, they look smaller.

### 18.3.4 Nail products

The nails belong to the so-called cutaneous appendages. Like hair, they are regarded as inert in contrast to skin. The nail plate consists of cornified cells, like the hair. But these cells are flat. Table 18.15 shows the water content and the water absorption capacity of nail plate and stratum corneum. The nail plate itself cannot be plasticized well by water and therefore also does not swell like stratum corneum.

The amino-acid composition from nail and hair also differs to skin. Due to the rigid keratin network nails are regarded as better barriers toward penetration than skin. The nail is primarily composed of a highly cross-linked keratin network that contains several disulfide linkages. This unique structure results in a highly effective permeability barrier [59]. Table 18.16 compares the amino-acid content of different keratinous substances. The amino acids with the biggest differences from nail to hair or to stratum corneum are depicted in this comparison [58].

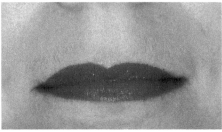

**Figure 18.16:** Comparison of different lip colors from top to bottom: light pink pearlescent, classic red, very dark red.

**Table 18.15:** Water content and absorption capacity of nail plate compared to stratum corneum [58].

|  | Water content | Water absorption capacity |
|---|---|---|
| Nail | 7–12% | Up to 25% of the dry matter weight |
| Stratum corneum | 5–15% | A multitude of the dry matter weight |

The nails are nourished by the nailbed and the nail fold, where the nail origi-nates. Traumata to those areas often result in damaged nails. Nails also betray a lot of the health status, for instance psoriasis also can be seen in the nails when small holes occur.

Nails are a very important feature of the hands. Evenly smooth nails convey health and youth, whereas long nails are proof of wealth and a life of leisure. Also, nail color is a topic of fashion and lifestyle. Therefore, it makes sense to treat nail colors together with facial makeup.

**Table 18.16:** Comparison of the amino-acid content (in %) of different keratinous substances [59].

| Amino acid | Nail | Hair | Stratum corneum |
|---|---|---|---|
| Histidine | 1.0 | 0.9 | 1.5 |
| Arginine | 6.4 | 6.5 | 3.8 |
| Aspartinic acid | 7.0 | 5.4 | 7.9 |
| Threonine | 6.1 | 7.6 | 3.0 |
| Proline | 5.9 | 8.4 | 3.0 |
| Glycine | 7.9 | 5.8 | 24.5 |
| Valine | 4.2 | 5.5 | 3.0 |
| Leucine | 8.3 | 6.1 | 6.9 |
| Hemicysteine | 10.6 | 15.9 | 1.2 |
| Organically bound sulfur | 3.2 | 4.5 | 1.4 |

Also finger nails are a very interesting part of the body. They are developed in the 9$^{th}$ week of gestation (at the same time as teeth), and convey information about genetic health. Considering that makeup should also improve the mating success, nails are an important factor.

Interesting is, that nail lacquers are grouped under makeup together with facial makeup, in contrast to hair colorants. The idea behind that might be, that in coloring of the hair chemicals are used that react on the hair whereas for coloring of the nails products are used that stay on top. Therefore, the application of more colorants and other substances are allowed which are restricted on normal skin.

### 18.3.4.1 Nail lacquers

Nail lacquers generally consist of nitrocellulose which is dispersed in a mixture of alcohols, mainly isopropanol and other solvents. Nitrocellulose is highly explosive and therefore production of nail colors is done under strict safety conditions which are not necessary for ordinary cosmetic products. The general composition of nail lacquers is depicted in Table 18.17.

Nitrocellulose is still the best film former for the formulation of lacquers, despite its time of usage. As it is highly explosive when dry, it is stored and used in a predispersed form. Typically used solvents are isopropanol, butyl-, ethyl- and propyl acetate. Further substances are added like toluenesulfonamide/formaldehyde resin or epoxy resins in order to improve gloss and adhesion of the film to the nails.

As gloss and elasticity of films generally are opposing characteristics, a good mixture of additional film formers like polyacrylic acid esters or polymethacrylic

**Table 18.17:** General composition of nail lacquers [60, 61].

| Substance | Concentration |
|---|---|
| Nitrocellulose | 10–20% |
| Solvents | 35–40% |
| Plasticizers | 5–10% |
| Thickeners | 0,5–5% |
| Filmformers and resins | 0.5–5% |
| Pigments, fillers, UV filters | q.s. |

acid esters and plasticizers such as camphor, dibutylphthalate, acetyl tributyl citrate, castor oil, diisobutyl adipate or butyl octyl adipate has to be found.

As thickening agents for those systems, bentonite gels frequently are used like stearalkonium hectorite. They are important for the stabilization of the system as they slow down syneresis, the sedimentation of the pigments. Generally, UV-absorbers are added to prevent the yellowing of the nitrocellulose.

For the production of nail lacquers generally the solvents are placed in the production vessel first. Predispersed nitrocellulose then is added. Finally, film formers, resins, plasticizers and thickeners are then added to give the base. That base can be stored and used to produce the final color. To that base the pigments are added according to the desired color. The pigments used for nail lacquers most often are preprocessed and used in chip format to reduce dusting. At the end the residual ingredients are added [62].

### 18.3.4.2 Gel nails

In order to overcome classical disadvantages of nail lacquers like chipping and poor lastingness, so-called gel lacquers have been developed. Derived from nail modeling, this trend has settled in the market of professional and non-professional manicure. These nails can either be achieved by modeling while using acrylates or urethane acrylates (a monomer and a starter) or by using so-called gel lacquers (where everything already is mixed together). The use of acrylates usually is limited in the professional applications as experience and certain working conditions are necessary. The curing of the polymers takes place at room temperature and is initiated by a so-called liquid, where monomers are incorporated [63]. Urethane acrylates usually come in premixed formats. They are cured using light: UV, LED or daylight) [64]. It is noteworthy that also in this area the innovation is not in the pigments but in the formula, which is derived from coatings industry [65].

## 18.4 Outlook

Even though pigments are the active ingredients in makeup formulations, they are only partially important for the product. At least the same impact is achieved by the right choice of fillers, waxes, polymers and oils. Together they achieve the desired properties. At the moment we can technically incorporate (almost) every pigment into every cosmetic formulation. Especially in the case of light cured nail varnishes, so-called gels, the pigments have to withstand polymerization reaction and UV or blue light irradiation.

As consumer perspectives are changing toward "sustainability" meaning that consumers prefer natural ingredients over synthetically derived molecules, they expect their cosmetic products also to deliver these values. Legislation additionally adds further restraints for instance for the use of volatile organic compounds like cyclomethicones. The industry meanwhile offers a lot of oils which can be used as silicone alternatives, but in the end give a completely different product. As most organic fillers are classified as microplastic, they are not further used in new developments. The problem arises for gel formers on acrylate or polyurethane basis, for which often no alternative is available. Mineral waxes such as Ceresin, Ozokerite or Microcrystalline Wax, which have been widely used in lipstick formulations have been replaced. From the standpoint of a formulator, these actions not always make sense, but honor the changing demands of consumers. It should be noted, that also dispersion aids like castor oil are restricted. This leads to the use of other dispersing agents.

Natural products such as talc, mica and kaolin also are scrutinized. Besides heavy metal content the consumers also ask for the areas where the minerals are mined or the working conditions of the laborers. These factors therefore influence the selection of the materials and suppliers, respectively. The same is true for natural oils, which should be GMO free and grown sustainably.

Packaging and applicator are relevant for makeup. They are essential parts of the cosmetic product. But the changing expectation of consumers will lead also here to more "refills" or reuse of certain formats, wherever that makes sense from a microbiological standpoint.

Traditional pigments will be used for all the known cosmetic applications. Further, inorganic pigments such as iron oxides and titanium dioxide will be favored over organic pigments. But for lipsticks and especially nail varnishes, these pigments will continue to be important. First trends emerge in natural makeup that strictly stick to mineral pigments. In order to achieve different colors, they are also relying on effect pigments.

As pigments and colorants are regulated separately in the cosmetic directives worldwide and the toxicological profile is examined closely, it is not expected much innovation to take place in that area. Innovation for cosmetic formulations therefore will come from application format and application characteristics.

# References

1.    Color cosmetics market by target market (Prestige Products and Mass Products) and application (Facial Make Up, Lip Products, Eye Make Up, and Nail Products) - global opportunity analysis and industry forecast. 2017–2023. https://www.alliedmarketresearch.com/color-cosmetics-market. Accessed: 11 Jun 2020.
2.    https://www.statista.com/statistics/297070/growth-rate-of-the-global-cosmetics-market/ Accessed: 11 Jun 2020.
3.    Dodson D. Putting the color in cosmetic sales, global cosmetic industries. 2008. https://www.gcimagazine.com.marketstrends/segments/cosmetics/16256206.html?prodrefresh=y. Accessed: 11 Jun 2020.
4.    Melage C. Global color cosmetic market assessment. San Francisco: Color Cosmetic Summit, 2006.
5.    Grabenhofer RL. Who is leading the color cosmetics race? Cosmetics & Toiletries (2016) e-Formulatory Directory. http://www.bnaeopc.com/Color-2016-ebook-CosmeticsAndToiletries.pdf. Accessed: 11 Jun 2020.
6.    https://www.ikw.org/ikw/der-ikw/fakten-zahlen/marktzahlen/. Accessed: 06 Jun 2020.
7.    https://www.loreal-finance.com/en/annual-report-2017/cosmetics-market. Accessed: 11 Jun 2020.
8.    https://marketpublishers.com/report/consumers_goods/cosmetics/cosmetics-market-by-category-skin-n-sun-care.html. Accessed: 11 Jun 2020.
9.    Singh B. Cosmetic Market: Size, Trends, Forecasts, 2016-2025: Variant Market Research. https://www.slideshare.net/karanSingh950/cosmetics-market-global-scenario visited 11 Jun 2020. Accessed: 11 Jun 2020.
10.   https://www.ikw.org/fileadmin/ikw/z-IKW-ENGLISCH/IKW_AnnualReport-2017-2018.pdf. Accessed: 11 Jun 2020.
11.   Cosmetics Europe. Annual report. 2017. https://www.cosmeticseurope.eu/files/1615/2872/3398/CE_Annual_Report_17.pdf. Accessed: 11 Jun 2020.
12.   https://makeup-in.com/05-trends-en/makeup-continues-to-pull-the-global-cosmetic-market-upwards-2/. Accessed: 06 Jun 2020.
13.   Cosmetics Directive 76/768/EEC, Article 1. https://eur-lex.europa.eu/ legal-content/EN/TXT/HTML/?uri=CELEX:31976L0768&from=EN.
14.   https://www.fda.gov/forindustry/coloradditives/coloradditivesinspecificproducts/incosmetics/ucm110032.htm. Accessed: 11 Jun 2020.
15.   https://colour-index.com/introduction-to-the-colour-index. Accessed: 09 Jun 2020.
16.   O'Lenick AJ. Comparatively speaking: natural- vs. mineral-based colorants. Cosmet Toiletries. 2011;3:e-version. https://www.cosmeticsandtoiletries.com/formulating/function/pigment/118610674.html. Accessed: 11 Jun 2020.
17.   https://sensientfoodcolors.com/en-eu/. Accessed: 11 Jun 2020.
18.   https://www.pinkmelon.de/magazin/geheimnis-kosmetik/tattoo-dermographische-impressionen-ein-bild-fur-die-ewigkeit.html. Accessed: 11 Jun 2020.
19.   Tätowiermittelverordnung. https://www.gesetze-im-internet.de/ttov/BJNR221500008.html. Accessed: 11 Jun 2020.
20.   Bäumler W. The fate of Tattoo pigments in the skin. BfR symposium, 2013. https://mobil.bfr.bund.de/cm/343/the-fate-of-tattoo-pigments-in-the-skin.pdf. Accessed: 11 Jun 2020.
21.   Petersen H, Roth K. To Tattoo or Not to Tattoo? part 1-4, Chemie in unserer Zeit. 2016. https://www.chemistryviews.org/details/ezine/10162381/ToTattoooorNottoTattooPart_1.html.
22.   Gieske S, editor. Lippenstift - Ein kulturhistorischer Streifzug über den Mund. Marburg: Jonas Verlag, 1996.

23. https://www.pinkmelon.de/magazin/geheimnis-kosmetik/vom-zauberstab-des-eros-zum-modernen-lippenstift.html. Accessed: 11 Jun 2020.
24. Deutscher G. Through the language glass. Metropolitan Books. 2010.
25. Heller E. Wie Farben wirken, 7th Aufl. Rowohlt Verlag, 2013.
26. https://www.pinkmelon.de/magazin/geheimnis-kosmetik/rot-rot-rot.html. Accessed: 11 Jun 2020.
27. https://www.pinkmelon.de/magazin/geheimnis-kosmetik/lippen-schoen-schminken.html. Accessed: 11 Jun 2020.
28. Little AC, Jones BC, DeBruine LM. Facial attractiveness: evolutionary based research. Philos Trans R Soc Lond B Biol Sci. 2011;366(1571):1638–1659. DOI:10.1098/rstb.2010.0404. https://www.ncbi.nlm.nih.gov/ pmc/articles/PMC3130383/#!po=16.6667.
29. Etcoff NL. Beauty and the beholder. Nature. 1994;368:186.
30. Henss R. Perceiving age and attractiveness in facial photographs 1. J Appl Social Psychol. 1991;21:933.
31. Rhodes G, Sumich A, Byatt G. Are average facial configurations attractive only because of their symmetry? Psychol Sci. 1999;10:52. DOI:10.1111/1467-9280.00106.
32. Ibáñez-Berganza M, Amico A, Loreto V. Subjectivity and complexity of facial attractiveness. Sci Rep. 2019;9:8364. DOI:10.1038/s41598-019-44655-9. https://www.nature.com/articles/ s41598-019-44655-9
33. Cunningham MR. Measuring the physical in physical attractiveness: Quasi-experiments on the sociobiology of female facial beauty. J Pers Soc Psychol. 1986;50:925.
34. Cunningham MR, Roberts AR, Barbee AP, Druen PB, Wu CH. "Their ideas of beauty are, on the whole, the same as ours": consistency and variability in the cross-cultural perception of female physical attractiveness. J Pers Soc Psychol. 1995;68:261. DOI:10.1037/0022-3514.68.2.261.
35. Henss R personal communication.
36. Cash TF, Dawson K, Davis P, Bowen M, Galumbeck C. Effects of cosmetics use on the physical attractiveness and body image of American College Women. J Soc Psychol. 1989;129:349.
37. Presentation of Seismo Market Research for Schwan Cosmetics. May 2004.
38. Pfaff G. Optical principles, manufacture, properties and types of special effect pigments. In: Special effect pigments. 2nd ed. Hannover: Vincentz Verlag, 2008:16–91.
39. Hollenberg J. Standardized color evaluation methods for cosmetic pigments. San Francisco: Color Cosmetic Summit, 2006.
40. https://gardnerlaboratories.com/2014/07/11/byko-chart-overview-variety-drawdown-charts-variety-applications/. Accessed: 11 Jun 2020.
41. Nofi M. Understanding color measurement geometry. PCI Magazine, 2016. https://www.pci mag.com/articles/102356-understanding-color-measurement-geometry. Accessed: 11 Jun 2020.
42. https://www.happi.com/issues/2018-09-03/view_formulary/you-niversal-foundation-spf-30/. Accessed: 11 Jun 2020.
43. Dow Corning Formulation Information. 2006.
44. https://www.pinkmelon.de/magazin/geheimnis-kosmetik/concealer-die-kunst-des-abdeck ens.html. Accessed: 11 Jun 2020.
45. https://de.kryolan.com/blog/posts/dermacolor-camouflage-creme-0. Accessed: 11 Jun 2020.
46. https://global.kryolan.com/ product/liquid-body-makeup#100ml-lakealtrot. Accessed: 11 Jun 2020.
47. https://www.pinkmelon.de/magazin/geheimnis-kosmetik/cushion-craze.html. Accessed: 11 Jun 2020.

48. Augen auf. Stiftung Warentest. 2005. https://www.test.de/Wimperntusche-Augen-auf-1288765-1288768/. Accessed: 11 Jun 2020.
49. https://www.pinkmelon.de/magazin/geheimnis-kosmetik/augenbrauen-geformt-dicht-wild.html. Accessed: 11 Jun 2020.
50. Lanzendörfer G. From deity to goddess, the magic of the eyeliner. San Francisco: Color Cosmetic Summit, 2006.
51. Zarling C, Lebok S.Decorative pencil cosmetics. SÖFW. 2005;131(11):3.
52. Patent DE. 94 21936 (Schwan Stabilo Cosmetics GmbH). 1994.
53. Patent DE. 94 22069 (Schwan Stabilo Cosmetics GmbH). 1994.
54. Patent EP. 1457192 (Beiersdorf AG). 2004.
55. Patent EP. 1080713 (Beiersdorf AG). 2001.
56. Floratech Formulary. Pot Lip Gloss. https://www.floratech.com/PDFs/Formulary/ColorCosmetics/M001.pdf. Accessed: 11 Jun 2020.
57. Lochhead R, Lochhead M Two decades of transfer-resistant lipstick. Cosmetic & Toiletries. 2015. https://www.cosmeticsandtoiletries.com/research/chemistry/Two-Decades-of-Transfer-resistant-Lipstick-290207561.html. Accessed: 11 Jun 2020.
58. Walters KA, Flynn GL. Permeability characteristics of the human nail plate. Int J Cos Sci. 1983;231.
59. Gupchup GV, Zatz JL. Structural characteristic and permeability properties of the human nail: A review. J Cosmet Sci. 1999;50: 363. https://pdfs.semanticscholar.org/ 7767/ 465d944c5ea12161243eabff1a8e094baeb5.pdf?_ga=2.173816483.1639022167.1591470668-1315617545.1588516513.
60. https://www.compoundchem.com/2017/04/06/nail-polish/. Accessed: 11 Jun 2020.
61. https://chemistscorner.com/cosmetic-formulation-basics-nail-polish/. Accessed: 11 Jun 2020.
62. Abrutyn ES. Deciphering nail polish formulas. Cosmetics & Toiletries 2013 (7) e-version. https://www.cosmeticsandtoiletries.com/formulating/category/color/premium-Deciphering-Nail-Polish-Formulas-214187831.html.
63. https://www.acrylnaegel.eu/. Accessed: 11 Jun 2020.
64. Pagano FC. A review of gel nail technologies. Cosmetics & Toiletries. 2015, 40–49. https://www.cosmeticsandtoiletries.com/formulating/category/color/A-Review-of-Gel-Nail-Technologies-290228861.html.
65. https://www.allnex.com/en/markets-applications/performance-applications/personal-care-encapsulation/gel-nails. Accessed: 11 Jun 2020.

Gerhard Pfaff

# 19 Colorants in plastic applications

**Abstract:** This review article is a summary of the current knowledge in the field of plastic coloring. Plastics belong as well as paints, coatings, printing inks, and cosmetic formulations to the most important application systems for colorants, both for pigments and dyes. Colorants have to meet increasing demands in plastic applications due to the growing number of polymers with specific properties. Crucial factors besides the plastic type are the processing method and the required fastness level. Among the most important polymers for coloring with pigments and dyes are polyolefins, polyvinyl chloride, polyurethane, polyamide, polycarbonate, polyester, and elastomers. Different processing methods are used for coloring of the individual plastics. The coloring processes need to be coordinated in accordance with the steps of the plastics processing leading to the final product.

**Keywords:** colorant, dye, pigment, plastics processing

## 19.1 General aspects

Plastics belong to the most important application systems for colorants, both for pigments and dyes, whereby pigments have the much higher importance for coloring of plastic materials. Due to the growing number of different plastics on the market, colorants have to meet increasing demands in this application field [1,2,3,4].

In making a sensible choice of pigments and dyes, the crucial factors besides the plastics are the processing method and the required fastness level. The intrinsic color of the plastic used is besides the colorant an important factor, which has to be regarded for the desired coloration. A wide variety of additives such as antioxidants, UV absorbers, HALS stabilizers (Hindered Amine Light Stabilizers), plasticizers, slip agents, antislip agents, blowing agents, antistatic agents, flame retardants, fillers and laser marking substances are added to plastics depending on the polymer type and the application.

As chemicals, colorants can synergize or interfere with needed reactions during the compounding process of polymers. Chemical reactions with some of the additives may happen to stabilize the polymer during the various process steps, especially during thermal treatment. Certain colorants can interfere with these reactions. Furthermore, pigments and dyes often influence the final thermal and UV stability of the material blend.

This article has previously been published in the journal Physical Sciences Reviews. Please cite as: G. Pfaff, Colorants in Plastic Applications *Physical Sciences Reviews* [Online] 2020, 5. DOI: 10.1515/psr-2020-0104

https://doi.org/10.1515/9783110588071-019

Mechanical properties like the impact strength can be altered depending of the chemical nature of the colorant used. The addition of titanium dioxide pigments, for example, can severely affect the impact strength of many products. In cases where the base polymer requires very high fabrication temperatures, the colorants also must be stable under these conditions. Temperatures up to 315 °C are possible during some of the manufacturing processes.

Most plastics e. g. polyolefins and polystyrenes and their derivatives such as ABS (acrylonitrile butadiene styrene) and SAN (styrene acrylonitrile) are supplied by the manufacturers in ready-to-use form. Most of the necessary additives are already incorporated in the polymer before it is sold. On the other hand, in the case of other polymers, e. g. PVC, the end user has to add the additives and colorants. Fluid and high-speed mixers are suitable devices for the incorporation of the additives. Gravity mixers or tumble mixers are also possible to use. The mixtures are homogenized on mixing rolls, kneaders, planetary extruders, twin-screw kneaders, or twin-screw extruders and further processed.

The processing methods are very much governed by the desired end product. Injection molding, injection blow molding, extrusion blow molding, calendaring, foaming, and coating as well as production of flat films, blown films, profiles, and fibers by extrusion are typical processes used. Injection-molding machines or single-screw extruders do normally not have the shear forces necessary for the dispersion of a colorant in the polymer melt. This step is done very often by specialized producers of masterbatches. Specially equipped twin-screw extruders, kneaders, triple-roll mills, or bead mills in a suitable carrier material are used for this purpose. The colorant concentration of masterbatches and other preparations depends on the end product, the wall thickness and the desired concentration in use. In preparations, the colorant is present in a well-dispersed form with the appropriate choice of polymer. Uniformly, homogeneously, and reproducibly colored plastics are therefore possible to produce under normal processing conditions.

Pigment and dye preparations are offered in solid form as granules and powders or as pastes. They normally contain between 5% and 50% colorant. Such preparations are typically suitable for only a few plastics depending on the type of carrier material used. General-purpose preparations are possible to produce, but are of minor importance compared to the preparations tailored to specific plastics.

Dispersion of the colorant particles in the polymer plays an extremely important role. Good dispersion of the colorants in a preparation and subsequently good distribution in the end product needs an increase of the flowability of the preparation above that of the plastic to be colored. This can be achieved by using carrier materials with a low molecular weight (wax), a copolymer, or an increased plasticizer content.

Color concentrates are of major importance for the coloration of plastics. These concentrates are defined as color preparations that contain two or more colorants and produce a defined shade when mixed with a specific amount of the uncolored polymer. Colorant mixtures that have been mixed previously in the correct ratio to give desired shades are used for this purpose in some cases. High-speed mixers are mostly used to produce such mixtures.

As already stated, pigments have the higher relevance for the coloration of plastics compared with dyes. Therefore, the focus is placed in the following on the use of pigments in plastics. Both inorganic and organic pigments are used in plastic materials. Most inorganic pigments have larger particles than organic pigments. They cause significantly less dispersion problems compared with organic pigments. Mixtures with organic pigments are possible without any negative influence on the overall properties. Organic pigments have grown in popularity and application since the 1990s due to the desired replacement of cadmium and lead containing inorganic pigments, which came more and more under environmental scrutiny. Inorganic pigments are limited in many cases to the coloring of opaque materials. Organic pigments, however, can be applied in either opaque or transparent polymer systems.

Whether it is better to use organic or inorganic pigments in a particular coloration case depends on technical and economic considerations and also on regulations in individual countries. Organic pigments are preferred if a transparent coloration is desired and high color intensity, especially for thin-walled articles such as films and fibers, is needed. Organic pigments are also usually applied for multicolor printing on films. If high opacity, high light and weathering fastness are required, inorganic pigments are the colorants of choice. The advantages of inorganic and high-quality organic pigments are often combined. The inorganic pigment with the lower color strength is usually present in excess in such cases in order to achieve the desired opacity, whereas the organic pigment used in smaller amounts produces primarily the color intensity of the pigment mixture and the brilliance of the shade.

Dyes are generally used with transparent plastics. They are usually added at low levels for tinting purposes. Since dyes are primarily used in transparent materials, their formulation systems are typically very simple. As with pigments, dyes must be selected to match the polymer and additives in connection with the processing temperature required.

Table 19.1 contains a comparison of the three different types of colorant playing a role for coloring of plastics.

In doing the right choice of pigments and dyes for coloring of plastics, crucial factors besides the polymer material itself and the dispersion behavior of the colorant are the processing method and the required fastness level.

Plastic raw materials are offered and processed in powder or granulated form. Several processing methods have been introduced for the successful coloring with pigments or pigment preparations. Plastics in powder form are often premixed with pigment powders in fluid mixers. The pigments are dispersed by this method while the plastic is being processed.

Typically, concentrates or masterbatches will be used instead of raw inorganic or organic pigments or dyes during the compounding procedure. Advantages of a

**Table 19.1:** Characteristics of different colorant types.

| Colorant | Loading level in the polymer system | Solubility in the polymer system | Coloring effects | Special challenges |
|---|---|---|---|---|
| Inorganic pigments | High | Insoluble | Opaque | Dispersion of the pigment particles in the polymer |
| Organic pigments | Medium | Insoluble | Opaque, transparent | Dispersion of the pigment particles in the polymer, thermal stability, cost |
| Dyes | Low | Soluble | Transparent | Compatibility with the polymer system, thermal stability |

masterbatch can include improved color development, higher productivity, and faster process cleanout. These concentrates are produced in most cases by specialized manufacturers. This approach allows the use of a simpler and more economical compounding process.

Masterbatches are typically colored using granular pigment preparations or paste preparations, e. g. pigment plasticizer pastes. Granular pigment preparations are advantageous for automatic metering by volumetric or gravimetric metering units and metering pumps. Such granules are, however, of only limited suitability for coloring of plastics in powder form. The polymer-pigment distribution in the final plastic is often insufficient in such cases. In addition, there is a risk of separation particularly with pneumatic conveying.

A disadvantage of the masterbatch approach is the difficulty to adjust specific colorants if color adjustments are required to meet quality standards. A possibility to overcome this problem is the use of several individual single-pigment concentrates.

A manufacturer of a polymer compound can select from several methods when blending pigments, dyes, polymers and additives to get a masterbatch. The decisive factor here is that the final product meets the color needs of the end-use application and/or the corporate color design requirements [5].

A first method of compounding is adding color via concentrates or a masterbatch at the related processing step (molding or extrusion). A two-component approach is used, whereby a natural polymer and a concentrated additive masterbatch is metered and blended at the press. The two components must be blended carefully to achieve a homogeneous mixture. Performing checks and adjustments are necessary to achieve the desired final color quality.

A second method requires that a supplier blends the base polymer and the color concentrate together before supplying the mixture to a fabricator for plastics. It is important to prevent any segregation of the two components prior to molding and extrusion steps.

A third method is the use of a precolored product, where the needed colorants and additives are precompounded into the base polymer through a melting process. The so obtained compounds are sold as single-component products. The precolored approach requires that the supplier provides a product that is already formulated and color matched to meet the demands of the end user. This is achieved by using the coloring additives in a melt-mixing process to guarantee final appearance and consistency. It is necessary that the fabricator has an acceptable and consistent processing window to achieve the final coloristic quality.

Instrumental methods for measuring color quality of plastic parts are available, which are accepted by plastic compounders and final users. On the basis of an approved light source, ΔE values are often declared in specifications as the color control target. Upper and lower limits are fixed on key values, such as red/green value (A value), blue/yellow value (B value), or lightness value (L value). Unacceptable color excursions can be avoided by this means. Injection or compression molded chips of suitable thickness are preferably used as samples for testing. Such color chips may be given to customers to confirm or document color as well as instrument values. Blown films can be used to check acceptable dispersion levels. The frequency of sampling depends on the quality systems and statistical methods used by the compounder. Certificates of analysis are often required by the final users in order to prove that agreed specifications are met. However, the best method to assure the quality of a plastic compound is the combination of visual evaluation by a trained human eye with sophisticated instrumental techniques.

Solvent fastness of the pigments in the plastic, particularly under the processing conditions, is of high importance. Migration problems such as blooming and bleeding are based on the complete or partial "dissolving" of the pigment in the polymer at its processing temperature. Recrystallization is also attributable to pigments, which have a certain "solubility" in the plastic. As a result, a change in transparency or opacity in transparent colorations and in the depth of shade in white reductions is observed.

As mentioned above, the use of pigments in a number of plastics is restricted by high processing temperatures. The residence time (duration of thermal stress) is an important parameter here. Important factors are the pigment concentration as a function of the heat resistance and the ratio of colored pigments to white pigments. The limit concentration, i. e. the ratio of lowest possible amount of a colored pigment to titanium dioxide, has to be determined in a test for the thermal stability.

The light fastness of a pigment depends strongly on the polymer used. It can be low for a certain plastic after exposure to light. In other polymers, however, the pigments are almost not affected by light and the fastness is high. The light fastness can always be stated only for the entire pigmented system. Standards exist for determining the light fastness of colored pigmented plastics in daylight and in xenon arc light, i. e. in accelerated exposure equipment, and for determining the weathering fastness. Due to their strong inherent scattering, lead-containing stabilizers or antimony trioxide (used as flame retardant in PVC) can influence the light fastness of the plastic system as strongly as titanium dioxide in white reductions.

Inorganic pigments and other colorants applied in plastic articles used for packaging food or cosmetics must be not only migration-fast and extraction-resistant but also physiologically safe. Higher fastness requirements can only be met in many cases by more expensive pigments. A compromise between the fastness requirements and the price of a specific pigment often needs to be found.

In pigment preparations, the pigment is present in dispersed form, which with appropriate choice of carrier material enables the plastic to be colored uniformly, homogeneously and reproducibly under normal processing conditions. Dispersibility of the pigments for the plastics industry is tested by the filter test according to EN 13,900-5 [6]. In this test, a specific amount of pigment in prepared form is forced through a fine screen pack by an extruder and the pressure rise upstream of the screen pack is measured. The calculation is done by the formula

$$P_{final} = \frac{(\text{final pressure } P_2[\text{bar}] - \text{initial pressure } P_1[\text{bar}])}{(\text{weight of pigment}[\text{g}])}$$

$P_1$ in this formula is the melt pressure of the polymer without pigment, $P_2$ is the melt pressure after the pigment is added. The tolerance in this test is 2 bar/g pigment.

Recrystallization of the polymer can be attributed to the pigment used with its specific solubility in the plastic. It can be seen primarily in a change of transparency or opacity in transparent colorations and in the depth of shade in white reductions. Lack of recrystallization stability is a problem, e. g. for the manufacture and processing of pigment-plasticizer pastes and in various polymers at elevated processing temperatures.

Some organic pigments can cause warping of certain thick-walled, large-area, non-axially symmetrical injection-molded parts such as bottle crates, where they act as nucleating agents for partially crystalline polymers. The light fastness requirements of plastic colorations are met by many organic pigments, but are similarly dependent on the ratio of colored pigment to titanium dioxide. The stability of the plastic used should also be taken into consideration here.

The compatibility of the pigments with the polymer system being colored must be considered in all application cases. Furthermore, the influence of the pigments on the physical and mechanical properties of the plastics used must be eliminated.

There are only a few colorants that meet all requirements mentioned here, as well as any additional ones imposed, and which are also particularly cost-effective. In general, higher fastness requirements can be met only by more expensive pigments. In the plastics sector inorganic pigments are included prominently in these considerations. Whether it is better to use organic or inorganic pigments in a particular case depends on technical and economic considerations and regulations in the individual countries.

## 19.2 Coloring of polyolefins

Polyolefins are quantitatively the largest group of plastics and their importance for coloration with pigments and dyes is accordingly high.

Polyethylene (PE) was firstly discovered accidentally in 1898 by the German chemist H. Pechmann [7]. The industrial history of polyethylene started in the early 1930s in Great Britain. Two researchers, E. W. Fawcett and R. O. Gibson, began in the laboratories of I.C.I. in Winnington to investigate the phenomena occurring in the field of high pressures including effects of high pressure on chemical reactions. Fifty reactions were investigated without any success, but one failure resulted, through a series of coincidences, in the discovery of polyethylene. The decisive experiment was carried out with ethylene and benzaldehyde under conditions that were not fully controlled. A white waxy coating on the inside of the autoclave was found at the end of the experiment – polyethylene. It took until 1939 that PE was manufactured industrially on a small pilot plant [8].

Polyethylene and polypropylene (PP) are the main representatives of the polyolefins. Other polyolefins with an industrial importance are polymethylpenten (PMP), polyisobutylene (PIB), and polybutylene (PB). Polyolefins belong to the semi-crystalline thermoplastics. They are characterized by easy workability, good chemical resistance, and electrically insulating properties. PE and PP may be divided depending on the starting materials (monomers) and the density (processing temperature) in the following types:
LDPE = Low Density Polyethylene
LLDPE = Linear Low Density Polyethylene
HDPE = High Density Polyethylene
PP = Polypropylene

The respective processing temperatures, which determine the criteria for pigment selection, are as follows:
LDPE = 160–220 °C
LLDPE = 220–240 °C
HDPE = 190–300 °C
PP = 200–300 °C

The flow properties of a polymer at a certain temperature are defined by the density and the molecular weight of the material. They are characterized by the melt flow index. Polyolefins scatter very little light when they are partially crystalline. They therefore tend to appear lighter on coloration, depending on the processing temperature. Another lightening effect is observed in stretched and foamed polyolefins. Injection molding and extrusion are the typical techniques used for the processing of polyolefins.

Polyethylene and polypropylene have the highest technical importance among the copolymers. They are elastomers and are produced in a number of varieties. HDPE and PP are manufactured as powders, while LDPE is generated from the melt preferably in the form of lenticular granules. All types, however, are supplied primarily as granules. Any thermoplastic transformation of a polymer powder or granulate is performed in the presence of additives.

Polyolefins have a low glass transition temperature. Pigments, which are partially soluble in polyolefins tend therefore to migration. Like in other media, this tendency is dependent on pigment concentration and temperature. The type of the polyolefin, especially its density and molecular weight, has a significant influence on the migration behavior of a pigment. Pigment migration occurs in LDPE to a much larger extent than in other plastics like HDPE or PP [9]. HDPE and PP may therefore be colored problem-free by organic pigments, which migrate in LDPE. The trend among pigments to migrate in LDPE, which includes both bleeding and blooming, increases with increasing melt flow index and decreasing molecular weight of the polymer, respectively. Additives such as lubricants or antistatic agents may also play a role.

Some organic pigments cause distortion in certain types of polyolefins, especially in HDPE. Such pigments can act as nucleating agents in the semi-crystalline HDPE. They initiate crystallization, which creates stress within the polymeric product. These pigments may also enhance the shrinkage of polyolefins, particularly in the direction of the flow.

Thermal stability of a polyolefin can be influenced by the added colorants, additives, and impurities. Heavy metal ions for example, especially copper, manganese, and iron adversely affect the thermal stability of polyolefins, while sulfide-containing inorganic pigments have a distinctly improving effect. The response of a system to additives or impurities is tested by measuring the loss of tensile strength after exposing the polymer to heat. The polyolefins are stored for that purpose at temperatures close to the melting point of the crystallites. Pigmented and pigment-free samples are tested under these conditions and compared for brittleness.

Light fastness of pigments in a polyolefin or in other plastics is measured using the entire pigmented polymer system with all its additives. Blending of plastics, especially of PP and HDPE, with HALS stabilizers has increased the importance of such measurements in regard of the protection of plastic materials against light and weather influences. Attention to the combination of pigments and stabilizers in polymers is necessary as well, because the effectivity of these agents is inferior in the presence of certain pigments.

The dispersion behavior of a pigment in a polyolefin is of great importance. This is particularly true for the coloration of extrusion films and of HDPE or PP ribbons made from stretched blow films or sheeted extrusion films.

High processing temperatures associated with a high degree of softening are the reason why only limited shearing forces are necessary for the dispersion of pigments in polyolefins. Pigment preparations, usually in combination with polyolefins as a carrier, are often used for coloring polyolefin plastics. Poor dispersion causes filler specks, holes in films and other problems, which can be avoided by using pigment preparations.

Pigment powders are still in use for special applications, e. g. in thick-walled articles, such as extruded sheets, hollow objects, or in injection-molded products, but color concentrates gain more and more importance also in these areas.

Paste preparations consisting of pigments and polyolefins are also used. They contain 20–70% pigment. Three-roll mills, agitated ball mills, or dissolvers are used by the manufacturers of such preparations. The preparations are employed primarily to produce bottles, injection-molded articles, or extrusion sheets.

## 19.3 Coloring of polyvinyl chloride

Vinyl chloride was discovered by H. V. Regnault in 1838 and its polymerization to polyvinyl chloride (PVC) succeeded for the first time in 1912 by F. Klatte. Industrial-scale production of PVC started in 1927 [10, 11]. PVC belongs to the most versatile plastics. Several methods for the manufacture of PVC are available. PVC can be thermoformed on all conventional processing machines if a slight thermal damage is taken into account. Processing is easy and the material can be bonded, bent, welded, printed, and thermoformed. A distinction is made between bulk, suspension and emulsion PVC on the basis of different polymerization methods.

Three methods are mainly used for the manufacture of polyvinyl chloride. The suspension polymerization is used for high-volume standard grades. High-quality products such as graft polymers and copolymers, pastes and paste extender resins are produced by this method. The emulsion polymerization is used primarily to manufacture special paste-making grades. The mass or bulk polymerization is a solvent-free two-stage process. The PVC produced using this polymerization technique cannot be processed unless it is compounded. Mass or bulk polymerization is becoming increasingly less important and no new capacities for this method are being created.

The dispersion behavior of pigments in PVC, as in other plastics, is a property that determines its use. Processing parameters are changing for economic reasons and are less and less suitable for achieving a good degree of pigment dispersion. This concerns in particular the processing temperature, which is increasing up to 200 °C and above.

Exact and reproducible coloration of PVC in the laboratory and in the plant is anything but simple. It can be achieved only if specified working conditions are kept and special equipment is used.

Pigment-plasticizer pastes are used in many cases for coloring of PVC. Because of the mixing or embrittlement gap in the PVC-plasticizer system, such pastes have little or no suitability for un-plasticized PVC compounds.

Organic pigments can be dispersed readily in plasticizers, for example with the aid of triple-roll mills. The throughput here is low because of the lack of smoothness of such pastes. This is the reason why attrition mills with a higher efficiency are used more and more.

The pigment content in plasticizer pastes is typically between 20% and 40%. The addition of special dispersing agents to some of the pigments in the manufacture of such pastes has brought benefits. Such additives promote accelerated wetting of the pigment and enable the pigment content of the pastes to be raised and the dispersing process to be shortened. Since some pigments differ considerably in their dispersion properties from others, when using pigment-plasticizer pastes, it is highly advisable not to disperse together mixtures of pigments necessary for adjusting the shade. Pastes containing only one pigment should be mixed homogeneously to adjust the shade.

Diluents, mainly volatile aliphatic or aromatic hydrocarbons are often added to PVC pastes to lower the viscosity. They do not have a gelling effect and are evaporated before the gelation starts. Pigments applied together with such diluents are required to have adequate resistance to the solvents used at the processing temperatures.

PVC spread-coating pastes are normally manufactured in high-speed planetary mixers that can be evacuated. The best procedure consists in adding the plasticizer first, stirring in the pigments or pigment pastes and adding finally the solid components in portions. For the pigmentation of plasticized PVC compounds, on the other hand, it is necessary to mix the pigments with the PVC before adding the plasticizer. Pigment-plasticizer pastes can however be added to the PVC at the same time as the plasticizer.

Pigments for articles that are permanently used outdoors, such as garden furniture, facade claddings or roller shutter profiles, need to have high weathering fastness. There is no problem for inorganic pigments regarding this requirement, but only a few organic pigments have sufficient fastness to long-term weathering. The way to stabilize the PVC plays an important role in this context.

Coloring the wide variety of PVC grades requires the compatibility of the pigments with the polymer and all its additives. Chemical reactions with inorganic or organic pigments may not occur. Exceptions are some lake pigments, which might decompose in the emulsifier-containing emulsion and release the organic dye as a result of hydrolysis. Such a process is often accompanied by changes in shade and fastness properties.

Another prerequisite is that the pigments used have little or no effect on the physical and mechanical properties of the polymer. A change in rheological properties of PVC plastisol or of PVC melts during processing must be kept in mind.

The influence of organic pigments on the electrical resistance of PVC cable insulations is another aspect that needs to be considered. Such an influence is caused not by the pigment itself but by ethoxylated surfactants, which are added as auxiliaries in the manufacture of these pigments, especially of azo pigments. Some pigment manufacturers offer special product ranges with verified dielectric properties to avoid possible problems.

Pigment preparations are frequently used in the PVC cable industry for coloration. The pigment content is often selected in such a way that one part by weight of the preparation is used to color 100 parts by weight of the polymer compound. The shades and color codes for cables and insulated lines are specified in standards in various European countries.

## 19.4 Coloring of polyurethane

Polyurethane (PUR) belongs to the most versatile plastics. Reasons are the wide variation of the starting materials and the possibility to use practically all processing methods known in the plastics sector.

PUR was firstly discovered by the German chemist O. Bayer in 1937 by polyaddition of isocyanates and alcohol [12]. A third basic principle for structuring of macromolecules was found therewith after the already known polymerization and polycondensation. It took nearly 10 years before the discovery of PUR could be used for practical purposes. One of the first applications was PUR foam followed by hard (insulating materials) and semi-hard variations of PUR.

Virtually the same pigments as for plasticized PVC are recommended for coloring of thermoplastic PUR. Pigment migration processes are comparable in both cases. This kind of knowledge is particularly important for coloring of PUR leathercloth, for which a high pigment concentration is normally used.

Pigment preparations are also marketed for coloring of thermoplastic PUR. Their carrier materials range from vinyl chloride or vinyl acetate copolymers to low-molecular polyethylene and to PUR itself.

PUR foams, especially for integral foamed parts for vehicles and the furniture industry, are also important for coloring with pigments. The required heat resistance of the pigments for this sector is in some cases very high. For PUR foaming by the high-pressure method, in which isocyanate and polyol are sprayed through narrow nozzles under high pressure, it is necessary for the pigments used to be fully dispersed in order to prevent production problems due to blocked nozzles. Normally pigment preparations are used here, too. Their carrier material is in most cases either involved in the reaction of the isocyanate in PUR formation or it takes part in foam formation.

## 19.5 Coloring of polyamide, polycarbonate, polyester, and polyoxymethylene

Other plastics with an important market size are polyamide (PA), polycarbonate (PC), polyethylene terephthalate (PET), and polyoxymethylene (POM). The pigmentation of these polymers takes place similar to that described above. In the case of PA, high-temperature-resistant pigments are required for injection molding and extrusion.

A further limitation concerning the choice of pigments for PA is the weakly alkaline and reducing character of the polymer melt. If organic pigments are to be used, preliminary testing of the pigments in the polyamide grade should be performed. The use of inorganic pigments in this group of plastics is comparably unproblematic.

The range of organic pigments suitable for coloring of PA, PC, PETP, and POM varies according to the requirements. From a comprehensive point of view, however, it is necessary to find a compromise between price and performance of the pigment to be used.

## 19.6 Coloring of polystyrene, styrene-copolymers, and polymethyl methacrylate

Polystyrene (PS), a highly rigid and surface-hardened thermoplastic, is a transparent and almost colorless polymer. PS has a typical slight yellow tinge, which is easy to compensate by adding transparent blue colorants. Polystyrene becomes soft in the temperature range between 80 °C and 100 °C. It is processed between 170 °C and 280 °C, up to a maximum of 300 °C, without color change. PS is processed by the methods which are typically recommended for thermoplastics. The list of products consisting of colored polystyrene includes extrusion made sheets, profiles, and films. Foaming plays a role for some of the applications.

The mechanical properties of PS may be improved significantly by copolymerizing the monomer with one or more out of a variety of rubber-like materials (graft polymers). Impact resistant PS contains mostly between 5% and 25% natural rubber, which is not dissolved but dispersed in PS. Natural rubber and other dispersed additives scatter much of the incident light due to a difference in the refractive indices between the polystyrene and the added material. As a result, opaque products may be obtained. The degree of opacity in the individual cases depends on the amount of rubber added. Impact resistant PS types are processed in the range from 170 °C to 260 °C. Copolymers with acrylonitrile and butadiene have a high impact strength and excellent fastness to aging.

Polymethyl methacrylate (PMMA) is a plastic material, which is extremely stable to aging and to weathering. It is transparent and has a comparatively high hardness.

The transparency respectively the opacity of a polymer, e. g. polystyrene and styrene/acrylonitrile copolymer as transparent materials or ABS as an opaque material, has significant influence on its coloration. In discussing the coloration of polymers whose glass transition temperature is far above room temperature, special rules have to be considered. It is rarely possible for dissolved molecules to migrate in such a case, thus these polymers may be colored mostly with pigments. Nevertheless, also a few soluble dyes can be used for coloring of these polymers. Some of the dyes are even acceptably lightfast and afford brilliant shades, especially in combination with opaque inorganic or organic pigments.

The requirements regarding heat stability are stringent for colorants which are used to color these plastic materials. Not all organic pigments tolerate the high end of the processing temperature range between 280 °C and 300 °C. Some types, however, withstand moderate temperatures between 220 °C and 260 °C. This also applies to a limited number of dyes.

Depending on the temperature, some organic pigments dissolve more or less in these types of plastics. The color almost invariably changes as the pigment dissolves, frequently accompanied by a change in the fastness properties, especially in the response of the system to light.

Completely dissolved pigments should be considered as dyes and be tested as such. This concerns features such as the extraction properties in the finished article. In PS, SAN, and other transparent plastics with a high glass transition temperature, they afford transparent, glass clear colorations.

Pigments dissolving reasonably well at low concentration and at the temperature at which their medium is processed, are frequently employed to advantage at higher concentrations. This phenomenon can be explained by considering the influence of the dissolved versus the undissolved portion of pigment. The coloristic properties of the polymer will be enhanced if the undissolved pigment particles dominate over the dissolved portion in determining the coloristics of the system, often resulting in high brilliance. According to the laws of physical chemistry, organic pigments (and dyes) dissolve at a rate which is largely dependent not only on the pigment particle size but also on the available time. PS and its copolymers are normally processed as pellets, which may be colored by colorants in form of powders, granulates, or pastes.

Pigments and pigment blends in powder form are incorporated in their medium by means of slow mixers. Organic pigments are often insufficiently wetted by molten PS or its copolymers and accordingly difficult to disperse. It is possible, however, to facilitate dispersion by initially applying an adhesion agent, i. e. a wetting agent, in concentrations up to 0.3% relative to the granulated plastic. PS, being a brittle and hard material, may cause metal abrasion after a certain mixing time, which gives a dull effect to otherwise clean, brilliant shades. Fluorescent colorants may even lose their fluorescence.

The advantage of using pigment preparations in paste form, which afford easy color matching by blending the corresponding pastes, is affected by the fact that a higher content of liquid component may have an influence on the mechanical properties of the plastic and can lead to stress corrosion cracking. In addition, it should be noted that the applicability of pastes is restricted by physiological reasons.

Polystyrene shows a certain yellowing behavior in exterior exposure due to UV radiation. In order to shield this plastic from degradation under the influence of UV light, it is also supplied in form of blends with UV absorbers. This extends the lifetime of the final products noticeable. Grades containing UV absorbers are slightly yellowish, a disadvantage, which may be corrected by the addition of transparent blue colorants such as soluble dyes or ultramarine blue.

## 19.7 Coloring of elastomers

Elastomers are characterized by their rubber-elastic behavior. The softening temperature of elastomers lies below room temperature. Elastomers in the unvulcanized state, i. e. without cross linking of the molecular chains, are plastic and thermoformable. In the vulcanized state, they deform elastically within a certain temperature range. Vulcanization transforms natural rubber into the elastic state. A large number of synthetic rubber types and elastomers are known and offered on the market. They have a number of improved properties compared to crude rubber. Some of them have substantially improved elasticity, heat resistance, low-temperature stability, weathering and oxidation resistance, wear resistance, resistance to different chemicals, oils, etc.

Pigments for coloring rubber compounds must fulfill a number of requirements. In particular, they must not contain rubber poisons. Even small amounts of copper and manganese may impair vulcanization of the rubber significantly and cause accelerated aging of the vulcanized rubber. Pigments for coloring rubber may therefore not have a content of more than 0.01% of these two heavy metals. In the case of copper phthalocyanine blue pigments, somewhat higher quantities of non-complexed metal are accepted but not more than 0.015%.

Pigments used in elastomers are also required to have a specific heat resistance. This property is investigated on five colored test samples with different pigment concentrations in the range from 0.01% to 1% together with 10 times the amount of chalk. The colored samples placed side by side are vulcanized hot for 15 minutes at 140 °C and evaluated coloristically against untreated comparative probes.

For many application cases, the pigments used are also required to be migration-resistant. The suitability test is done as before with 5 different pigment concentrations. To test the fastness to bleeding, the unvulcanized colored samples are brought into defined contact with a white milled sheet of specific composition and vulcanized wet for 20 min in open steam at 140 °C. Half of the samples is often covered with a wet cotton cloth during this process to determine whether the cloth, the rubber or both are stained by bleeding.

Blooming of colored samples is tested by rubbing the colorations of different concentrations with a white cloth immediately after they are produced, after 6 months' storage at room temperature and, if necessary, also in an accelerated test after 24 h storage at 70 °C.

Weathering fastness of pigments for natural rubber is of virtually no significance because the material itself has poor weathering fastness. On the other hand, some high-quality synthetic elastomers with high weathering fastness require a correspondingly stable pigment.

Rubber compounds are colored with pigments in powdered form or with granulated pigment preparations, referred to as masterbatch in the rubber industry.

## 19.8 Coloring of thermosets (thermosetting plastics)

Thermosets are formed by cross linking, respectively, curing of reactive linear and branched macromolecules. They can be manufactured by polycondensation, polymerization or polyaddition. Thermosets can therefore be processed once only with the application of heat and pressure to form semi-finished products or finished articles and cannot be recovered. This means that their processing is irreversible. Combinations of formaldehyde with phenol, resorcinol, etc. (phenolics), urea, aniline, melamine and similar combinations (aminoplastics) belong to the most familiar thermosets.

Phenolics have a dark intrinsic color. They are processed predominantly with a high filler content (up to 80%) leading to limits for their coloration.

Thermosets are processed by compression molding, transfer molding and injection molding or by extrusion, depending on their structure. The pressing temperatures are between 150 °C and 190 °C. The pigments used have to be stable under these conditions. Dyes are also used for the coloration of thermosets.

The colorants are incorporated into the resins before cross linking to ensure homogeneous coloration. This process step can be done in the molten resin, for example in a kneader at about 90 °C, or in dissolved or liquid resins by the liquid resin method. Ball mills are typically used for coloring pre-wetted powder molding compounds that have not yet been cured.

Casting resins based on epoxy resins, methacrylate or unsaturated polyester are also thermosets. Epoxy resins are cured with amines or phthalic anhydride. Curing is not influenced by organic pigments. Moisture, on the other hand, delays curing. The pigments used must therefore be largely dry.

Unsaturated polyester and methacrylate casting resins are normally cured with organic peroxides. The polyester casting resins are dissolved in monostyrene. Temperatures up to 200 °C are reached in their polymerization. Thick-walled articles in particular are often exposed to heat for a long period of time. The heat resistance of organic pigments used has to correspond to these conditions.

Methacrylate resins are produced from monomeric methacrylic acid methyl ester. A significantly lower temperature is employed during their polymerization. Heat resistance is therefore only of minor importance for the pigments in this medium.

High light and weathering fastness are frequently required, e. g. for car bodies. These fastness properties can be severely impaired by the type and amount of the peroxide catalysts. Therefore peroxide-resistant pigments have to be used in such cases. At the same time, they must not affect the curing process, i. e. they must neither accelerate nor delay it. Organic pigments can behave completely different under the necessary process conditions. Their behavior depends on the method of curing and type and amount of peroxide used. Disazo yellow pigments for example, such as Pigment Yellow 17 and 83, have no effect at all, whereas copper phthalocyanine green delays curing severely and copper phthalocyanine blue prevents it altogether.

## 19.9 Spin dyeing

Spin dyeing of chemical fibers is a technique that may be classified somewhere between the areas of textiles and plastics. In contrast to textile coloration, the material which has to be extruded is colored before the fiber is produced. The requirements to be met for pigments in this process are therefore comparable to those known for the coloration of plastics. Heat stability is the foremost concern in connection with pigment selection for spin dyeing. A prerequisite for pigments processed with this technique is a high insolubility in the solvents used.

The requirements with respect to pigment dispersibility are high. Special care should be taken to ensure that the size of remaining pigment agglomerates does not exceed 2–3 µm. Particles, which are larger affect adversely the tensile strength of the fiber and frequently cause failure through breakage, especially as the fiber is stretched. In practice, pigment powders rarely afford such a high degree of dispersion. There is accordingly no guarantee for the quality of such a dispersion. It is therefore often inevitable to replace pigments by pigment preparations in spin dyeing processes.

Three different methods are available for spin dyeing of polymers:

*Melt spinning*: This technique is used for thermoplastic materials, such as polyester, polyamide, or polypropylene. The polymer is melted in the extruder and then pressed through a spinneret. It solidifies by cooling as it falls vertically through a shaft to the bottom of the extruder. Pigments used must have an excellent heat stability. The spinning temperatures are in relation to the melting points of the polymers (Table 19.2). As a rule, pigment preparations for this application are based on a carrier material, which is identical or similar to the polymer which is to be extruded. Figure 19.1 shows the principle of the melt spinning technique.

*Wet spinning*: This technique is characterized by spinning a filtered viscous polymer mass, dissolved in a suitable solvent, into a precipitation or coagulation bath. Polyacrylonitrile, polyvinyl acetate, cellulose acetate, and other polymers are processed by this method. Thermal requirements for pigments used are less stringent than for melt spinning. A precondition for the pigments is that they are fast to the solvents and chemicals used.

*Dry spinning*: The polymer, dissolved in a suitable solvent and filtered, is pressed through spinnerets and, in an oxygen-free atmosphere, pulled by vacuum through a heated shaft, where polymer solidifies as the solvent evaporates. The requirements of this process regarding the heat stability of pigments are much less stringent than in melt spinning. Similar to wet spinning, pigments must be fast to the solvents and chemicals used. Examples for polymers processed by this method are polyacrylonitrile, triacetate, and polycarbonate.

Special coloration techniques have been introduced for a number of thermoplastics such as polyester. Instead of using pigments, dyes are employed, which dissolve in the polymer melt. The dyes used are sufficiently heat stable and sublimation-fast

**Table 19.2:** Melting points and spinning temperatures of different polymers.

|  | Melting point (°C) | Spinning temperature (°C) |
|---|---|---|
| Polyester | 255 | 285 |
| PA 6 (6-polyamide) | 220 | 250–280 |
| PA 66 (nylon) | 245 | 275 |
| Polypropylene | 175 | 240–300 |

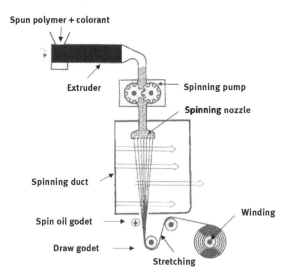

**Figure 19.1:** Scheme of the melt spinning process.

to tolerate the processing conditions. These techniques have certain advantages over traditional methods based on organic pigments. They avoid, for example, fiber failure through breakage during stretching, plugging or the filter, and other faults.

### 19.9.1 Polyacrylonitrile

Polyacrylonitrile (PAC) decomposes already at 220 °C, which is below its melting point (about 290 °C). PAC can therefore not be processed by melt extrusion. Dry and wet spinning techniques are used for this polymer. The list of appropriate solvents includes dimethyl acetamide, dimethyl formamide, dimethyl sulfoxide, and aqueous solutions of inorganic salts. Methods of primary importance for spin dyeing are those based on solutions of inorganic salts. Pigments used in special applications such as window blinds, awnings, and tents, should be dispersable in the appropriate solvents and meet the respective demands regarding light fastness and weathering fastness. Less lightfast pigments are acceptable for the use in clothing, decorative, and home textiles. Examples for such applications are upholstery, curtains, and carpeting. Polyacrylonitrile is commonly used for these purposes because it is by far the most weatherfast of all synthetic and natural fibers. Pigment preparations are also supplied for the coloration of PAC.

### 19.9.2 Polyethylene terephthalate

The thermal requirements for pigments, which are targeted for the melt extrusion of polyethylene terephthalate (PETP) are particularly severe. The individual conditions at the various stages of polymer coloration have to be considered. Pigments, which are added during the so-called condensation process in a glycol dispersion prior to trans-esterification or condensation in the autoclave, are exposed to temperatures between 240 °C and 290 °C for 5–6 h. These harsh conditions are only tolerated by inorganic pigments and very few organic pigments with a polycyclic structure. Representatives of these organic pigments are quinacridone, copper phthalocyanine, naphthalene tetracarboxylic acid, and perylene tetracarboxylic acids.

The choice of pigments which are to be added to a ready-made polyester is much less restricted. At this stage, a pigment will only be exposed to heat for about 20–30 min in the mentioned temperature range. This is possible by mixing the granulated polyester with a pigment preparation, or by transferring the molten pigment concentrate to the melting zone of the spin extruder, for instance via a side-screw extruder. The carrier material of the pigment preparation is equally subject to compatibility restrictions. The use of dyes for the coloration of polyesters by melt extrusion is possible as well. Dyes used have to be melt-soluble, sufficiently heat resistant, and sublimation proof during spinning.

### 19.9.3 Polyamide

Pigments, which are targeted for polyamide spin dyeing need an extreme heat stability and must also be chemically fast to the highly reducing medium of the polymer melt. Spinning temperatures are between 250 °C and 290 °C, depending on the type of polyamide. As in the case of polyester extrusion, only very few polycyclic pigments are suitable for this application. Pigment preparations are also available for the spin dying of polyamide.

### 19.9.4 Viscose

Viscose, the alkaline solution of sodium cellulose xanthate, is produced by treating cellulose with sodium hydroxide solution and carbon disulfide. Viscose is colored in the form of cellulose xanthate prior to extrusion. Aqueous pigment preparations in paste form are used. The pigments are expected to fulfill the typical fastness requirements. In addition, they should be fast to strong acids, alkali, and reducing agents. Pigments, which affect the so-called maturation of the viscose or the coagulation and regeneration in the spin bath or in one of the after-treatment baths cannot be used.

## 19.10 Other manufacturing methods for colored plastic objects

### 19.10.1 Injection molding

Injection molding is a prototype technique for polymer processing [13, 14]. The respective material is liquefied with an injection molding machine. The thus created melting is injected under pressure in a form called injection tool. The material passes in the tool from the liquid back to the solid state by cooling or cross-linking. After opening the tool, the finished part can be removed. The cavity of the tool determines the form and the surface structure of the finished part. Today, parts in the weight range from a few tenths of grams to a scale of 150 kg are possible to produce. Injection molding allows the cost-effective production of large numbers of directly usable polymer parts. An advantage besides the economic production is that it requires almost no finishing. It is used typically for mass production and achieves highest cost efficiency on long-time production.

During the injection-molding process, the colored or not colored thermoplastic polymer in form of a granulate trickles into the aisles of a rotating screw. The granular parts are transported towards the screw tip and melts through the heat of the cylinder and the frictional heat from cutting and shearing of the material. The melted polymer collects at the screw tip. In the following injection phase, the melt is pressed under pressures of 500 to 2000 bar through an open nozzle into the forming cavity. The tool is with 20–120 °C cooler than the plastic mass (200–300 °C). The polymer melt is therefore cooled down in the tool and solidifies if freezing point is reached. At the end of the cooling time the mold opens and the final article will be removed. Figure 19.2 shows schematically the process of injection molding.

### 19.10.2 Extrusion blow molding

Blow molding is a method for the manufacture of hollow articles from thermoplastic polymers. It is a special variant of the injection molding process [15]. The plastic melt prepared in an extruder is discharged in this case through a die gap. A two-part mold in the open position closes to the pipe form, cutting the top and bottom ends. At the same time compressed air is blown through the bottle neck and expands the soft material against the cooled walls of the mold. When the cooling time

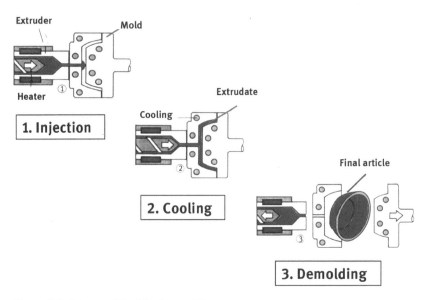

**Figure 19.2:** Scheme of the injection molding process.

is over the mold opens and the bottle or container is removed. Figure 19.3 illustrates the process of extrusion blow molding in a schematic way.

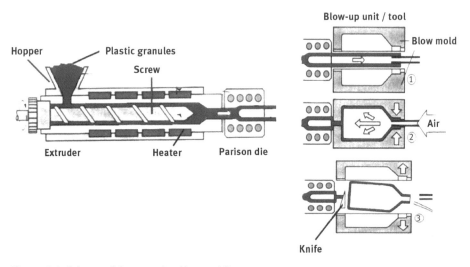

**Figure 19.3:** Scheme of the extrusion blow molding process.

### 19.10.3 Calendaring

The calendaring process is mainly used for the production of PVC, PP, and rubber films [16]. The investment costs for this polymer technology are relatively high. It is used therefore only if the produced volumes are high. Shaping of the polymer takes place through a series of rolls. Calendars with four or five heated rolls, fed by a kneader, a twin-screw extruder and/or mixing rolls are used to manufacture the films. Lamination of substrates is likewise possible. Typical examples here are plastic films, linoleum, textiles, or paper.

### 19.10.4 Extrusion

Extrusion is a continuous or semi-continuous process used to produce objects of a fixed cross-sectional profile. The extrusion technology can basically be applied for metals, ceramics, and plastics. In all these cases, the starting material in form of solid to viscous masses is pushed under pressure continuously through a die of the desired cross-section. Objects are created having the profile of the die. Different variants of the extrusion process are used: hot extrusion, cold extrusion, warm extrusion, friction extrusion, and micro-extrusion. The production of thermoplastic objects through extrusion is technically of high importance [17].

Extrusion is used for the production of many different plastic articles, e. g. window frame profiles, other profiles, pipes, films, tapes, wires, cables, or laminates. The rotation screw takes polymer in powdered or pelletized form out of a hopper, melts the material on the way through the heated barrel, homogenizes it and finally presses it through the die. There are many different extruders like single-screw, twin-screw, or planetary extruders on the market. Figure 19.4 shows schematically the process of plastics extrusion.

### 19.10.5 Blowing of films

Blowing of films, also referred to as blown film extrusion, is a technology used for the manufacture of plastic films for the packaging industry. The process consists of extruding a tube of molten polymer through a die and inflating the material to several times of its initial diameter to form a thin film bubble. The bubble collapses in the following to form a lay-flat film or a bag [18].

Films made from polyethylene are normally produced by the blown film extrusion process. The usual variations of this process differ in the extrusion direction of the blown tube (upwards, downwards, or horizontal) and in the way the formed bubble is collapsed. A blown film plant consists of an extruder with cross or straight head, cooling ring, bubble guides, pinch and collapsing rolls, spreader rolls and finally a wind-up unit.

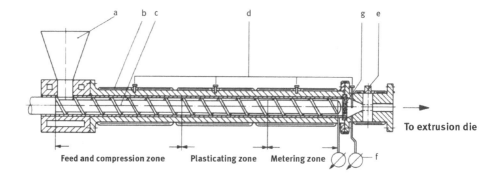

a Feed hopper
b Heaters
c Screw
d Thermocouples
e Back pressure regulating valve
f Pressure gauges
g Breaker plate and screen pack

**Figure 19.4:** Design of an extruder plasticating unit.

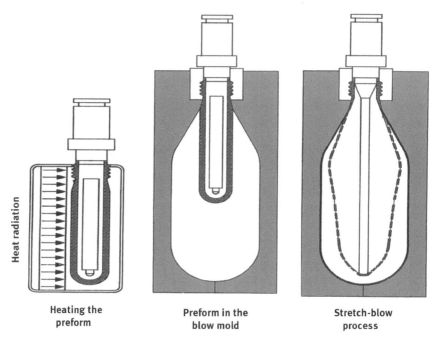

**Figure 19.5:** Scheme of the injection stretch blow-molding process.

### 19.10.6 Injection stretch blow molding

Injection stretch blow molding is a special variation of the blow molding process mainly used for the production of plastic bottles. In a first step, preforms are produced with an injection molding machine (Figure 19.5). In a second step, these preforms are blow molded to cavities (bottles). The preforms are stretched vertically by a mechanical stamp during this step. At the same time, the preforms are blown horizontally to the full volume. The process is highly suitable for low volumes and short runs. There is only limited restriction on bottle design. Cylindrical, rectangular, or oval bottles are possible to produce [19].

# References

1.   Charvat RA. Coloring of plastics, 2nded. Hoboken: John Wiley & Sons, 2004.
2.   Müller A. Coloring of plastics. München: Carl Hanser Verlag GmbH & Co. KG, 2003.
3.   Harris RM, editor. Coloring technology for plastics. Norwich: Plastics Design Library, 1999.
4.   Christensen IN. Development in Colorants for Plastics (Rapra Review Report 157). Rapra, 2003.
5.   Russell S D. Color Compounding. In: Charvat RA, editor. Coloring of plastics, 2nd ed. John Wiley & Sons: Hoboken, 2004:268-75.
6.   EN 13900-5 (2005). Pigments and extenders - Methods of dispersion and assessment of dispersibility in plastics - Part 5: determination by filter pressure value test.
7.   v. Pechmann H. Über Diazomethan und Nitrosoacylamine. Ber Dtsch Chem Ges. 1898;31:2640.
8.   Gibson RO. The discovery of polyethylene. The Royal Institute of Chemistry, Lecture Series 1964, Number 1, 1–30.
9.   Herbst W, Hunger K. Industrial organic pigments, 3rd ed. Weinheim: Wiley VCH Verlag GmbH & Co. KGaA, 2004.
10.  Kaufmann M. The history of PVC. MacLaren, London: The chemistry and industrial production of polyvinyl chloride, 1969.
11.  Mulder K, Knot M. PVC plastic: a history of systems development and entrenchment. Technol. Soc. 2001; 23:265.
12.  Seymour R. B. Reinforced plastics: properties and applications. Asm Intl., 1991.
13.  Stitz S, Keller W. Spritzgiesstechnik. München: Carl Hanser Verlag, 2001.
14.  Pötsch G, Michaeli W. Injection molding, 2nd ed. München: Carl Hanser Verlag, 2007.
15.  Thielen M, Hartwig K, Gust P. Blasformen von Kunststoffhohlkörpern. München: Carl Hanser Verlag, 2006.
16.  Hopmann C, Michaeli W. Einführung in die Kunststoffverarbeitung, 6. Aufl. München: Carl Hanser Verlag, 2010.
17.  Greif H, Limper A, Fattmann G, Seibel S. Technologie der extrusion. München: Carl Hanser Verlag, 2004.
18.  Chanda M, Roy SK. Plastics technology handbook. 4th ed. Florida: CRC Press, 2007.
19.  Brandau O. Stretch blow molding. 2nd ed. Oxford: Elsevier Inc, 2012.

Thomas Rathschlag

# 20 Colorants in printing applications

**Abstract:** This review article is a summary of the current knowledge in the field of colorants in printing applications. Printing inks belong as well as paints, coatings, plastics, and cosmetic formulations to the most important application systems for colorants, both for pigments and dyes. Colorants have to meet increasing demands in printing applications due to the considerable number of printing methods and consequently of a large number of specific printing formulations. Crucial factors besides the specific properties of a certain printing ink are the processing method and the required quality of the final printed product. Amongst the most important printing methods are letterpress printing, offset printing, flexographic printing, gravure printing, screen printing, and digital printing. Different processing methods are used for coloring of the individual printing inks. The coloring processes need to be coordinated in accordance with the steps of the printing processes leading to the final product.

**Keywords:** colorant, dye, pigment, printing inks, printing methods

## 20.1 General aspects

Printing inks are liquid or pasty preparations consisting of colorants, binders, solvents and additives, which are processed in high-speed printing machines. Printing inks are similar to lacquers and paints, but dry typically faster and have a substantially smaller layer thickness after the application. Table 20.1 contains the composition of printing inks. The relation of the printing process, the ink viscosity, the dried ink film thickness and ink consumption is shown in Table 20.2.

Printing inks are used in all printing processes such as letterpress, offset printing, flexographic printing, gravure printing, screen printing or digital printing. While sheet-fed as well as rotary printing presses were the most common printing equipment in the last 50 years, digital printing processes got nowadays more and more importance.

Typical sample applications are books, newspapers, magazines, catalogues, packing materials, wallpapers, cash notes, stamps, and labels. Printing inks were delivered in the past by printing ink producers exclusively to printers and to specialized dealers, but not to the final consumers. Today, digital printing inks can be bought by electronical medias in several open purchase portals as well.

This article has previously been published in the journal Physical Sciences Reviews. Please cite as: T. Rathschlag, Colorants in Printing Applications *Physical Sciences Reviews* [Online] 2021, 6. DOI: 10.1515/psr-2020-0162

https://doi.org/10.1515/9783110588071-020

**Table 20.1:** Composition of printing inks.

| Components | Content in the printing ink (%) |
|---|---|
| Binder | 10–40 |
| Pigments, dyestuffs and fillers | 1–40 |
| Solvents | 10–80 |
| Additives | 0.01–10 |

**Table 20.2:** Relation of printing process, ink viscosity, film thickness and ink consumption.

| Process | Viscosity (Pa*s) | Dried ink film thickness (µm) | Ink consumption (g/m$^2$) |
|---|---|---|---|
| Letterpress | 3–50 | 2–4 | 1.5–3.5 |
| Offset lithography | 4–80 | 1–2 | 0.7–1.3 |
| Flexography | 0.050–0.500 | 1–3 | 1–2.5 |
| Gravure | 0.040–0.300 | 2–8 | 2–7 |

In Germany the per capita consumption of printing inks and varnishes in 2016 was approximately 3,7 kg [1]. The market for printing inks and printing aids in Germany is estimated to approximately 294.000 tons for 2017 with a turnover of approx. 790 million EUR. Germany represents the largest single market in Europe [2].

## 20.2 Printing inks

### 20.2.1 Composition of printing inks

In order to print on flexible and rigid substrates for packaging and goods for the daily life, it is necessary to adapt the printing ink to the structure of the substrates. Therefore, various ink systems with different formulations are available at the market.

Scale printing inks are highly transparent printing inks, particularly designed for the multicolor printing process (three-, four-, seven-colors printing process). The colors are selected in that way, that they can work effectively in the multicolor printing process with the chosen printing application to perform the desired color effect. In order to achieve a maximum of print quality, ink suppliers and printing houses have to work according to standardized processes. Industrialized countries standardized the optical features of the three colors yellow, magenta and cyan in so-called color scales for offset and letterset printing.

On the basis of the theory of colors, colored images are achieved with the multi-color printing process using the standardized basic colors yellow, magenta (bluish red) and cyan (greenish blue). The four-color printing process is supplemented by black and the also used seven-color printing process by orange, violet and green. The color scale series are summarized in Table 20.3 [3].

**Table 20.3:** Existing color scale series.

| Series | Description |
| --- | --- |
| European Colour Scale (Euro-Scale, European Colour Standard for Multicolour Printing) | Multicolor printing CiE 30–89, Tl.1–3 DIN EN ISO 11,664–4:2011–07 |
| European Colour Scale for Letterpress (set of printing inks for letterpress) | Letterpress CIE 30–89, Tl.1 + CIE 12–66; ISO 2846–1; |
| European Colour Scale for Offset printing (set of printing inks for offset printing) | Offset printing especially sheet fed offset CEI 30–89, Tl.1; CIE 13–67; ISO 2846; DIN 16539:1971–10 |
| European Colour Scale for Sheet-fed Offset Printing (European Colour Standard for Sheet-Fed Offset Printing) | Sheet-fed offset printing CIE 30–89, Tl.2, Novelty CIE-67 |
| European Colour Scale for Heat-Set Web Offset Printing (European Colour Standard for Heat-Set Web-Offset Printing) | Heat-set web offset printing CIE 30–89, Tl.3 |
| European Colour Scale for Newspaper Offset Printing (European Colour Standard for Newspaper Web-Offset Printing) | Newspaper offset printing CIE 30–89, Tl.4 |

## 20.2.2 Binders and solvents

### 20.2.2.1 Binders

Suitable binders for printing inks are
- in general, a mixture of resins, solvents and additives.
- mostly transparent
- compositions especially designed for the needs of the different printing application processes.

Table 20.4 gives an overview on the different binder systems for the printing processes including resin bases, solvents and application areas.

**Table 20.4:** Overview on binders for printing processes.

| Printing process | Resin base | Solvents | Application area |
|---|---|---|---|
| *Offset* | | | |
| Letterset/Offset, conventional | (a) fatty, drying oils of vegetables<br>(b) alkyd resins of fatty drying oils of vegetables | mineral oil-fractions<br><br>230–260 °C (heat-set oils), 260–320 °C (cold-set oils), vegetable oils | book-, packaging-, label-, magazine-printing |
| UV-offset | (a) polyester-AY<br>(b) epoxy-AY<br>(c) PU-AY<br>(d) polyether-AY<br>(e) full-acrylated-types<br>(e) mixtures of a – e | di-, tri-, tetra-, hexa-functional AY | book-, packaging-, label-, magazine-printing |
| UV-OPV´s<br>– free-radical<br>– cationic | free-radical:<br>(a) polyester AY,<br>(b) epoxy AY,<br>(c) PU AY,<br>(d) full-acrylated types<br>cationic:<br>(a) cycloaliphatic di-epoxide,<br>(b) caprolactones | free-radical:<br>di-, tri-, tetra-, hexa-functional AY<br>cationic:<br>– vinylether monomers,<br>– alcohols/glycols | protection and gloss improvement of displays, folded carton boxes, book covers, labels |
| *Flexo* | | | |
| UV-flexo | (a) polyester-AY<br>(b) epoxy-AY<br>(c) PU-AY<br>(d) full-acrylated-types<br>(e) mixtures of (a) – (d) | mono-, di-, tri-, tetra-, hexa-functional AY | flexible packaging on foil and paper/board, food and label applications |
| Water-based | (a) STY- AY or AY disp.,<br>(b) PU disp.,<br>(c) alkyd disp.,<br>(d) mixtures of (a) – (c) | water, water-alcohol mixtures, water-glycol mixtures | flexible packaging on foil and paper/board, food and label applications |
| Solvent-based | (a) NC,<br>(b) NC + plasticizer or/and aldehyde/-ketone resins,<br>(c) polyvinyl-butyrale,<br>(d) PA, mixtures of a – b | ethanol, ethylacetate, methoxy-/ ethoxypropanol | flexible packaging on foil and paper/board, food and label applications |

**Table 20.4** (continued)

| Printing process | Resin base | Solvents | Application area |
|---|---|---|---|
| *Gravure* | | | |
| Water-based | (a) STY- AY or AY-disp., (b) PU disp., (c) alkyd disp., (d) mixtures of (a) – (c) | water, water-alcohol mixtures, water-glycol mixtures | flexible food packaging on foil or paper substrates |
| Solvent-based | (a) NC, (b) NC + plasticizer or/and aldehyde/-ketone resins, (c) polyvinyl-butyrale, (d) PA, mixtures of a – b | ethanol, ethylacetate, methoxy-/ ethoxypropanol (most used quantities) | flexible food packaging on foil or paper substrates |
| Publication | pentaerythritol ester of rosin | toluene | magazines, advertisements |
| *Screen* | | | |
| Water-based | (a) AY-emulsions/disp., (b) PU-AY-emulsions/disp., (c) PU-emulsions/disp. | water, water-glycol mixtures | plastics, paper/ board, textile |
| Solvent-based | (a) alkyd Resin, (b) CAP/AY, (c) AY, (d) pentaerythritol ester of rosin | (a) alcohols, ester, ketones, (b) aromatic solvents, (c) mineral spirits, (d) glycol ethers | plastics, paper/ board, textile |
| UV | (a) polyester-AY (b) epoxy-AY (c) PU-AY (d) polyether-AY (e) full-acrylated-types mixtures of a – e | mono-, di-, tri-, tetra-, hexa-functional AY | plastics, paper/ board, textile |

AY = acrylates; STY = styrene; OPV's = overprint varnishes; PA = polyamide; PU = polyurethane; Disp. = dispersion; CAP = cellulose acetate propionate

Binders are important for
- covering the pigments during the grinding process to achieve a narrow particle size distribution or to dissolve dyes
- ink transfer
- protection of the printed ink film against chemical or mechanical influences.

## 20.2.2.2 Solvents

There is a differentiation for solvent-based ink systems between real solvents, able to dilute the resins, which will be used in a binder and non-solvents, which are in use for blending the real solvents. To perform a constant ink drying without problems, it is necessary that the real solvent for the resins in a binder will evaporate later than the solvents used for blending. Solvents are also necessary for the adjustment of the viscosity/efflux time of the printing ink. Solvents, which are used for physical drying inks in gravure-, flexo-, screen-, or tampon-printing determine the drying of the inks by their evaporation time. Solvent-mixtures (co-solvents) are often used for the improvement of the solubility strength.

Solvents used in printing processes are summarized in Table 20.5.

**Table 20.5:** Overview on solvents used in printing processes.

| Printing process | Used solvent | Application |
|---|---|---|
| Gravure | Volatile | Food packaging |
| Flexo | Volatile | Food packaging |
| Publication | Aromatic, sometimes aliphatic | Magazines |
| Screen | Medium volatile | Displays |
| Tampon | Medium volatile | PP-cups |
| Letterset | Light and medium mineral oils | Books |
| Offset | Light and medium mineral oils | Labels, folded carton boxes |

In addition to organic solvents in printing inks containing only solvents, water is also used in water-based and reactive thinners in UV-curable printing inks.

## 20.2.3 Pigments

Inorganic pigments find only limited application in printing processes because of their weaker color strength and in most cases smaller luster effect in printing ink applications [3, 4]. On the other hand, organic pigments are used in a broad manner. Nowadays, more than 50 different organic pigments are in use for high quality printing purposes. Among these, special pigments with a high light fastness and resistance towards chemical substances are preferred. The most expensive organic pigments based on chinacridones, dioxazines, indanthrenes or isoindolinones belong to the class of polycyclic pigments.

To use pigments in printing inks, they have to be ground down to primary particles, which are usually significantly smaller than 10 μm, in order to achieve best brilliance and color strength. Because of the broad variety of possible pigments, the focus in this chapter will be on those pigments with the highest sales potential and the usefulness according to the so-called European Colour Standard for Multicolor Printing Scale.

A standardization of the color tones for the process colors Cyan, Magenta, and Yellow, and in addition Black for the offset printing, was already specified in the DIN 16539 in 1956. These colors are also called scale colors or Euro colors. With various mixtures of these colors, most different colors can be technically achieved.

Pigments are offered to the market as powders, solids or liquid pigment preparations and slurries. Table 20.6 contains the color indices of inorganic and organic pigments suitable for high quality printing inks. The most important properties of pigments used in the ink industry are shade, color strength and fastness, but also dispersibility, rheological behavior and transparency. These properties have their origin in the chemical composition and particle size distribution, the particle shape and the surface characteristics.

**Table 20.6:** Color indices of inorganic and organic pigments suitable for high quality printing inks.

| Shade | Color indices | Name |
| --- | --- | --- |
| White | P.W. 6 | Titanium dioxide |
| Black | P.BL. 7 | Carbon black |
| Yellow | P.Y. 1 | Monoazo yellow |
| | P.Y. 3 | Monoazo yellow |
| | P.Y. 12 | Diarylide yellow |
| | P.Y. 13 | Diarylide yellow |
| | P.Y. 14 | Diarylide yellow |
| | P.Y. 74 | Monoazo yellow |
| | P.Y. 81 | Diarylide yellow |
| | P.Y. 83 | Diarylide yellow |
| | P.Y. 95 | Diazo-condensation |
| | P.Y. 97 | Monoazo yellow |
| | P.Y. 111 | Monoazo yellow |
| | P.Y. 155 | Bisacetessigarylid |
| | P.Y. 181 | Benzimidazolone |
| Red/Magenta | P.R. 2 | Naphthol AS |
| | P.R. 48:1 | BONS:Ba |
| | P.R. 48:2 | BONS:Ca |
| | P.R. 53:1 | β-Naphthol:Ba |
| | P.R. 57:1 | BONS:Ca |
| | P.R. 81 | Triarcylcarbonium |
| | P.R. 112 | Naphthol AS |
| | P.R. 122 | Chinacridone |

**Table 20.6** (continued)

| Shade | Color indices | Name |
|---|---|---|
| | P.R. 146 | Naphthol AS |
| | P.R. 166 | Diazo-condensation |
| | P.R. 169 | Triarcylcarbonium |
| | P.R. 176 | Benzimidazolone |
| | P.R. 184 | Naphthol AS |
| | P.R. 185 | Benzimidazolone |
| | P.R.188 | Naphthol AS |
| | P.R.254 | Di-keto-pyrrolo pyrrole |
| | P.R. 264 | Di-keto-pyrrolo pyrrole |
| | P.R.266 | Naphthol AS |
| Orange | P.R.268 | Monoazo |
| | P.R.269 | Naphthol AS |
| | P.O. 5 | β-Naphthol |
| | P.O. 13 | Diazopyrazolone |
| | P.O. 16 | Diarylide yellow |
| | P.O. 34 | Diazopyrazolone |
| | P.O. 43 | Perinone |
| Purple | P.V 3 | Triarcylcarbonium |
| | P.V.19 | Chinacridone |
| | PV 23 | Dioxazine |
| | PV 27 | Triarcylcarbonium |
| Blue | P.B.1 | Triarcylcarbonium |
| | P.B.15:3 | Cu-phthalocyanine blue; β-modification |
| | P.B.15:4 | Cu-phthalocyanine blue; β-modification |
| | P.B.16 | Phthalocyanine blue |
| | P.B.60 | Indanthrone |
| | P.B.61 | Triarcylcarbonium |
| | P.B.79 | Phthalocyanine blue |
| Green | P.G. 7 | Cu-phthalocyanine green |
| Brown | PB 5 | BONS:Cu |
| Fillers | PW 24 | Aluminum hydrate |
| | PW 19 | Aluminum silicate, kaolin |
| | PW 21 | Blanc fixe |
| | PW 18 | $CaCO_3$, chalk |
| | PW 26 | Magnesium silicate, talc |
| | PW 27 | $SiO_2$, silica |
| Metal effect pigments | Metal 1 | Aluminum based |
| | Metal 2 | Copper/zinc based |

### 20.2.3.1 Inorganic pigments

#### 20.2.3.1.1 White pigments
Pigment White 6 (titanium dioxide) is the most common white pigment for printing applications showing the highest hiding power of all white pigments. Two crystalline modifications of $TiO_2$, anatase and rutile, are used depending on the demands in the application.

*Examples:*
Rutile: Kronos 2066 (Kronos International), Billions TR 52 (Lomon Billions)
Anatas: Hombitan A 300 (Venator Germany GmbH), A-100 (Pangang Group)

#### 20.2.3.1.2 Black pigments
Among the black pigments, carbon black (Pigment Black 7) has the greatest significance for printing inks. Carbon black pigments are produced in most cases according to the channel black or the furnace black process.

*Examples:*
Channel Black: Special Black 4 (Orion Engineered Carbons)
Furnace Black: Regal 330 R (Cabot), Black Pearls E (Cabot), Raven 820 (Birla Carbon), Special Black 350 Black (Orion Engineered Carbons)

### 20.2.3.2 Organic pigments

#### 20.2.3.2.1 Yellow pigments
Diarylide yellow pigments (Pigment Yellow 12 and Pigment Yellow 13) are intensely colored, transparent colorants showing high resistance to solvents and ink components. These pigments are used in all technically relevant printing inks. They are particularly important for the yellow process ink in the multicolor printing process.

*Examples:*
Pigment Yellow 12: Permanent Yellow DHG (Clariant), Sunbrite Yellow 12 (Sun Chemical),

Vibfast Yellow 4004-T (Vibfast Pigments PVT,LTD).

Pigment Yellow 13: Permanent Yellow GR01 (Clariant), Diacetanil Yellow GR 1314 C (Ferro Performance Pigments Belgium NV), Sunbrite Yellow 13 (Sun Chemical)

Pigment Yellow 74 as an arylamide yellow (monoazo yellow) pigment, is sold in form of fine particle-sized, non-laked grades. It offers a light fastness, which is 2–3 times higher than that of the coloristically comparable Pigment Yellow 12. The good light fastness is the reason why Pigment Yellow 74 is often used for package printing. As a yellow pigment according the European Colour Standard for Multicolour Printing, the shade of Pigment Yellow 74 is a little too greenish and has to be tinted with suitable more reddish shaded yellow pigments.

Pigment Yellow 83 as a diarylide pigment, is a reddish shaded yellow pigment with high color strength combined with excellent light fastness (pure tone 6–7 on wool scale). Pigment Yellow 83 is established as standard pigment in the reddish-yellowish shaded segment.

### 20.2.3.2.2 Red pigments
The pigment class of Pigment Red 57 covers the laked process of different organic acids with Ca-, Ba- or Mn-salts. The common characteristic of so-called BONS pigments is the use of beta-oxynaphthol acid as coupling agent. The most popular pigments of this class are the litholrubines.

*Examples:*
Pigment Red 57:1 belongs to the laked BONS pigments. Ca-laked pigments are also known by their trivial name "4B-toner". The light fastness of the pure tone on the wool-scale is 4–5. The color shade corresponds to the Magenta of the European Color Standard for Multicolor Printing for the 3- and 4-color set.

Pigment Red 184 (P.R.184) belongs to the Naphtol AS pigments. It is used when alkali-, acid- or soap- fastness does not permit the employment of P.R. 57:1. The color shade corresponds also to the Magenta of the European Color Standard for Multicolor Printing for the 3- and 4-color set according to the European Color Scale 12–66. The light fastness for the pure tone is around 5–6 on the wool-scale.

### 20.2.3.2.3 Blue pigments
Pigment Blue 15:3 belongs to the Cu-phthalocyanine-blue pigments($\beta$-modification) and offers a pure turquoise shade. For a better dispersibility in printing ink binders, resin-modified grades are available. Low-priced pigments of excellent light fastness and resistance against alkali and sour chemicals are also available. The light fastness for the pure tone is 7–8 on the wool-scale. Pigment Blue 15:3 corresponds to the Cyan of the European Color Standard for Multicolour Printing for the 3- and 4-color set according to the European Color Scale 12–66.

Pigment Blue 15:4 belongs also to the Cu-phthalocyanine-blue pigments($\beta$-modification). The major advantages are the better flow properties in comparison to Pigment Blue 15:3, e.g. in the publication printing process [5].

### 20.2.3.3 Miscellaneous colorants

#### 20.2.3.3.1 Dyes

Dyes are used only in subordinated quantities due to a lower fastness compared with organic colored pigments.

#### 20.2.3.3.2 Organic soluble dyes

Due to the fact, that the light fastness is poor, the use of organic soluble dyes in liquid printing inks is limited. The "laked process" is carried out with tannin or phenolic resins.

Salts of basic dyes (Fanal® types from BASF): The salts are manufactured by formation of cationic tri-aryl-methane or xanthen coloring dyes with inorganic complex acids. They offer moderate light fastness and brilliant color shades with high color strength. The range of Fanal® types includes blue, violet and pink color shades.

Alkali types (Reflexblue® from Clariant): the internal salts of basic dyes are produced by the installation of sulfonium groups in the dye molecules.

#### 20.2.3.3.3 Solid dyes

Viktoria Blue-B-Base and Nigrosin-B-Base can be cracked through fatty – or rosin acids. Solid dyes are in use as oil- or toluene-soluble oleates or resinates in black news inks and, in smaller extents, for illustration printing inks, as a tinting dye.

#### 20.2.3.3.4 Metal-complex-dyes

These dyes are known as Zapon- or Neo-Zapon, predominantly chromium complex pigments, with solubility in alcohols, esters and ketones. In water they are almost insoluble [3].

### 20.2.3.4 Effect pigments

#### 20.2.3.4.1 Metal effect pigments

Metal effect pigments are usually either aluminum or zinc/copper flakes and vary in shades from silver to gold, depending on the composition of the metal used. The silver colored powders can also be toned with organic pigments to produce gold or copper shades using transparent yellow or red pigments [6].

#### 20.2.3.4.2 Special effect pigments (pearl luster pigments)

Special effect pigments (pearl luster pigments, pearlescent pigments) consisting of thin transparent platelets and containing at least one high refractive layer are used in printing processes due to their possibility to supply multiple reflection and iridescent effects [7].

### 20.2.3.5 Luminescence pigments

#### 20.2.3.5.1 Daylight fluorescence pigments
Daylight fluorescence pigments are used in all printing inks where brilliant fluorescent shades are required. The main applications are in packaging, greeting cards, wrapping paper, and posters. They are generally not very fast to light because they consist of dyed resins, which are then pulverized to resemble a pigment.

#### 20.2.3.5.2 Fluorescence pigments
Fluorescence pigments are organic or inorganic compounds, which could be accelerated by the exposure of energy and emit the absorbed energy as "light". The acceleration is possible via electrons, X-Rays, ionic or UV radiation and is also possible via the short-wave part of the daylight. The accelerated radiation is always of more energy power than the emitted light. The emission of light is possible in the visible, as well as, in the UV and IR spectrum of the light. Fluorescence pigments are used for marking of goods or security coding [8].

### 20.2.3.6 Functional pigments

#### 20.2.3.6.1 Conductive pigments
The most important electro-conductive pigment is carbon black. The specific resistivity measured at 300 bar pressed test specimen lies between $10^{-1}$ and $10^{-2}$ $\Omega$cm. Silver is the classic pigment for conductive varnishes for printed circuit boards. Silver coated copper powder is more and more used because of the high price for silver.

#### 20.2.3.6.2 Magnetic pigments
The most important magnetic pigments are $\gamma$-$Fe_2O_3$, cobalt-containing $\gamma$-$Fe_2O_3$, $Fe_3O_4$, cobalt-containing $Fe_3O_4$ and $CrO_2$. The magnetic particles are needle-shaped and have an average length of 200–600 nm [9].

### 20.2.3.7 Fillers
Fillers (or extenders) are commonly used in inks to reduce the overall cost or to achieve special properties. Extenders are sometimes used to reduce the color strength and change the rheology and printability of the inks. Important raw materials are aluminum hydroxide (aluminum hydrate), magnesium carbonate, calcium carbonate, blanc fixe (precipitated barium sulfate), talc and clay [10].

Aluminum hydroxide (Al(OH)$_3$, PW 24) is used as a filler mainly in letterpress and offset inks. It also finds application as a white filler for tinting of printing inks when ground with binders. Blanc fixe (PW 21), which has a better hiding power than aluminum hydroxide can also be used for this application.

Calcium carbonate (chalk, CaCO$_3$, PW 18) is a widely used filler, which is applied in form of finely ground powders or precipitates. In some cases, e.g., in combination with waxes or fatty acids, calcium carbonate powders with a hydrophobic coating are used. Specific mat effects can be achieved with the addition of chalk to paste printing inks.

Silicon dioxide (silica, SiO$_2$, PW 27) is used in many printing ink formulations. It is applied as an amorphous, highly dispersed powder. The surface of the silica particles is hydrophobically coated for special applications.

Bentonite (PW 19) with the main component montmorillonite is used in form of uncoated and coated powders. It is characterized by special thickening properties, which are applied in different formulations, even in small quantities. China clays also belong to this group.

Magnesium silicate (talc, PW 26) influences the dispersibility of pigments as well as the rheological properties. It works as a tack-reducer in offset inks.

Testing methods for fillers in printing inks are summarized in Table 20.7.

**Table 20.7:** Testing methods for fillers in printing inks.

| Testing method | Remarks | Test conditions |
|---|---|---|
| Sieve analysis | Mesh size 0.063 and 0.045 mm | DIN EN ISO 787–7: 2010–02 |
| Particle size analysis | Sieve analysis | DIN 66,165–1: 2016–08 |
| Specific surface determination method | BET method | DIN ISO 9277: 2014–0 |
| Specific surface determination method | Nitrogen adsorption according the single point determination method after Haul and Dümbgen | DIN 66132: 1975–07 |
| Oil value (g/100 g) | | DIN EN ISO 787–5: 1995–10 |
| pH-value | aqueous suspension | DIN EN ISO 787–9: 2018–03 |

## 20.2.4 Additives

Additives are defined as substances, which are added to paints, lacquers or printing inks in small quantities in order to achieve specific properties for the liquid application media or for the final coatings and prints. Additives are used for the manufacture of liquid coatings, powder coatings and printing inks as well as for their application [11].

### 20.2.4.1 Waxes

Waxes are used to impart an improved slip, scuff and block resistance to ink films. Dispersions and emulsions of polyethylene, polypropylene, paraffin, vegetable waxes or fatty acid amides are dispersed in a vehicle or solvent. PTFE or polyolefin waxes are also available as powders, which can be directly mixed into inks.

### 20.2.4.2 Plasticizers

Plasticizers are products of natural or synthetic origin. They are used for the improvement of
  – elasticity by plasticizing of the dried ink-film
  – adhesive strength properties, especially on metal surfaces

In solvent-borne nitrocellulose-based food packaging inks, plasticizers like citric acid ester or lactic acid ester are in use. Further product classes in use are esters of benzoic acid (benzoates), esters of the adipic and sebacic acid (adipates), phosphoric acid esters (phosphates), chlorine containing and soft resins (e.g. alkyd resins). Phthalic acid ester like DOP (dioctyl phthalate) or DBP (dibutyl phthalate) are no longer in use, because of their mutagenic risk potential.

### 20.2.4.3 Driers

Driers are soaps of cobalt, manganese and other metals formed with organic acids such as linoleic or naphthenic acid. They catalyze oxidation of drying oils and thus are used in inks that dry by oxidation. The heavy metal ions contained in most driers (cations of Pb,Mn;Co; Zn) involve a toxicological risk.

### 20.2.4.4 Antioxidants

Antioxidants are based on oximes, substituted phenols, aromatic amines and naphthol. They prevent or retard
  – pre-mature oxidation procedures by admission of oxygen in the ink container
  – skin formation and gelling in the ink container

### 20.2.4.5 Wetting and leveling agents

Wetting and leveling agents work on the base of non-ionic, anionic, cationic, or amphoteric compounds to pre-wet the surface of the substrate, which has to be printed. An overview of wetting agents used in printing inks is given in Table 20.8.

Effective interfacial substances based on silicones are used as leveling agents. Fluor additives are the most effective wetting agents. Their disadvantage is the tendency to stabilize foam in waterborne applications.

**Table 20.8:** Wetting agents used in printing inks.

| Class | Chemical substance |
|---|---|
| non-ionic | Alkyl-poly-glucoside |
| anionic | Alkyl-carboxylate |
| cationic | quarterneric ammonia compounds |
| amphoteric | alkyl-dimethyl betaine |

### 20.2.4.6 Neutralizing agents
A salt forming reaction with functional groups like -COOH in acrylates is usual to create a water-soluble or water-dilutable resin-solution; emulsion or dispersion. Ammonia, amines or amino-alcohols are widely used for this purpose.

### 20.2.4.7 Bactericides
Fungi or bacteria preventing auxiliaries are used for protection against degeneration of the binder system in water-based formulations and the odor, which can be formed during storage.

### 20.2.4.8 Crosslinker
Crosslinker are used to increase the resistance properties in 2-component solvent- or water-based ink formulations. Polyisocyanates are working in solvent-based as well as in water-based systems, while polyaziridines or carbodiimides are used in water-based application systems.

### 20.2.4.9 Photoinitiators
The task of photoinitiators is the acceleration of free-radical or cationic polymerization of acrylic-or cycloaliphatic epoxidized oligomers and resins in UV-inks in combination with reactive diluents. The market is looking for liquid and pulverized products with good incorporation properties. After polymerization, the photoinitiators should form only fragments of low residual odor. In general, two major photoinitiator types exist for free-radical polymerization.

Norrish-I-Typ

Photoinitiators of this type have the ability to form radicals after initiation. There is no need to add co-initiators. The combination of Norrish I and Norrish-II-photoinitiators is probably possible.

Norrish-II-Typ

There is the need of a combination of benzophenone with an amine functional co-initiator to accelerate the complete curing of free-radical UV-systems.

Cationic photoinitiators may consist of onium salts, ferrocenium salts, or diazonium salts. UV cationic polymerization of cycloaliphatic epoxy resins is the result of the formation of a strong acid. The polymerization of the epoxy monomer (initiation, chain reaction) works via a ring-opening of the epoxy moiety to form a reactive cationic species, which attacks and opens the next epoxide monomer.

### 20.2.4.10 Defoamers

Defoamers are based on interfacial substances like silicones, mineral oils, and polyglycols. They have to fulfill different tasks, especially activity against
– micro foaming
– macro foaming
– instability during storage (in can)

The printed surface often has to be over-printable. Mixtures of defoamers are common. An ideal defoamer for all cases does not exist.

### 20.2.4.11 Complexing agents

The task of complexing agents consisting of EDTA or tartrates is the formation of insoluble Ca-salts. EDTA works very well in buffer systems within a fountain solution.

### 20.2.4.12 Miscellaneous additives

Important miscellaneous additives are lubricants, thickeners, gelling agents and preservatives.

## 20.3 Application methods

### 20.3.1 Printing ink sequence

The printing ink sequence is chosen according to the printing forms and the associated printing inks, for example, in the Multicolor Printing Process of scale printing inks (e.g. black, magenta, cyan, yellow). A suitable sequence is very important, especially in the wet-on-wet printing (e.g. in two or four-color sheetfed offset presses). If required, the printing inks must be adjusted to a certain printing sequence.

In practice, different sequences are used. Cyan, yellow, magenta and black are typically printed one after the other. In the wet-on-wet printing process on two-color offset machines, the first printing process sequence might be black and yellow and the second one magenta and cyan in order to avoid a Moiré effect. In the gravure and flexographic printing process the printing sequence is always yellow, magenta, cyan and black.

## 20.3.2 Letterpress printing

Johannes Gutenberg was in the period around 1450 the inventor of a casting equipment for movable types performed by a lead alloy, the so-called letterpress. Letterpress printing is based on hard printing elements. These can be uniformly inked with rubber rollers. Pasty printing inks are rolled uniformly to layer thicknesses of a few micrometers, typically to 1–3 µm. Long ink devices are used together with 10 to 20 rubber and metal rollers to achieve suitable print qualities. The ink composition has to be adapted to the equipment in order to attain a relatively slow drying process and to avoid hardening in the inking device [3, 12].

Letterpress was the dominant method for printing on paper until 1970, but since that time it has been more and more replaced by offset printing. The main factors for this replacement were the relatively high manufacturing costs for the letterpress forms, the laborious preparation of the printing press and certain difficulties with ink trapping in multicolor wet-on-wet printing. Applications in which letterpress printing is still used are stamping, embossing, hot embossing, perforation, and numbering. Special equipment is needed for all these procedures, particularly raised patterns.

Letterpress printing typically uses offset printing inks. To achieve optimal printing results, these inks usually have to be modified somewhat by adding printing aids.

### 20.3.2.1 Newspaper web offset printing

Newspaper offset printing (cold-set printing) is more and more used instead of newspaper letterpress printing due to quick and cost-effective plate making processes and improved image reproduction. The printing inks used for newspaper web offset printing method dry by penetration. They contain hard resins dissolved in high boiling mineral oils.

Ink additives for black newspaper offset printing are dark viscous mineral oils, bitumen and asphalt. Infrared dryers are used to enhance the drying step allowing thicker print layers and leading to more brilliant printing results [13].

High-grade colored printing inks contain organic pigments and often lighter mineral oils. Cold set printing machines, which are in use to produce newspapers and paperback books can achieve more than 65.000 cylinder revolutions per hour. The components of high-grade black and colored printing inks include organic pigments and sometimes also lighter mineral oils and hard resins.

As with sheet-fed offset, there are usually four inking systems set up consecutively in web offset, while the paper web, unlike the individual sheets used in sheet-fed offset, is always printed on both sides. This perfecting printing process means that the paper web runs through the rollers and impression cylinders at speeds of up to 50 km/h.

To avoid deposition of partially dried ink on the metal parts of the equipment, the ink should not yet be dry when it is succeeded on the former of the printing device.

Newspaper offset inks do not contain binder components undergoing oxidative cross-linking. As Offset newspaper inks are free of such substances they can be used in regard of recycling without any problems. De-inking is used to remove the inks from waste paper.

Further demands for cost-effective newsprints, e.g. for color pictures, have led to the use of alternative processes, like water-based flexographic newspaper printing.

### 20.3.2.2 Indirect letterpress and dry offset printing

Dry offset (Letterset) is an offset printing process which combines the attributes of letterpress and offset. With this technology a special plate prints directly onto the blanket of an offset press. The blanket then transfers the image onto the paper surface. The whole process is called dry offset because the plate is not dampened as it usually would be in the offset lithography process.

For the printing of plastic tubes, plastic capsules and other preformed containers (e.g. plastic beakers), UV-hardening inks are in use.

### 20.3.3 Flexographic printing

Flexographic printing, also known as flexography or flexo, is mainly used for packaging prints. The process goes back to the older aniline-dye rubber printing. A breakthrough of the flexographic process was achieved in the 1960s when photopolymer plates came in use and pigmented inks could be applied. Inks for flexographic printing are characterized by fast drying [3, 12].

Modern flexographic printing presses have a high-speed rotary functionality, which can be used for printing on almost any type of substrate, including metals, plastic, cellophane, paper, board and other packaging materials. The process is also applied for printing on non-porous substrates for food packaging.

The printing process utilizes raised, inked elements of a rubber or photopolymer printing plate, which presses directly against the material that is to be printed. This part of the process is comparable to letterpress printing.

The printing form consists of flexible rubber or elastic photopolymers. Inking-up of the printing form proceeds with an etched or engraved anilox roller consisting of steel. Its surface is either chromium-plated (chromium oxide) or of ceramic material, whose cells are filled with printing ink.

The ink-excess is squeezed off the anilox roller by blade or (in older machines) with a rubber roller. The flexographic printing process is schematically shown in Figure 20.1.

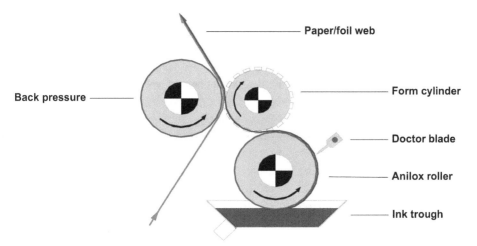

**Figure 20.1:** Scheme of the flexographic printing process.
(source: Coatema Coating Machinery Gmbh).

### 20.3.3.1 Solvent-Containing flexographic printing inks

Low viscosity printing inks are applied for solvent-containing flexographic printing. These inks dry quickly by a fast evaporation of solvents. Flexographic printing presses as a part of a compact inking system use today often laser-engraved ceramic anilox rollers transferring a well-defined quantity of ink to the flexible printing form. Chambered doctor blades have replaced the open ink fountains of former times. Printing speeds of about 500 m/min are achieved depending on the equipment and ink formulation.

Flexographic printing inks and packaging gravure inks consist of very similar components. They differ in the solvent amount, which has an influence on the drying step. Different distances from the doctor blades to the printing nips are therefore used for the two processes. The quantity of solvents with lower evaporation rate is higher in the case of flexographic inks. Hot air dryers are commonly used during the flexographic drying process.

There is also a price advantage over gravure printing. Among other things, cliché production is cheaper than prepress. This is particularly important for bulk goods such as packaging. The flexographic printing process is therefore becoming increasingly important in this area.

### 20.3.3.2 Water-Based flexographic printing inks

Solvents in printing inks are more and more replaced by water due to environmental aspects. The replacement has proceeded already completely for porous, absorbent substrates like paper or board. The compositions of water-based systems contain only very small quantities of solvents, in some cases smaller than 1%. Excellent printing results can be achieved even with ink systems without any solvent. Printing with water-based inks on non-absorbing substrates such as polyethylene and polypropylene has made significant progress. The solvent content is below 5% even for these inks. Printing of wallpaper is possible likewise when using optimized adapted inks. Printing rates of 200–400 m/min are possible, depending on the substrate type.

Water-based and solvent-based inks are dried using a similar equipment. In order to increase the speed of drying, infrared dryers combined with a cold-air blower can be used. This is practiced especially during printing of non-absorbing films. The joint application of mainly acrylic solutions and dispersions, which are solubilized by ammonia or amines, has a beneficial effect on the drying process and on the fastness properties of the finished print. The neutralizing agents are evaporating together with the water. Table 20.9 shows typical formulations of water-based flexographic inks.

**Table 20.9:** Typical formulations for water-based flexographic inks, suitable for paper, corrugated board and foil.

|  | Component | Percentage |
| --- | --- | --- |
| (A) Pigment concentrate | Joncryl SCX 8078 (BASF) | 29.1 |
|  | Novoperm-Yellow HR 02 (Cariant), PY 83 | 34.7 |
|  | Tego Foamex 805 (Tego Chemie) | 1.0 |
|  | Water | 35.2 |
| (B) Ink | Pigment concentrate A | 43.2 |
|  | Joncryl 617 (BASF) | 44.0 |
|  | Tego Foamex 805 (Tego Chemie) | 0.5 |
|  | Water | 12.3 |

Water-based flexographic compositions can also be used for newspaper printing. The achievable quality is lower in most cases compared with newspaper letterpress printing.

### 20.3.3.3 UV-Curing flexographic printing inks

In the field of UV-curing flexographic inks, two major classes are distinguished – free-radical and cationic printing inks (Table 20.10). Mixtures of components of both reaction mechanisms are known as hybrid formulations. UV curing flexographic inks show a superior printing behavior on paper and films compared with water-based inks. Higher gloss and better fastness can be achieved. Similar compositions for UV-curing offset inks are used, but the printing viscosity of UV-curing flexographic inks is generally lower (Table 20.11).

**Table 20.10:** Comparison of free radical and cationic curing.

| Feature | Free radical | Cationic |
|---|---|---|
| Cure speed | High | Moderate to high |
| Initiation | Light | Light and heat |
| $O_2$ sensitivity | Yes | No |
| Shrinkage | Large | Negligible |
| Adhesion | Moderate to good | Excellent |
| Post cure | Limited | Strong |
| Chemical resistance | Good | Moderate to good |
| Humidity sensitivity | No | Yes |
| Acid/base sensitivity | No | Yes |

**Table 20.11:** Free-radical systems – composition of a pearl luster UV-flexo ink.

| Weight (g) | Component |
|---|---|
| 15–30 | UV-flexo resin binder, e.g., Ebecryl 810 (Allnex) |
| 15–25 | Glyceryl propoxy triacrylate, e.g., OTA 480 (Allnex) |
| 10–20 | Dipropylene glykol diacrylate, e.g., DPGDA; also possible 1,6 Hexandiol diacrylate, e.g., HDDA |
| 5–15 | Hexafunctional aliphatic urethane acrylate, e.g., Ebecryl 5129 (Allnex) |
| 0.5–1 | Silicone additive, e.g., Dowsil 57 (Dow) |
| 20–25 | Iriodin® pigment, e.g., Iriodin 123 (Merck KGaA) |
| 5–15 | Photoinitiators (mixing) |

### 20.3.3.4 Offset coating

The finishing of surfaces, which were printed with offset inks, is nowadays quite important. Therefore, so-called chamber doctor blade systems are in use to transfer either water-based offset – or UV-varnishes. These systems are also suitable for the even transfer of effect varnishes pigmented with silver or bronze pigments as well as with pearl luster pigments.

Advantages of chamber doctor blade systems are:
- The ink amount in the chamber is extremely small. There is an ink circulation between an ink reservoir and the chamber.
- The chamber blade system is very easy to clean because of the circulation system. In the chamber blade system, the anilox roller did not dip from the top into an ink tray. The chamber is assembled to the side of the anilox roller. This is beneficial for replacing the air in the anilox roller cells by the ink/varnish. The ink/varnish flow is more smoothly.

The most advanced system for overprint varnish (OPV) the ink tray is encapsulated by two doctor blades and the anilox roller.

### 20.3.4 Gravure printing

Gravure printing is a widely used printing process especially applied to print large volumes of magazines and catalogues. Special printing forms are used, into which the printing areas have been engraved [3, 12, 14].

#### 20.3.4.1 Rotogravure

Rotogravure goes back to copper plate engraving and etching. Printing forms used for this process consist typically of copper cylinders plated with chromium. The printing areas are composed of tiny cells. These cells are filled with a low viscous printing ink. An excess of printing ink is wiped off from the cells before the cylinder hits the surface of the substrate. Gravure printing is suitable for the print of large copy numbers, usually more than 500.000 copies. Cylinder preparation is cost-intensive, but it is compensated by high printing speeds because the drying step runs very fast. A rapid evaporation of solvents in combination with the use of suitable binders is necessary to transport the ink from the ink tray to the substrate surface.

#### 20.3.4.2 Publication gravure

Publication gravure is the method of choice when a large number of magazines and catalogues are to be printed. Most of these prints are done in multicolor technology. Typical printing speeds are in the range of 15 m/s with paper widths of 4,32 m [3, 15].

Binders for publication gravure printing are often based on hard resins dissolved in toluene. The toluene used in publication gravure printing is recovered by adsorption on activated carbon and desorption with hot steam. The ink formulations contain pigments (8–15 wt-%), resins (15–20 wt-%), toluene (60–70 wt-%), and additives (0.5–5 wt-%).

#### 20.3.4.3 Packaging gravure

Packaging gravure involves the use of different printing materials. Typical requirements are a high degree of flexibility in relation to varying quality demands. Printing speeds of up to 400 m/min are employed depending on the size of the print job and the surface to be printed.

Ethyl acetate and alcohols are used as solvents instead of toluene and aliphatic solvents [14]. The process for packaging gravure as it is used for paper, plastic films, aluminum foils or laminated stock is shown schematically in Figure 20.2.

Binder compositions vary widely and have to be adapted to the different requirements. Cellulose nitrates, polyvinyl butyral, polyurethanes and other binders are used together with plasticizers, alcohols and aliphatic esters to achieve optimized printing results. The most important binder components used for packaging gravure are based

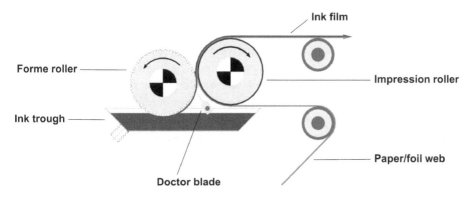

**Figure 20.2:** Scheme of the gravure printing process. (source: Coatema Coating Machinery Gmbh).

on cellulose nitrate. They contain specific plasticizers, acrylate resins, polyurethane resins, and polyamide resins with ethanol and ethyl acetate as solvents.

Solvent-free or low-solvent compositions are more and more used for environmental reasons and occupational health. Water-based inks are used for printing of polyolefin and aluminum foils as well as for wallpapers.

### 20.3.4.4 Gravure transfer inks

Transfer inks are used in a favorable manner in gravure printing. Water-based inks containing selected disperse textile dyes are printed on special papers. The printed image is transferred to the textile material by pressing. Temperatures of about 200 °C are necessary for this step. The dyes are absorbed by the textile fibers. Production and storage of the paper prints are relatively cost-effective. Another advantage of this process is the possibility to achieve high-quality textiles without greater time expenditure.

Typical formulations for gravure transfer inks contain dispersed dyes (10–12 wt-%), acid acrylates together with neutralizing agents (15 wt-%) and water (73–75 wt-%).

### 20.3.4.5 Special gravure inks

There are special applications where the printed product must be stable over the long term against light and other influences. Examples for this are printed wallpapers, decorative wood surfaces, and PVC floor coverings. These materials, when printed, must withstand high process temperatures in many cases. Especially water-based printing inks are gaining more and more importance in these applications. Stabilized pigments and special binders are used to fulfill the demands for light and solvent interaction. Inorganic pigments have clear advantages concerning stability properties in these applications.

### 20.3.4.6 Intaglio printing

Intaglio printing, also known as line intaglio or steel-plate gravure, uses the same principles as traditional copper plate engraving. Cylindrical polished printing forms made from Ni-Cr alloys and plated with chromium are applied. The intaglio process can be used advantageously for the mass manufacture of high-quality printed products, e.g., for stamps and banknotes. During the printing process, the printing ink is transferred from a cutout rubber cylinder via a roller inking device to the surface to be printed on. A reverse-rotating plastic roller is used to wipe off the excess ink.

The printing ink is directly placed on the paper using a high contact pressure. An aqueous, alkaline solution is applied for cleaning the wiping cylinder after each revolution. Brushes and doctor blades are used in addition to achieve optimum cleaning. Air drying is the last step of cleaning before the next revolution can start. Selected pigments are necessary to achieve high printing speeds together with the adapted binder systems. The tack of the ink is further adjusted by the addition of fillers leading also to an optimized printing viscosity. Drying of the printing ink on the substrate proceeds by oxidation and evaporation of the solvents. The dried ink remains raised on the paper, which is different to gravure printing.

### 20.3.5 Screen printing

Screen printing is used for difficult surfaces such as glass, ceramics, metals, and several plastics. Three different types of devices are commonly in use-
(a) roll against roll
(b) flat against flat
(c) flat against roll

The technique for screen printing uses woven mesh to support an ink blocking stencil. The attached stencil forms open areas of mesh allowing the transfer of the ink as a sharp-edged image onto a substrate. A roller or squeegee is moved across the screen stencil during the printing process in order to pump the ink paste through the threads of the woven mesh in the open areas. Most mesh is produced of manmade materials such as steel, nylon, and polyester. Multi-color designs on textile items are often generated using a wet-on-wet technique. The screen can be re-used several times after cleaning or reclaimed if the design is no longer needed.

Prints achieved with screen printing have relatively good weather stability. Examples are plastic sheets with sizes up to several square meters, transparencies, posters made of paper as well as labels for glass bottles, bottle crates, ampoules, candies, and tablets.

Screen printing is also used in the manufacture of printed electrical circuits. Typical examples of this are track conductors and discharge resists. Conical or cylindrical objects can be printed using special devices. Rotary screen-printing devices work with a screen that is stretched over a special cylinder (web printing). Printing speeds of 100 prints/h (manual printing) up to 4000 prints/h (fully automatic rotary screen devices) can be achieved (Figure 20.3).

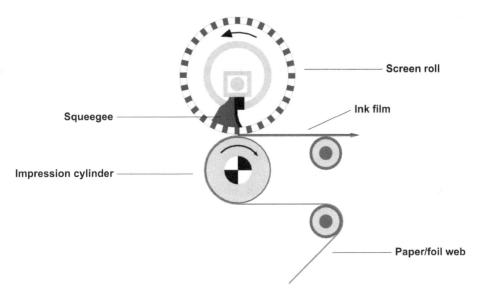

**Figure 20.3:** Scheme of the rotary screen-printing process.
(source: Coatema Coating Machinery Gmbh).

The composition of the printing ink depends on the substrate, the process and the required stability. The printing inks flow easy and produce thicknesses on the printed material greater than those achieved with other techniques. They are based on physical and thermal-hardening compositions. Oxidative, UV-hardening, or two-component inks are in use.

Plastic surfaces are often printed using acrylates as binders. Combinations of acrylates with other resins or polymers are possible likewise. Difficult surfaces can be printed with vinyl chloride copolymers, as in the case of gravure printing. The used organic solvents are commonly higher-boiling compared with those typically used for gravure printing. Screen printing inks that are compatible with water are getting more and more importance and offer comparable weather stabilities in most cases.

### 20.3.6 Offset printing

Offset printing is a commonly used printing process where the inked image is transferred from a plate to a rubber blanket and then to the printing surface. Mutual repulsion between water and fatty printing inks is a basic principle of this process. The printing areas on the plate take on the fatty oleophilic inks. On the other hand, the nonprinting areas on the same plate are wetted by water. This leads to the rejection of the ink in these regions. A special situation exists with waterless offset printing, where a layer of silicone rubber is used instead of the ink-repelling aqueous wetting agent.

Offset inks are characterized by viscous and tacky properties. It must be avoided that they dry already in the inking device. The sensitive printing plate is protected by the elastic rubber blanket. The blanket also adheres to rough surfaces permitting the offset print of even embossed and rough materials.

Pigments used for offset printing are dispersed and ground before they are added to the printing ink. Devices used for these process steps are dissolvers, bead mills, and three-roll mills. Fine pigment dispersions are obtained in this way. It is also possible to use finely divided pigment particles precipitated from aqueous solutions. They are directly inserted in the oleophilic phase by a flushing procedure without intermediate drying. This is how the formation of agglomerates can be prevented.

Brochures, calendars, posters, and many business papers are printed in most cases using sheet-fed offset printing. 20.000 sheets/h can be produced on modern devices.

Printing of absorptive surfaces is using the penetration of the thin oils into the paper where drying proceeds. Final hardness of the print is achieved after several hours.

Materials to be printed possess different absorptive abilities. Unsuitable components in the wetting agent or strong water absorption through the ink should be avoided to achieve perfect printing results. Sticking of the drying prints can be excluded by dusting them with finely powdered chalk or plant starch [3].

Formulations for sheet-fed offset inks typically contain organic pigments (12–20 wt-%), hard resins (20–25 wt-%), soft resins and drying oils (20–30 wt-%), mineral oils with a boiling point of 250–300 °C (20–30 wt-%), and additives (5–10 wt-%).

Printing inks based on a similar base are used for sheetfed offset and letterpress printing.

Types and quantities of raw materials are selected in accordance with the interactions to be expected in the offset ink. Hard resins are beneficial to achieve the tack for a good ink transfer in the equipment and sharp print images. Their use also leads to prints with an appealing gloss on the surface. Phenol- and aldehyde-containing resins have the greatest significance among the hard resins. Using such adapted resins offers mineral oil solubility in the desired range.

The mineral oils are important for dissolving the resins and keeping the ink in a liquid state in the printing device. It is important that these oils are quickly separated from the resin on printing materials with high absorption ability and penetrate into the surface of the material to be printed. Of great importance are also the solution properties of the oils. They are dependent on the content of aromatics, naphthenes, and paraffin.

Special sheet-fed inks are also applied, in which vegetable oils or their derivatives are used instead of mineral oils. Comparable printing results can be achieved when using such renewable raw materials. Those inks are actual preferably in use for tobacco and food packages.

Soft resins and drying oils can be used as dissolution chemicals for hard resins. These substances are linked under the influence of dryers and oxygen. Fatty-acid-modified isophthalic acid polyesters are commonly used as soft resins. Linseed, soybean, tung, castor, and safflower oils find application for the drying oils. Dryers in form of oxidation catalysts and waxy lubricants are used as additives to achieve a sufficient mechanical resistance. Figure 20.4 shows a scheme of the offset printing process.

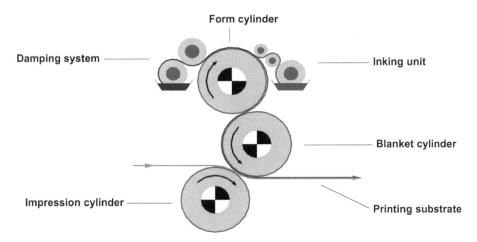

**Figure 20.4:** Scheme of the offset printing process. (source: Coatema Coating Machinery Gmbh).

### 20.3.7 Miscellaneous

### 20.3.7.1 Sheet-Fed metal decoration printing

Printing of metal plates and other non-absorbing materials is difficult because the solvents of the binder cannot penetrate into the surface. Resin-dissolving agents are therefore added to perform the oxidative drying of the printed surface. The drying step is usually carried out within a few minutes in a special oven. A more recent development is the use of UV-LED curable printing inks in a printing process called sheet-fed metal decoration printing.

### 20.3.7.2 UV-Curing offset printing + UV-OPV

UV printing ink systems commonly contain unsaturated acrylates, photoinitiators, pigments and additives. Exposure to UV radiation is the final step of the printing process. UV-curing offset inks are also used in continuous form printing when the prints are further processed in a laser printer.

A lithographic ink is most often prepared in two steps, the preparation of a pigment paste followed by mixing of the paste (25 wt-%) with free-radical hardening UV-varnish (75 wt-%) (Tables 20.12 and 20.13).

**Table 20.12:** Example for a free-radical hardening blue flexo ink.

| Weight (g) | Component |
| --- | --- |
| 30 | Polyester-Tetraacrylate, e.g., Ebecryl 657 (Allnex) |
| 25 | Aromatic hexafunctional urethane acrylate, e.g., Ebecryl 220 (Allnex) |
| 0.15 | Genorad 16 (Rahn AG); Stabilizer |
| 1.5 | Clay(BASF) |
| 25 | Heliogen Blue D 7081D (BASF) |
| 0.5 | Genorad 16 (Rahn AG); Stabilizer |
| 4 | Omnirad 500 (IGM) |
| 4 | Benzophenone Flakes (IGM) |
| 3 | Omnirad EDB (IGM) |
| 1 | Omnirad ITX (IGM) |
| 1 | Glyceryl-Propoxy-Triacrylate, e.g., OTA 480 (Allnex) (to adjust final viscosity) |

**Table 20.13:** Example for a free-radical hardening UV-varnish.

| Weight (g) | Component |
| --- | --- |
| 42 | UV-flexo resin binder, e.g., Ebecryl 657 (Allnex) |
| 15 | Aromatic hexafunctional urethane acrylate, e.g., Ebecryl 220 (Allnex) |
| 10 | Ebecryl 1608 (epoxy-acrylate) (Allnex) |
| 5 | Ebecryl 860 (epoxidized soybean oil-acrylate) (Allnex) |
| 0.1 | Stabilizer |
| 13 | $CaCO_3$ |
| 5 | Omnirad 500 (IGM) |
| 2.5 | Benzophenone (Rahn AG) |
| 2 | Omnirad EDB (IGM) |
| (5) | Glyceryl-Propoxy-Triacrylate, e.g., OTA 480 (Allnex) (to adjust final viscosity) |

UV-curing printing OPV´s (overprint varnishes) are of interest for the mechanical protection of printed surfaces, printed by UV-offset, conventional offset and letterpress inks, because of their high stability and good gloss properties [15].

### 20.3.7.3 Heat-Set web offset printing

Double-sided advertising brochures, catalogs and magazines in four- or five-color printing are often produced using heat-set web offset printing. The printing machines can achieve more than 65.000 cylinder revolutions per hour [3].

Heat-set web offset printing uses a horizontal web guide and heat-drying printing inks. After the last printing unit, the paper web is fed into the dryer. There are temperatures of approx. 250 °C, which causes the paper web to heat up to approx. 120 °C. In the subsequent cooling roller unit, the paper is suddenly cooled to 20 to 30 °C on bright chrome-plated roller surfaces, which causes the inks to harden.

Volatile compounds such as mineral oils are removed from the exhaust air by after-burning. High-gloss printing images are achieved after this step.

Formulations used for heat-set web offset inks typically consist of organic pigments (15–25 wt-%), hard resins (25–35 wt-%), soft resins and drying oils (5–15 wt-%), mineral oils with a boiling point of 200–300 °C (25–40 wt-%), and additives (5–10 wt-%). The compositions used are similar to those of sheet-fed offset inks. However, the ratio of oxidatively drying resins and oils is lower in the case of heat-set web offset inks.

### 20.3.7.4 Fountain solutions

In the conventional offset printing, the fountain solution serves for separating the printing and not printing portions on the pressure plate. It consists to a predominant part of water. The pH value should vary between 4.8 and 5.5. The fountain solution usually contains printing plate protective agents, wetting agents, isopropanol and other antimicrobial effective materials. As printing plate protective agent gum arabic is used. The wetting agents and the isopropanol serve for the reduction of the surface tension. Isopropanol-free, the fountain solution contains frequently glycol as alcohol substitute [16].

Typical formulations for wetting agents or concentrates consist of the buffer system (4–8 w-%), alcohols (5–20 wt-%), surfactants (0–1 wt-%), hydrophilic polymers (1–7 wt-%), complexing agents (0–2 wt-%), preservatives (0–4 wt-%), anticorrosion additives (0–2 wt-%), and water (56–90 wt-%).

An effective regeneration of the printing plate can be achieved by using buffer systems producing a slightly acidic pH-value. Such buffers also protect the plates against alkaline media. The wetting power is improved by the addition of alcohols and surfactants. Hydrophilic polymers with alcoholic groups are used to render the printing plate permanently hydrophobic. Ethylene diamine tetra acetic acid and other complexing agents are part of the ink composition in order to deactivate interfering ions. Fungal growth is inhibited by preservatives in the fountain solution. Metallic parts of the printing equipment are prevented from corrosion by the addition of anticorrosion additives in the formulation.

### 20.3.7.5 Digital printing

The digital printing process makes it possible, that print images could be transferred from a file or data stream directly from a computer to a printing press. The use of static printing forms is not necessary. Laser and ink-jet technology are the most common applications. Digital printing inks could be solvent-based, water-based or UV-based.

Beside dye-based inks, pigment-based inks are in use. Pigment-based inks meet meanwhile highest resistances and therefore permanent printing onto substrates based on paper/board, textiles or plastics is possible without problems [17]. A scheme of the digital printing process is shown in Figure 20.5.

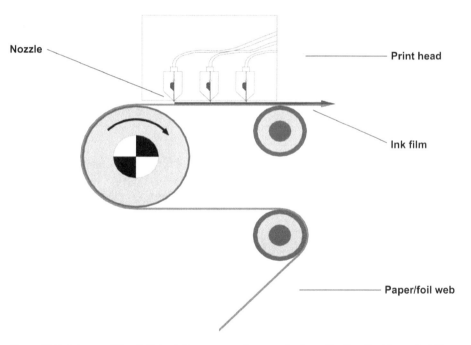

**Figure 20.5:** Scheme of the digital printing process. (source: Coatema Coating Machinery Gmbh).

### 20.3.7.6 Security printing

The intention to prevent items like passports, cheque-cards or banknotes against any kind of forgery is the reason, why security printing often combines different printing methods like intaglio-screen-or offset-printing in a product to be manufactured.

When used as the final layer, UV-curing lacquers offer a high mechanical and chemical protection. The use of special printing stocks and of specific pigments prevents undesired copying of sensible features.

## 20.4 Quality control and test methods

Quality control of prints plays an important role in the printing industry. Several test methods are used for the characterization of the ink and the printed films:
- Scanning electron microscope: topography of the printed film
- Quartant abrasion tester: abrasion resistance
- Brookfield viscosimeter: viscosity of the ink
- Krüss system: contact angle measurement
- Static and dynamic coefficient of friction: slip and anti-slip properties
- Fourier transmission infrared measurement: functional groups
- Gravimetric measurement: determination of water vapor permeability

Fastness properties play an outstanding role. They are strongly influenced by the pigments and binders used in the ink.

Fastness standards are described in national and international regulations, e.g. in ISO 2835:1974 (Prints and printing inks-assessment of light fastness) and 2836:2005–10 (Prints and printing inks-assessment of resistance of prints to various agents).

Important fastness properties that are measured include the fastnesses to water, to solvents, to alkali, to soap, to detergents, to wax, to spices, to edible fats and oils, and to cheeses

Light fastness is measured in a quick-exposure device described in ISO 2835:1974. One half of the sample is covered whereas the other half is exposed to filtered xenon light until a definite change in color can be observed. As a reference, an eight-level blue wool scale is exposed simultaneously. The following system of leveling is used for the characterization:

8 outstanding
7 excellent
6 very good
5 good
4 quite good
3 moderately
2 poor
1 very poor

## 20.5 Regulatory compliance

Important laws and Regulations for chemicals in printing applications are:

1. Regulation (EEC) 1907/2006 for the registration, evaluation, permission and restriction of chemical materials (REACH), to the creation of a European agency for chemical materials, for the change of the guideline 1999/45/EG and for the abolition of the regulation (EEC) 793/93 of the advice, the regulation (EEC) 1488/94 of the commission, the guideline 76/769/EWG of the advice as well as that guidelines 91/155/EWG, 93/67/EWG, 93/105/EG and 2000/21/EG of the commission.
2. Guideline (EEC) 88/378/EG for the security of toy.
3. Security of toys, European standard EN 71 part 3:2013 + A2:2017: Migration of certain elements.
4. Cosmetic guideline (EEC) 1223/2009.
5. (EG) 1935/2004 – EU framework for food contact materials (FCM).
6. GMP-Guidance (EG) 2023/2006 (in force since 1.8.2008) – Good manufacturing practice: "Materials and articles intended for food contact".
7. Toxic Substances control act (TSCA).
8. OSHA (US-Occupational Safety and Health Administration)
   The Hazard Communication Standard is aligned with the Globally Harmonized System of Labelling of Chemicals (GHS).
9. Colorants for polymer, 21 CFR parts 175,176,177,178.
10. Colorants for polymers: 21 CFR part 178.3297.
11. Swiss Ordinance of the FDHA on materials and articles intended to come into contact with foodstuffs SR 817.023.21.
12. Nestle guidance note on packaging inks (August 2016).

The use of pigments for finishing colored printed packaging, for example for foodstuff, should be in accordance with the physiology without hesitation. The pigment producers have to obey regulations and guilty laws.

## References

1.  Teaching material "Varnishes, paints and printing inks. What makes the world colorful" Fonds of the Chemical Industry within the Chemical Industry Association e. V. (FCI). Frankfurt/Main, 2018.
2.  Nicolay K-P. Druckindustrie: Stabiler Brachenumsatz. Markt & Zahlen (Branche). 2017;(8):22.
3.  Zorll U, editor. Römpp-Lexikon "Lacke und Druckfarben". Stuttgart/New York: Thieme Verlag, 1998.
4.  Sandos JD. Pigments for inkmakers, select. London: Industrial Training Association Ltd., 1989.
5.  Hunger K, Schmidt MU. Industrial organic pigments. 4th ed. Weinheim: Wiley-VCH Verlag GmbH & Co. KGaA, 2018.

6.  Wissling, P, Kiehl A. In metallic effect pigments, Wissling P, editor. Hannover: Vincentz Network, 2006:12.
7.  Pfaff G, Rathschlag T. Paintindia. 2002:65.
8.  Ronda C. In industrial inorganic pigments. Buxbaum G, Pfaff G, editors. 3rd ed. Weinheim: Wiley-VCH Verlag GmbH & Co. KGaA, 2005:269.
9.  Horishi N, Pitzer U. In industrial inorganic pigments. Buxbaum G, Pfaff G, editors. 3rd ed. Weinheim: Wiley-VCH Verlag GmbH & Co KGaA, 2005:195.
10. Pfaff G. Inorganic pigments. Berlin/Boston: Walter de Gruyter GmbH, 2017:286.
11. Kittel H. Lehrbuch der Lacke und Beschichtungen. vol. 3. 2nd ed. Stuttgart/Leipzig: S. Hirzel Verlag, 2001.
12. Kipphan H, editor. Handbuch der Printmedien. Berlin: Verlag Springer, 2000.
13. Owen DJ Printing inks for lithography: chemistry, formulation, application and trouble shooting, Sita technology series. New York: Scholium International Inc., 1991.
14. Bassemir RW, Bean A. Inks, in Kirk-Othmer encyclopedia of chemical technology. Hoboken: J. Wiley & Sons, 2004.
15. Leach RH, Pierce RJ The printing manual, 5th ed. Dordrecht: Kluwer Academic Publishers, 1999.
16. Ink world. vol. 11. Ramsey: Rodman Publishing, 2006.
17. Zapka W, editor. Handbook of industrial inkjet printing. Weinheim: Wiley-VCH Verlag GmbH & Co. KGaA, 2018.

Gerhard Pfaff

# 21 Colored pigments

**Abstract:** Colored pigments are inorganic or organic colorants, which are insoluble in the application media, where they are incorporated for the purpose of coloration. Their optical action is based on the selective light absorption together with light scattering. Colored pigments do not include pigments that are colored but which are not used because of their colored character but because of properties such as corrosion protection and magnetism. White pigments, black pigments, grey pigments, effect pigments, and luminescent pigments also do not belong to the colored pigments.

**Keywords:** colored pigments, Color Index, mineral colors, natural colored pigments, synthetic colored pigments

## 21.1 Classification

Colored pigments are inorganic or organic colorants, which are insoluble in the application media, where they are incorporated for the purpose of coloration. Their optical action is based on the selective light absorption together with light scattering. Colored pigments do not include pigments that are colored but which are not used because of their colored character but because of properties such as corrosion protection and magnetism. White pigments, black pigments, grey pigments, effect pigments, and luminescent pigments also do not belong to the colored pigments.

Inorganic pigments are in some cases also termed mineral colors or earth colors. There are historical names for natural inorganic pigments such as ocher, umbra and ultramarine, giving a hint on the mineral origin. Later these names were used also for all other inorganic pigments, however, their use is becoming uncommon more and more.

Each pigment is characterized by its Color Index (C.I.) Generic Name. This internationally accepted nomenclature system for pigments and dyes is edited by the Society of Dyers and Colorists, Bradford, England together with the American Association of Textile Chemists and Colorists.

Inorganic colored pigments have already been used in prehistoric times. Natural pigments were obtained from minerals or their fired products. The weathered products of iron oxide containing minerals and rocks were yellow, orange, red, brown and black powders, which could be used as pigments. In ancient times and in the Middle Ages, several mineral oxides, sulfides and carbonates of the metals lead, mercury, copper

This article has previously been published in the journal Physical Sciences Reviews. Please cite as: G. Pfaff, Colored Pigments *Physical Sciences Reviews* [Online] 2021, 6. DOI: 10.1515/psr-2020-0163

https://doi.org/10.1515/9783110588071-021

and arsenic were utilized to achieve red, blue or green color shades. Table 21.1 gives an overview on natural and older synthetic inorganic pigments [1].

**Table 21.1:** Natural and older synthetic inorganic colored pigments [1].

| Name | Chemical formula | C.I. Pigment | Historical significance |
|---|---|---|---|
| *Natural products* | | | |
| Egyptian blue | $CaCuSi_2O_5$ | C.I. Pigment Blue 31 | Most important blue pigment in ancient times |
| Auripigment | $As_2S_3$ | C.I. Pigment Yellow 39 | Used from Egyptians and Romans; book painting |
| Azurite | $2\ CuCO_3 \cdot Cu(OH)_2$ | C.I. Pigment Blue 30 | Used in Greece, Egypt; in Europe until seventeenth century one of the most important blue pigments |
| Iron oxide red | $Fe_2O_3$ | C.I. Pigment Red 101 | Used in prehistorical caves (Altamira, Lascaux); Roman paintings |
| Yellow ocher | $FeOOH$ | C.I. Pigment Yellow 42 | Most important yellow pigment in Egypt; in Europe used in the Middle Ages |
| Green earth (veronese green) | silicates with FeO | C.I. Pigment Green 23 | Belongs to the largest known mineral pigments |
| Malachite | $CuCO_3 \cdot Cu(OH)_2$ | C.I. Pigment Blue 30 | Used in Greece, Egypt; in Europe until seventeenth century one of the most important blue pigments |
| Realgar (ruby arsenic) | $As_4S_4$ | C.I. Pigment Yellow 39 | Yellow pigment in ancient times |
| **Burnt sienna (Terra di Sienna)** | $Fe_2O_3$ | C.I. Pigment Brown 7 | Used in the Middle Ages |
| Ultramarine (lazurite) | $[Na_8(Al_6Si_6O_{24})]S_{2\text{-}4}$ | C.I. Pigment Blue 29 | |
| Umber | $Fe_2O_3 \cdot x\ MnO_2$ | C.I. Pigment Brown 7 | Important pigment in Egypt; in Europe used in the Middle Ages |
| Cinnabar | $HgS$ | C.I. Pigment Red 106 | Most important red pigment beside iron and lead oxides; used for mural painting, book painting; first synthetic manufacture in the eighth century |
| *Synthetic products* | | | |
| Iron blue (Prussian blue) | $Fe_4[Fe(CN)_6]_3$ | C.I. Pigment Blue 27 | Since 1704 in Europe the most important blue pigment for artist paints and coating compounds |
| Burnt sienna | $Fe_2O_3$ | C.I. Pigment Red 102 | Used in the Middle Ages |
| Burnt umber | $Fe_2O_3 \cdot x\ MnO_2$ | C.I. Pigment Brown 7 | Used in the Middle Ages |

**Table 21.1** (continued)

| Name | Chemical formula | C.I. Pigment | Historical significance |
|------|------------------|--------------|--------------------------|
| Massicot (lead ocher) | PbO | C.I. Pigment Yellow 46 | Mural painting in Thracia in the fourth or third century B.C. |
| Minium (red lead) | $Pb_3O_4$ | C.I. Pigment Red 105 | Used in Egypt, Japan, China, India and Persia |

Natural inorganic colored pigments can be distinguished from each other with regard to their composition and therewith their properties in dependence on the finding location. Some of them have to be classified as hazardous substances from today's assessment. They contain environmentally relevant heavy metals. Some of them do not fulfill anymore the high technical demands of the present time. They were therefore replaced by synthetic colored pigments in modern times. On the other hand, the historically grown designations for the characterization of color shades are used further on, e. g., sienna, ocher or Naples yellow.

With the exception of the natural iron oxides, the other natural pigment types are only utilized in artist paints and for the purpose of restoration. Only a few companies are specialized today on the recovery and the supply of natural colored pigments.

The industrial manufacture of synthetic colored pigments started in the eighteenth and nineteenth century in order to enlarge the color variety by new product classes. The production of some of these newly developed pigments was meanwhile stopped again due to toxicological reasons. Examples for this are Schweinfurt green (cupric acetate arsenite), Scheele's green (copper(II) arsenite) and Naples yellow (lead antimonite).

Synthetic pigments, which are not industrially produced any longer for application technological or toxicological reasons or which have only a limited importance today are grouped in Table 21.2 [1].

**Table 21.2:** Inorganic colored pigments, which are insignificant today [1].

| Name | Chemical formula | C.I. Pigment | Color |
|------|------------------|--------------|-------|
| Barium yellow (lemon yellow) | $BaCrO_4$ | C.I. Pigment Yellow 31 | Very bright yellow |
| Coeline blue | $CoSnO_3$ | C.I. Pigment Blue 35 | Cyanic blue |
| Yellow ultramarine (lemon yellow) | $BaCrO_4/SrCrO_4$ | | Bright yellow |
| Cobalt violet | $Co_3(PO_4)_2$ | C.I. Pigment Violet 14 | Violet |
| Manganese blue | $BaMnO_4 \cdot BaSO_4$ | C.I. Pigment Blue 33 | Greenish bright blue |
| Manganese violet | $NH_4MnP_2O_7$ | C.I. Pigment Violet 16 | Reddish violet |

**Table 21.2** (continued)

| Name | Chemical formula | C.I. Pigment | Color |
|------|------------------|--------------|-------|
| Naples yellow | $Pb(SbO_3)_2$ | C.I. Pigment Yellow 41 | Bright to dark yellow |
| Scheele's green | $Cu(AsO_3)_2$ | C.I. Pigment Green 22 | Brilliant green |
| **Schweinfurt green** | $Cu(CH_3COO)_2 \cdot 3\ Cu(AsO_2)_2$ | C.I. Pigment Green 21 | Bright brilliant green |
| Zinc green | zinc chromate/iron blue | | Bright to dark green |
| Cinnabar | HgS | C.I. Pigment Red 106 | Bright red |

With the exception of some natural iron oxides, the predominant amount of the presently used inorganic colored pigments is today produced in the industrial scale under well-controlled conditions. The pigments are optimized in regard to their properties for their respective application purpose. The inorganic colored pigments are summarized and classified concerning their chemical composition in Table 21.3. Table 21.4 contains the classification of organic pigments, which are colored in all cases [2].

**Table 21.3:** Classification of inorganic colored pigments in regard to their chemical composition [2].

| Pigment | | Color Index | |
|---------|--|-------------|--|
| *Elements* | | | |
| Aluminum | Al | Pigment Metal 1 | 77,000 |
| Copper-zinc-alloy | Cu-Zn | Pigment Metal 2 | 77,400 |
| Carbon black | C | Pigment Black 7 | 77,266 |
| *Oxides/oxide hydrates* | | | |
| Iron oxide yellow | FeOOH | Pigment Yellow 42 | 77,492 |
| Iron oxide red | $Fe_2O_3$ | Pigment Red 101 | 77,491 |
| **Chromium oxide green** | $Cr_2O_3$ | Pigment Green 17 | 77,288 |
| Iron oxide black | $Fe_3O_4$ | Pigment Black 11 | 77,499 |
| *Mixed metal oxides* | | | |
| Bismuth molybdenum vanadium yellow | $(Bi,Mo,V)O_3$ | Pigment Yellow 184 | |
| Chromium titanium yellow | $(Ti,Cr,Sb)O_2$ | Pigment Yellow 53 | 77,788 |
| Spinel blue | $CoAl_2O_4$ | Pigment Blue 28 | 77,346 |
| Reddish blue | $Co(Al,Cr)_2O_4$ | Pigment Blue 36 | 77,343 |
| Greenish blue | | | |
| Iron manganese brown | $Mn_2O_3 \cdot Fe_2O_3, Fe(OH)_2$ | Pigment Brown 7 + 8 | 77,727 |
| Iron chromium brown | $(Fe,Cr)_2O_3$ | Pigment Brown 29 | 77,500 |
| Zinc iron brown | $ZnFe_2O_4$ | Pigment Yellow 119 | 77,496 |
| Iron manganese black | $(Fe,Mn)_3O_4$ | Pigment Black 26 | 77,494 |
| Spinel black | $CuCr_2O_4 \cdot Fe_2O_3$ | Pigment Black 28 | 77,428 |

**Table 21.3** (continued)

| Pigment | | Color Index | |
|---|---|---|---|
| *Cadmium sulfides/selenides* | | | |
| Cadmium yellow | $(Cd,Zn)S$ | Pigment Yellow 35 | 77,205 |
| | $CdS$ | Pigment Yellow 37 | 77,199 |
| Cadmium red | $Cd(S,Se)$ | Pigment Red 108 | 77,202 |
| *Chromates/molybdates* | | | |
| Chromium yellow | $Pb(Cr,S)O_4$ | Pigment Yellow 34 | 77,603 |
| Molybdate orange/red | $Pb(Cr,Mo,S)O_4$ | Pigment Red 104 | 77,605 |
| *Complex salts* | | | |
| Iron blue | $Fe_4[Fe(CN)_6]_3$ | Pigment Blue 27 | 77,510 |
| *Ultramarines* | | | |
| Ultramarine | $[Na_8(Al_6Si_6O_{24})]S_{2-4}$ | Pigment Blue 29 | 77,007 |
| Blue | | Pigment Violet 15 | 77,007 |
| Violet/red | | | |

**Table 21.4:** Classification of organic pigments in regard to their chemical composition [2].

| Pigment | Color Index | |
|---|---|---|
| *Monoazo pigments* | | |
| Acetoacetarylide | Pigment Yellow 1 | 11,680 |
| Benzimidazolon | Pigment Orange 36 | 11,780 |
| Naphth-2-ol | Pigment Red 3 | 12,120 |
| Naphthol AS | Pigment Red 112 | 12,370 |
| Pyrazolone | Pigment Yellow 10 | 12,710 |
| B-naphthol pigment lake | Pigment Red 57:1 | 15850:1 |
| B-naphthol pigment lake | Pigment Red 53:1 | 15585:1 |
| *Disazo pigments* | | |
| Azocondensation pigments | Pigment Red 144 | |
| Bisacetoacetarylide | Pigment Yellow 16 | 20,040 |
| **Diarylide** | Pigment Yellow 83 | 21,108 |
| Dipyrazolone | Pigment Orange 13 | 21,110 |
| *Polycyclic pigments* | | |
| Anthanthrone | Pigment Red 168 | 59,300 |
| Anthraquinone | Pigment Red 177 | 65,300 |
| Anthrapyrimidine | Pigment Yellow 108 | 68,420 |
| Azomethines (metal complex) | Pigment Yellow 129 | |
| Quinacridone | Pigment Violet 19 | 73,900 |
| Quinophthalone | Pigment Yellow 138 | |
| Dioxazine | Pigment Violet 23 | 51,319 |
| Flavanthrone | Pigment Yellow 24 | 70,600 |
| Indanthrene | Pigment Blue 60 | 69,800 |
| Isoindoline | Pigment Orange 66 | |

**Table 21.4** (continued)

| Pigment | Color Index | |
|---|---|---|
| Isoindolinone | Pigment Yellow 100 | 56,280 |
| Perinone | Pigment Orange 43 | 71,105 |
| Perylene | Pigment Red 149 | 71,137 |
| Phthalocyanine | Pigment Blue 15 | 74,160 |
| Pyranthrone | Pigment Orange 51 | |
| Thioindigo | Pigment Red 88 | 73,312 |
| Triphenylmethan | Pigment Violet 1 | 45170:2 |

New developments for industrial inorganic colored pigments are strongly limited by the availability of suitable chemistry. There are only a few new developments in the last years with relevance for the market. The most important of these are bismuth vanadate pigments. Cerium sulfides and rare earths containing oxonitrides did not achieve a breakthrough up to now [3].

There are on the other hand newly developed types of the existing pigments, which show special properties realized by using of various chemical and physical routes. The optimization of these pigments is driven from special purposes of application and from the target to improve the industrial safety [4].

## 21.2 Uses

Main application fields for colored inorganic and organic pigments are coatings, paints, plastics, artist paints, cosmetics, printing inks, leather, building materials, paper, glass, and ceramics. Most of the inorganic colored pigments show the following advantageous properties:
- strong hiding power
- highest light- and weather fastness
- very high color shade and temperature stability
- solvent fastness and non-bleeding behavior

Inorganic pigments have in many cases the better value-in-use compared with organic pigments with similar coloristic properties. A disadvantage compared with organic pigments is their weaker chroma, which is especially relevant for lightening. With the exception of lead chromate, cadmium pigments, bismuth vanadate and ultramarine pigments, inorganic pigments show a cloudier color shade. A solution for many coloration problems is the combined use of inorganic pigments with their strong hiding power and organic colored pigments with their high chroma.

# References

1. Endriss H. Aktuelle anorganische Buntpigmente. Zorll U, editor. Hannover: Curt R. Vincentz Verlag, 1997.
2. Spille J. In Lehrbuch der Lacke und Beschichtungen. In: Kittel H, editor. Pigmente, Füllstoffe und Farbmetrik. vol. 5. Stuttgart: S. Hirzel Verlag, 2003:20.
3. Buxbaum G, Pfaff G, editors. Industrial inorganic pigments. Weinheim: Wiley-VCH Verlag, 2005:99.
4. Pfaff G. Inorganic pigments. Berlin/Boston: Walter de Gruyter GmbH, 2017:92.

Werner Rudolf Cramer

# 22 Color fundamentals

**Abstract:** Colors are only created in the human brain. During this process, light rays from the environment reach the retina of the eye. There they trigger an optical stimulus, which is transmitted to the brain. And the brain translates this stimulus into colors. The perception of color is influenced by multiple factors, which is why it is highly subjective.

Originally, the light rays come from the sun. They are described by the color spectrum, which comprises rays visible to us from 400 to 700 nm. These light rays can be influenced by different pigments that manipulate the light rays. This manipulation can happen by absorption, reflection and refraction and interference.

**Keywords:** Sun, eye, brain, purple, manipulation of light, colors, color pigments, aluminum pigments, interference pigments

## 22.1 Fundamentals and properties

There is a world in front of the human eye, i.e. the environment, and a world behind the eye, i.e. in the brain. Both worlds are independent of each other and are connected by the eye as a bridge.

We always assume that the environment exists as we see it. But we should be aware that the brain translates and interprets the information it receives from the eye into a world of colors. In this respect, we "see" only an image of the environment, which is adapted, changed and manipulated in the brain by many different processes.

The processes in the environment, which ultimately lead to color sensations, begin with the rays of the sun, which take about eight minutes from the sun to the earth (Figure 22.1). Most of these rays are reflected by the earth's atmosphere; only a part penetrates the atmosphere. These are mainly so-called visible rays as well as ultraviolet and infrared rays (Figure 22.2).

Rays are described as waves and by their wavelengths: These are defined as the smallest distance between two equal points (phases). For example, if a sine wave starts at 0°, it reaches a maximum at 90° and passes through the zero line at 180° and through a minimum at 270°, only to return to the same starting point at 360°. Figure 22.3 shows the case that two sine curves are close together, which leads to the situation that the resulting curve is amplified. If two sine curves are far apart, as shown in Figure 22.4, the resulting curve is reduced.

This article has previously been published in the journal Physical Sciences Reviews. Please cite as: W. R. Cramer, Color Fundamentals *Physical Sciences Reviews* [Online] 2021, 6. DOI: 10.1515/psr-2020-0164

https://doi.org/10.1515/9783110588071-022

**Figure 22.1:** The sun is the source of life and light on earth. At midday it appears white and the sky is mostly blue.

The rays visible for the human eye have wavelengths between 400 and 700 nm (1 nm (nanometer) = $10^{-9}$ m). They cover the color range from blue–violet to blue, green, yellow and orange to red, as shown in Figure 22.5. Although rays of each individual wavelength trigger a color stimulus in the eye, objects in the environment reflect over the entire spectral range. There are no "holes" in the color spectrum where pigments do not reflect light rays. Figure 22.6 shows the coherence to the rainbow that begins with blue and continues through green and yellow to red. This corresponds to the diffraction of white light by a prism.

If the maxima of all reflected light waves are combined, a reflection curve is obtained that is typical for each pigment or object. It is so to speak the fingerprint. It should be noted that different reflections (reflection curves) give the same color impression with one type of light. With another kind of light, they lead to a different color impression. This phenomenon is called metamerism.

**Figure 22.2:** When the sun is low in the evening, the light rays need a much longer path through the atmosphere. Therefore, the sun appears red to us.

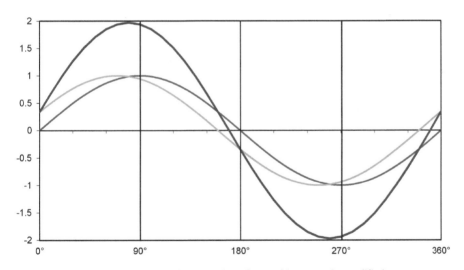

**Figure 22.3:** If two sine curves are close together, the resulting curve is amplified.

The explanations and representations are made with the help of wavelengths. These depend on the surrounding medium. And usually it is assumed that the surrounding medium is a vacuum. Colors are basically dependent on the frequency of the light beams! In different media – for example, water and air – the frequency remains the same, but the wavelengths change:

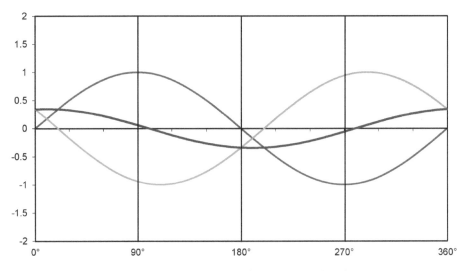

**Figure 22.4:** If two sine curves are far apart, the resulting curve is reduced.

400 nm          500 nm          600 nm          700 nm

**Figure 22.5:** Spectrum of colors visible for the human eye ranges from blue–violet to blue, green, yellow, orange and red.

$$f = \frac{c}{\lambda}$$

f = frequency
c = speed of light ($c_0$ = speed in a vacuum),
$\lambda$ = wavelength ($\lambda_0$ = wavelength in a vacuum) and
n = refractive index.

$$c = \frac{c_0}{n} \qquad \lambda = \frac{\lambda_0}{n}$$

When a light wave changes from one medium to another, its wavelength and speed change while the frequency remains the same.

The sun exists and shines for about 4.5 billion years and will continue to do so for another 4.5 billion years. As the light rays travel different distances through the atmosphere at noon and evening, further phenomena occur here: During the short journey through the atmosphere at noon or at the equator, light rays of shorter wavelengths are mainly scattered, and the sky appears blue to us (Rayleigh scattering).

**Figure 22.6:** Inside, the rainbow begins with blue and continues through green and yellow to red. This corresponds to the decomposition of white light by a prism.

On the long way through the atmosphere in the evening, mainly long-wave rays are scattered. The sunlight thus appears bluer at noon and redder in the evening. A simulation of the color shift through the atmosphere is shown in Figure 22.7. Looking at a white car or a white object at noon and in the evening, the color perception has hardly changed. A white car or a white object can be seen at both times. This is because the human perception is influenced by information stored in the brain. Since we know that the vehicle is white, it remains white despite new information from the eye. This phenomenon is called "changeover".

Sunlight has maximum intensity at 500 nm in the green spectral range. This is shown by measurements with the Hubble satellite and also by calculations according to Planck's radiation law. If these calculations are based on frequencies, a maximum in the red spectral range is shown. This is the reason for the green color of plants and leaves. They absorb red light with maximum frequency for chlorophyll synthesis and reflect in the green spectral range [1].

Isaac Newton described the decomposition of white light into spectral colors in 1704 in his famous treatise "Opticks or a treatise of the reflections, refractions, inflections and colors of light". He was also able to prove that he could reassemble the spectral colors back into white.

**Figure 22.7:** Simulation of color shift through the atmosphere.

Almost one hundred years later, in 1810, Johann Wolfgang Goethe published his theory of colors [2]. In it he dismissed Newton's experiments, because he came to different results with his experiments. For him yellow, red and blue were the basic colors from which all other colors could be mixed. Since he mixed green from yellow and blue, it was a mixed color for him.

Goethe's theory of colors is based on the mixing of colors/pigments and has nothing to do with the today's understanding of the perception. The eye does not care whether it sees a green pigment or a green color or a mixture of yellow and blue. In this respect Goethe's theory of colors is a theory of mixing. His corresponding color wheel is shown in Figure 22.8. It must be stressed at this point that everything that happens in front of the eye has nothing to do with perception. Colors only arise in observer's heads.

Goethe made experiments with rotating apparatus (flywheel) with discs. If a disc with yellow and blue is used there, the mixture does not turn green, but gray.

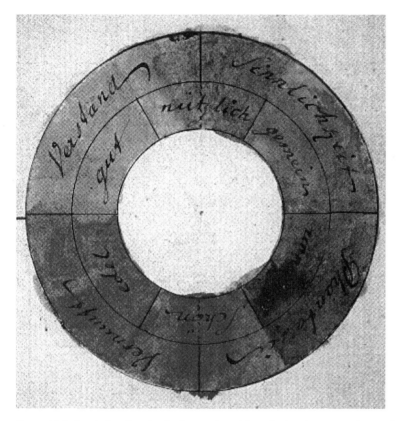

**Figure 22.8:** Goethe's color wheel refers to the mixing of color pigments. It has nothing to do with the today's understanding of the perception of color.

Since this result does not fit to his general explanations, he called it an "apparent mixture". If yellow and blue artist's colors or similar colorants are mixed, the result is green. This process happens in front of the eye and is physically described as a mixing process. If both colors are offered to the eye, for example on a fast-rotating disc, the eye cannot distinguish the two and mixes them additively to gray. This happens in the eye and in the brain.

Wilhelm Ostwald and Albert Henry Munsell have arranged colors according to physiological aspects. Thus, Wilhelm Ostwald used spinning tops to arrange colors into his color wheel, which initially consisted of 100 parts, then 24 parts. In this way, he obtained a color wheel that corresponds to the sensations and perceptions of the eye and the brain. With further gradations he darkened and lightened each color. Wilhelm Ostwald set up his color wheel, which is shown in Figure 22.9, based on sensory distances. Each color is defined by the number of the full color and by letters for the black and white portion. Otto Philipp Runge described the colors in a three-dimensional ball to include mixtures with white and black (Figure 22.10).

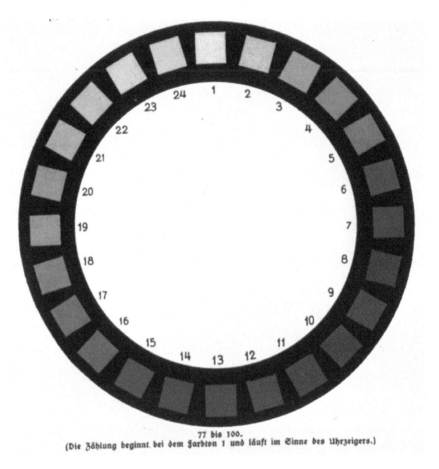

77 bis 100.
(Die Zählung beginnt bei dem Farbton 1 und läuft im Sinne des Uhrzeigers.)

**Figure 22.9:** Wilhelm Ostwald set up his color wheel based on sensory distances.

## 22.2 The eye

On one side, the sun is a light transmitter and on the other side, the eye are light receivers (Figure 22.11). Over the many years of the evolution, the eye has adjusted to the sun.

The eye corresponds to a simple optical system. Rays of light hit the lens of the eye, which is variable. The amount of light is regulated by the pupil. It reacts by reflection, i.e. it adapts automatically. The light rays penetrate the lens and hit the retina on the so-called macula. The lens adjusts itself to focus on the fovea centralis in the center of the macula. The laws of optics also apply here: short-wave light rays are refracted more strongly than long-wave ones. Here the lens adjusts itself when focusing. As a result, blue writing on a red background begins to flicker: The

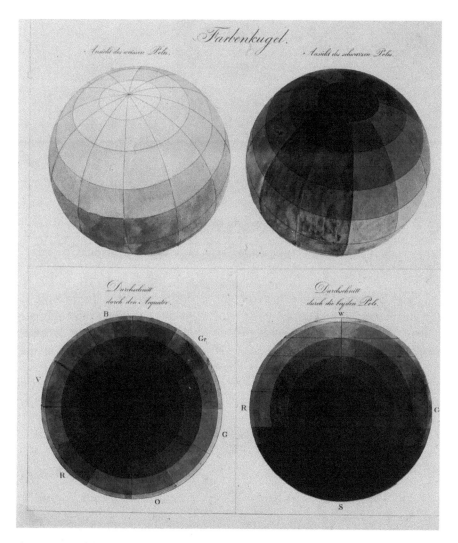

**Figure 22.10:** Philipp Otto Runge showed colors three-dimensionally for the first time.

lens moves between blue and red when focusing. These movements cause flickering. This phenomenon is known in optics as chromatic aberration.

The retina covers the inside of the eye. It contains two different types of light-sensitive cells: The so-called rods are responsible for light-dark vision and the cones for color vision. The rods are distributed over the entire retina, while the cones concentrate on the fovea. The light-sensitive cells transmit their information to the brain via the optic nerve. This nerve runs inside the eye and leaves the retina in the blind spot. At this point, it is impossible to see.

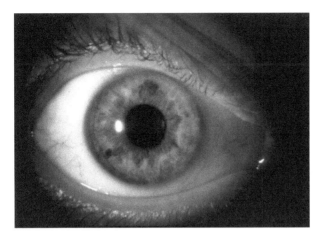

**Figure 22.11:** Human eye has adapted to sunlight over many millions of years. Its pupil regulates reflexively the incidence of light into the eyeball.

The cones differ in their sensitivity to light rays: One species is maximally sensitive to the blue spectral range, another to the green and the third to the red spectral range. These three types are responsible for the color vision of humans. Basically, all three cone types always react, i.e. the brain always receives a common reaction to incident light. No type of cone can be active alone (Figure 22.12).

Rod cells are more sensitive than cones. These photoreceptor cells run at lower light and are responsible for the night vision (scotopic vision) of humans (Figure 22.13).

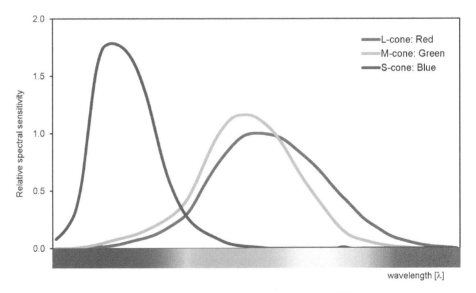

**Figure 22.12:** Human eye has three types of cones that are sensitive to different spectral ranges.

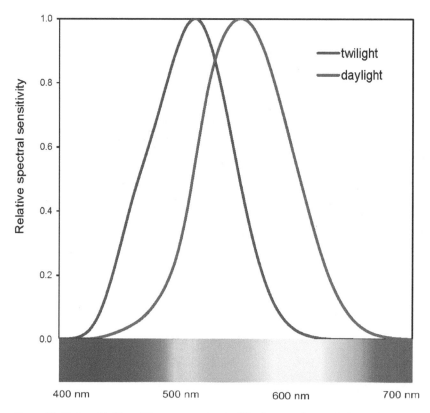

**Figure 22.13:** Sensitivities of the cones and rods differ by a maximum of about 50 nm.

Since only hard parts such as bones have survived millions of years, it can only be speculated about the evolution of soft parts such as the eye. The position of the cones has probably changed more often than the previously diurnal mammals became nocturnal because of the many predators. Today's blue cone could therefore have migrated into the UV range. In the visible spectrum, only one cone is likely to have remained. When the mammals became diurnal again, the blue cone migrated into the visible range. The green cone separated from the red one, as genetic engineering investigations have shown. This process happened only about thirty million years ago, which is a young process in evolution. This may also be the reason for the many color defects in red-green. Since color vision defects are inherited dominantly by men, they have a rate of 8–9%, while those of women are less than 1%. There are different types and strengths of color vision defects. With test charts, defective color vision can be determined. Figure 22.14 shows a red vehicle with a green door that a person with normal vision can clearly see. People with deficient color perception, on the other hand, cannot see the difference of the car paint shown in Figure 22.15.

Some marsupials have four cones at equal distances and probably can differentiate colors better. A crab species also has twelve cones, some of which are in the

**Figure 22.14:** This red vehicle has a green door that a person with normal vision can clearly see.

**Figure 22.15:** People with deficient color perception cannot see the difference of the car paint.

UV range. What and how these animals see can only be guessed. And whether they see colors as humans do, is also questionable.

The sensitivities of the red and green cones overlap. If both cones are excited, this leads to a yellow reaction. Because of the overlapping, humans cannot see dark yellow. They can see dark blue, dark green and dark red, but no dark yellow.

The optical information of both eyes is combined additively and results in a final image. If the eye and the brain are offered a blue (Figure 22.16) and a yellow image (Figure 22.17), these are combined additively to form a neutral image as shown in Figure 22.18. Mixing of colors outside the eye has nothing to do with the human perception.

**Figure 22.16:** Eye and the brain mix colors additively. A blue glass colors the surroundings blue.

## 22.3 The brain

The optical nerves run crosswise from the eyes to the brain. There the optical stimuli are converted into colors. The brain's greatest achievement is to combine the colors blue–violet and red at the two ends of the spectrum to form a new color, purple, as it is shown in Figure 22.19. Purple does not exist in the visible spectrum as a spectral color and cannot be assigned to a wavelength range as illustrated in Figure 22.20.

Because the brain makes this linkage possible, we see the colors in a color circle: Starting with yellow, the color circle runs from orange to red and violet. From

**Figure 22.17:** In the same way a yellow glass colors the surroundings yellow.

**Figure 22.18:** If the eye and the brain are offered a yellow and a blue image, these are combined to form a neutral image.

**Figure 22.19:** The colors at the respective end of the spectrum are connected by purple.

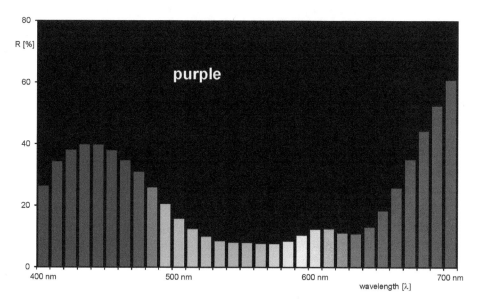

**Figure 22.20:** Purple does not exist in the color spectrum. A purple chromatic color is composed of red and blue reflective components.

there it continues toward blue and then via green back to yellow. This is how the color circle is created! The color circle exists only in the human brain (Figure 22.21).

When we open the eyes, we see colors as a matter of course. This makes us believe that the colors are properties of the objects in the environment. Since we do not see and recognize the reflected light rays as such, we associate the colors with the objects. Viewed purely objectively, the objects "only" reflect rays of light that reach the eye and trigger an optical stimulus there. This is translated into colors in the brain.

So: Colors only arise in the human brain!

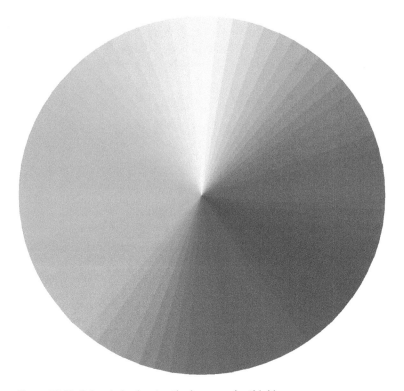

**Figure 22.21:** Color circle showing the human color thinking.

While the color spectrum represents the properties of light waves – which light wave triggers which color stimulus – the color wheel represents the brain's reaction to light waves and the relationships between colors. In the color wheel the colors are arranged in such a way that they have the same brightness and the same chromaticity (chroma).

The arrangement of the colors in the color wheel has already been described by Ewald Hering [3], a German physiologist, with his theory of opponent colors (Figure 22.22). According to this theory, yellow and blue as well as red and green are opposite each other in the color wheel: We don't know yellow which is bluish. A yellow can only be reddish or greenish. And a blue cannot be yellowish, but only greenish or reddish. It is the same with green and red: A green cannot be reddish, but only yellowish or bluish. And a red cannot be greenish. There is only a yellowish or bluish red. For this reason, in modern color science, yellow and blue are represented opposite each other in a coordinate system on the y-axis as shown in Figure 22.23. Perpendicular to this is the x-axis with the green and red components. Perpendicular to both is the brightness axis.

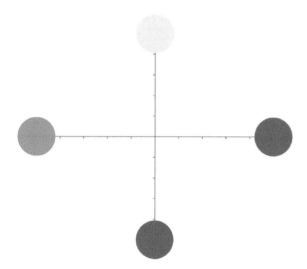

**Figure 22.22:** Ewald Hering postulated the theory of opponent colors.

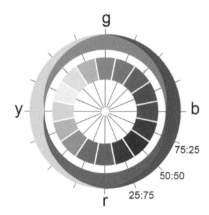

**Figure 22.23:** In the human perception of color, yellow and blue as well as red and green are opposite to each other. There is no reddish green and vice versa. Just as there is no bluish yellow and vice versa.

Colors that are opposite to each other in the color wheel are called complementary. If they are mixed, white or gray is generated. The eye and the brain mix colors additively, i.e. a mixture becomes whiter the more colors are added.

Colors are described by their hue (red, yellow, green etc.), their chroma (intense, less intense) and their brightness (light, dark). These three components characterize each color. This also applies to real color pigments (absorption pigments). It should be noted that mixing does not only change one component, but at least two. An addition of a white pigment provides more brightness, but also a change in chroma.

The colors appear different, which is not shown in the color wheel: For us, yellow is more colorful and brighter than blue.

Two special characteristics of the human brain are important in terms of colors: On the one hand, we cannot remember any color and "park" it in the brain. This makes it almost impossible to recognize a color (e.g. when shopping for clothes). On the other hand, we can differentiate colors very well. The smallest differences in color can be distinguished well.

Besides the direct translation of light rays into colors, there are psychological and psychological influences on color perception. For example, looking at a green against a yellow background, the green one appears more yellowish. If the same green is placed in front of a blue background, it appears more bluish. Another example is the placing of circular rings alternately in green and blue followed by the replacement of one of these by red. If the order of the green and blue rings is changed subsequently, the red also changes in the perception (Figure 22.24). Evaluation of colors should therefore always be done in a neutral environment!

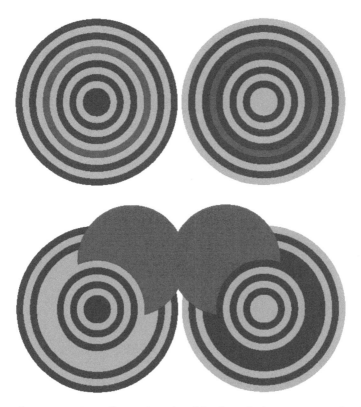

**Figure 22.24:** Depending on the order of the rings, the human eye perceives the red to be yellowish or bluish.

Many examples show how the human perception can be changed or even deceived. Since the human brain only reproduces a virtual image of the environment, the question also arises as to what we really see? This is where philosophy comes into play.

## 22.4 Manipulation of the light

Pigments can manipulate the incident light in different ways: They can absorb part of the light and reflect the rest in all directions. Or they reflect the incident light at a certain angle (angle of incidence = angle of reflection). The third type of pigment divides the light in such a way that both parts subsequently interfere, i.e. due to the displacement of the waves, they are attenuated or amplified.

These processes depend on light: If white light falls on a green pigment, mainly light waves from the green spectral range are reflected. If, on the other hand, red light falls on the same pigment, only little light is reflected, since the green pigment only has reflective properties in the green spectral range. The color is therefore not a property of a pigment and changes with the incident light. The here described situation is illustrated in the Figures 22.25–22.28.

**Figure 22.25:** In white light, the colors appear "normal" to the human eye.

**Figure 22.26:** If the same panels are illuminated with red light, yellow and red appear to us as light gray and green and blue as black.

**Figure 22.27:** Turquoise light turns yellow into yellow–green, red into black.

**Figure 22.28:** Violet light makes that yellow and red appear almost in the same red, while green becomes blue.

Basically, reflection occurs in the entire spectral range: A red pigment therefore also reflects in the yellow, green and blue range – only much less. There are no black holes in the reflections. The situation is similar with a blue pigment, for example: It reflects over the entire spectral range, especially strongly in the blue wavelength range.

## 22.5 Color pigments

Color pigments, respectively, absorption pigments belong to the classic pigments that were used in cave paintings as early as the Stone Age. Besides many earth colors, colorants made from plants or animals were also in ancient times. Figure 22.29 shows as an example a "paintbox" that belonged to the daughter of an Egyptian pharmacist who lived 3000 years ago. Even crushed mummies were used as "mumia" in painting, especially for painting shadows, as shown in Figure 22.30. Artificial pigments such as the Berliner Blue, which was produced from the beginning of the eighteenth century onward, also brought new colors into the world. A great leap forward was the discovery of "mauve", an aniline compound discovered by William Henry Perkin in 1856. His discovery led to a new class of organic colors. And due to his discovery, well-known chemical companies were founded.

**Figure 22.29:** This "paintbox" belonged to the daughter of an Egyptian pharmacist and is 3000 years old.

**Figure 22.30:** "Mumia" was a much sought-after color in paintings (photograph: Manfred Oppermann, Wuppertal, Germany).

Pigments are distinguished from dyes. These are primarily used for dyeing, as they are prepared in solutions. Their behavior toward white light corresponds largely to the behavior of pigments [4].

The optical properties shown below were established based on samples with car paints. Car paints require exact handling, which is why the results are scientifically sound. They can easily be transferred to other areas of application (plastics, industrial coatings, cosmetics, artists' paints, etc.). Real deviations are possible due to different binding agents, but in principle the statements are generally valid. This applies to color pigments, aluminum pigments and interference pigments [5].

When talking about colors in the following, it is the optical properties of the pigments and not spectral colors. The latter are based on individual light waves which are characteristic for the substance. For example, sodium has a double line at 589 nm. This produces a yellow color reaction in the human brain. A yellow pigment reflects more or less strongly in the entire spectral range (Figure 22.31). This distinguishes light colors (spectral colors) from body colors (pigments).

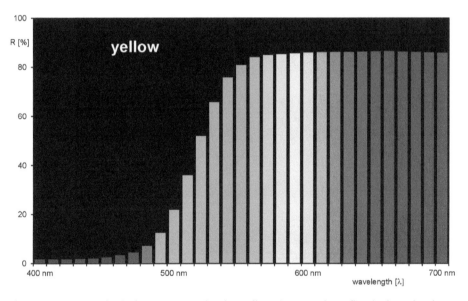

**Figure 22.31:** Human brain does not recognize that yellow pigments also reflect in the red and green spectral range. Both areas are mixed additively by the brain to yellow.

Color pigments mix subtractively (multiplicatively), whereby figuratively speaking, less and less light is reflected when two color pigments are mixed. It must be emphasized once again that a color wheel created by mixing colored pigments has nothing to do with the color wheel on which the human perception and color sensations are based. However, most people believe that the mixed color circle reflects their color sensations. This is wrong!

There are two ways of mixing color pigments: In subtractive mixing, the original colors act like two color filters connected in series, as shown in Figure 22.32. The first one only allows certain wavelengths to pass, e.g. those from the yellow range. This yellow filter blocks all other colors. The second color filter only receives the light waves that have been transmitted by the first one. In the example, light waves from the yellow range meet a blue filter. This filter only allows light waves from the blue spectral range to pass, i.e. none from the yellow range. Only some light waves from the green spectral range, which pass both color filters lead to the resulting color, as shown in Figure 22.33.

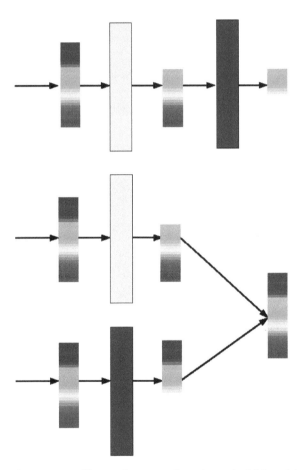

**Figure 22.32:** Difference between subtractive and additive mixing: In subtractive mixing, the color filters are placed one behind the other, in additive mixing, they are placed parallel.

If the two filter colors are far apart like red and blue, the entire spectral range is blocked. Since pigment colors, in contrast to filter colors, reflect throughout the entire spectral range, there are overlapping reflections for red and blue, but these do not lead to violet, but at most to brown, as shown in Figure 22.34.

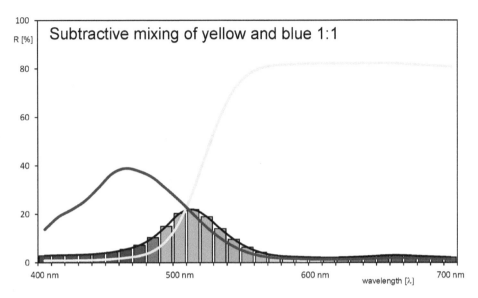

**Figure 22.33:** Blue and yellow mix to green because both source colors contain green. Green remains when mixing.

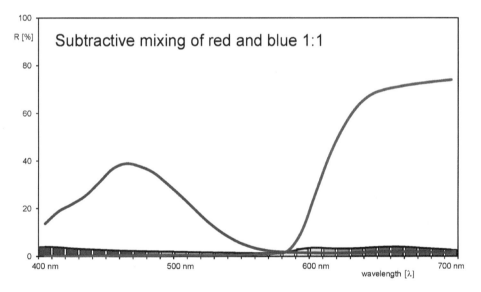

**Figure 22.34:** Subtractive mixing of red and blue in the ratio 1:1 to brown.

In additive mixing, as the name suggests, the ingredients add up: If a yellow and a blue interference color are mixed, they add up to a white. Here, both spectral ranges are combined, as shown in Figure 22.35. This corresponds to a parallel arrangement of color filters. After the light waves have passed through both filters

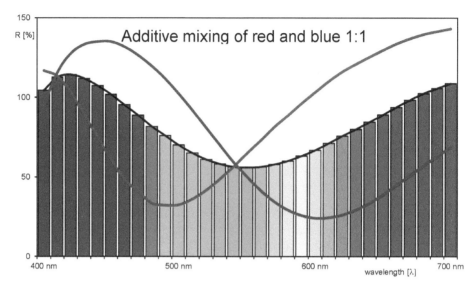

**Figure 22.35:** Additive mixing of red and blue in the ratio 1:1 mix to violet.

separately, they unite in the mixture. The appropriate mixture of yellow and blue leads to white (Figure 22.36).

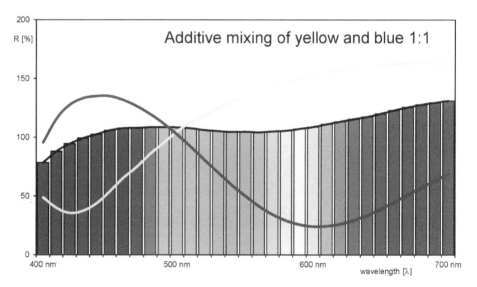

**Figure 22.36:** Appropriate mixture of yellow and blue leads to white.

If the common color pigments are put together or the color program of a manufacturer of artist's colors is taken, the colors do not arrange themselves in a circle. Sometimes there are also gaps between the different colors. As the pigments are produced in some way or other, one can only use the results of the production. The spectrum of colors is limited by the manufacturing process as shown in Figure 22.37. It is not possible to make an intense yellow more intense if it cannot be produced. To reproduce a color circle as good as possible, many pigments are necessary.

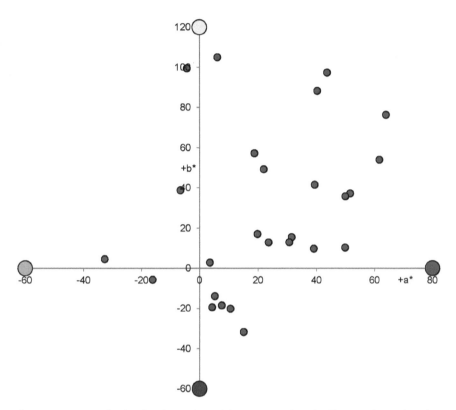

**Figure 22.37:** Typical paint distribution in a mixing system. The a*b* diagram shows the color values of the individual mixed paints.

## 22.5.1 Yellow

Yellow is one of the most interesting colors: Like all other body colors, it reflects in the entire spectral range, but especially strongly in the green, yellow, orange and red range. The high reflections in this range lead it in brightness close to white. Its colorfulness and chroma are also very high [6].

Interesting are the reflections: It reflects not only in the yellow, but also in the green and red areas. The eye and the brain combine both areas to yellow. The human eye cannot recognize the reflections; it transmits the optical stimulus of these reflections to the brain. There yellow is then perceived. The further the reflections reach into the green spectral range, the more greenish the yellow appears. The less it reaches into the green range, the redder it is!

Yellow can be mixed with red to orange. Here, the so-called color strength plays a major role: Small amounts of a red pigment are sufficient to change the yellow to an orange. Conversely, huge amounts of yellow would be needed to get an orange from a red: Therefore, when mixing an orange, it is best to start with yellow Figure 22.38 shows reflection curves of mixtures of yellow and red characterized by typical saddle shapes.

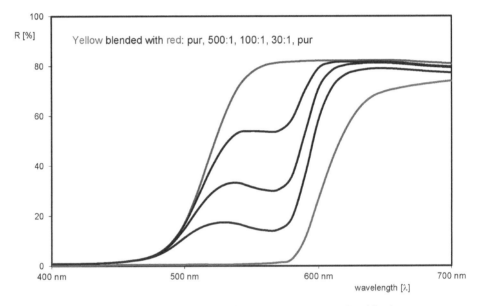

**Figure 22.38:** Reflection curves of mixtures of yellow and red show typical saddle shapes.

If the yellow is mixed with green, a gradation of yellow–green is obtained. There are no special features to be considered here.

### 22.5.2 Yellow ochre

Yellow ochre pigments "darken" less than a black pigment, but also have much less chroma than a yellow one. The addition of yellow ochre makes the mixture slightly darker and reduces the chroma. If yellow ochre is removed from the mixture, the brightness and chroma increase.

### 22.5.3 Orange

This color can be mixed from yellow and red as described. A manufactured orange pigment is much more color intensive (more brilliant), which is why it has advantages over mixing. A mixed orange can be easily recognized by its reflection curve. This has a typical saddle, which is missing in the orange produced (Figure 22.39).

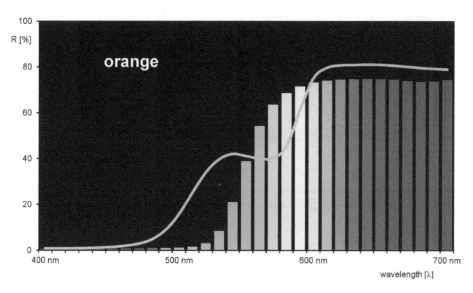

**Figure 22.39:** Comparison of the reflections of an orange (columns) with a mixture of yellow and red (line).

### 22.5.4 Red

Colors whose maximum reflections are in the red spectral range are perceived as red. The further these are in the yellow range, the more yellowish the red becomes. Conversely, humans recognize it as red, sometimes even as bluish, when reflections in the blue–violet range are added (Figure 22.40). A red pigment does not show the actual optical color character. Mixtures with white show this clearly. Figure 22.41 shows that red pigments react typically with a color shift when adding white. Blue and green pigments, on the other hand, increase their chroma up to a turning point when white is added. From this turning point, the chroma decreases while the lightness continues to increase.

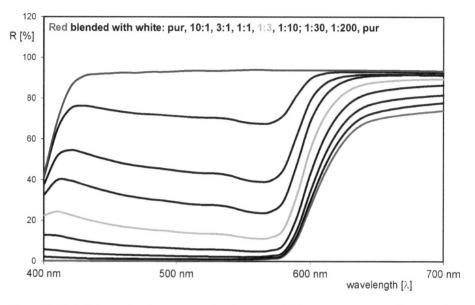

**Figure 22.40:** Red chromatic colors have a color character, which becomes apparent when mixed with white. Here, the blue reflective component is initially intensified when mixed with white.

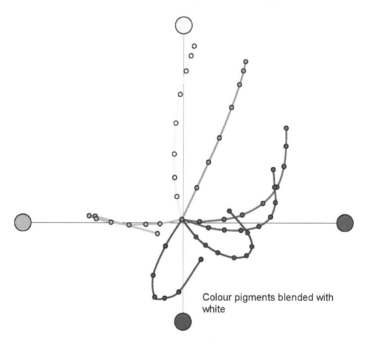

Colour pigments blended with white

**Figure 22.41:** Red pigments react typically with a color shift. Blue–violet, blue and green pigments increase their chroma and brightness up to a turning point when a white pigment, an aluminum pigment or an interference pigment is added. From this turning point, the chroma decreases while the brightness continues to increase.

### 22.5.5 Purple

The closest to red in the color wheel is violet. Depending on the proportion of red or blue, it can appear red-violet or blue–violet. By mixing of red and blue, a brown rather than a bright violet can be achieved. In watercolor painting one often believes that mixing violet from red and blue is possible. Firstly, a white background is used, secondly, transparent watercolors are used and thirdly, a produced violet is always more colorful and intense than a mixture.

Violet, which corresponds to this color direction, reflects in the red and blue–violet spectral range. As already mentioned, it has no spectral lines of its own in the visible spectrum. Only the brain combines the two reflected parts to a new color that does not exist in the spectrum.

### 22.5.6 Blue

This color is heavier and less brilliant than yellow for the human impression. Blue has a maximum of its reflection in the blue spectral range with flanks to the green and blue–violet range. The steeper these flanks are, the more intense the blue is. This also applies to the other colored pigments. Blue reflects the incident light in the entire spectral range. The further its maximum extends to the green spectral range, the more greenish the blue is. On the other hand, a blue can also appear reddish, the less it extends into the green range. Figure 22.42 shows that depending on the color character, a blue also reflects in the green and in the blue–violet color range.

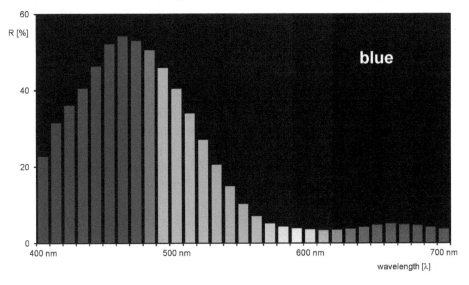

**Figure 22.42:** Depending on the color character, a blue also reflects in the green and the blue–violet color range.

### 22.5.7 Blue–green

If green and blue artist's paints are mixed, the corresponding mixed colors are obtained. In contrast to the mixing of yellow and red, there is little to consider here.

### 22.5.8 Green

Green is not far from blue in the spectrum and shows similar properties to this. This is especially apparent when mixed with white, which is described below. On the one hand, green can be bluish, on the other hand yellowish. Both alignments lead to corresponding mixed colors.

The addition of yellow pigments makes green colors brighter, yellowish and more colorful because yellow is a very bright and intensive color. Conversely, the brightness decreases when yellow is reduced. Mixtures become greener in this case.

Green mixed paints are very dark like blue ones and become more colorful (= green) by adding white. For this reason, the mixture becomes darker, more intense and greener when green is added.

### 22.5.9 Yellow–green

In order to get the desired mixing color, it is possible to start either with yellow or with green. It is more advantageous to use a yellowish green and a greenish yellow to prevent "contamination" of the mixture: Blue portions in green and red portions in yellow would cause undesired color reactions.

### 22.5.10 White

White, like the subsequent black, is an achromatic color, as shown in Figure 22.43. It is used alone or in mixtures with colored or black pigments. Since yellow itself is very bright and intense, the intensity cannot be enhanced by adding white. A mixed series of yellow and white shift directly from yellow to white.

Red pigments not only become brighter with white, but often change color. Therefore, a mixed series does not run directly from red to white, but may well make a turn to violet. Green and blue colored pigments are usually almost like black and not very colorful. Through white, both pigments unfold their chroma with increasing addition, as shown in Figure 22.41. At the same time, the brightness of the mixtures increases. From a turning point, however, the chroma of the mixtures decreases, while the brightness continues to increase until white is reached.

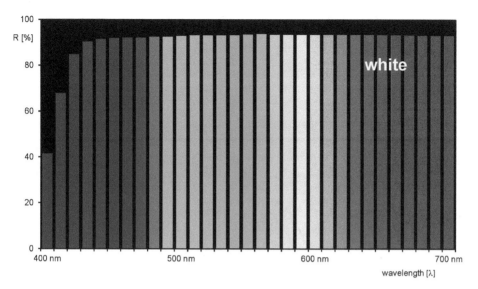

**Figure 22.43:** White reflects the incident light strongly in the entire spectral range. Differences occur when it is based on different materials.

### 22.5.11 Black

While mixtures of green and blue with black decrease in brightness, mixtures with yellow and red show additional color shifts. Especially with yellow and also with orange, black reacts to olive green.

Conversely, as with white, the addition of black leads to a reduction in brightness and chroma, as expected. Furthermore, the mixture becomes greener. A reduction causes the brightness and colorfulness to increase.

And red produces brown together with black. This situation is shown in the a*b* diagram in Figure 22.44.

In many applications, more than two color pigments are mixed to obtain the desired final color. Different interactions can lead to overlapping of effects. The example of a real color for the automotive paint sector in Figure 22.45 is chosen to illustrate the interactions during color adjustment. The green shown here is composed of the mixed paints green, yellow, yellow ochre, white and black.

## 22.6 Aluminum pigments

Aluminum pigments are tiny platelets that reflect incident light like a mirror. They were originally developed for protective coatings (e.g. for the roofs of freight cars).

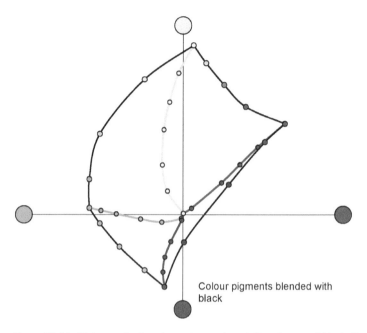

Colour pigments blended with
black

**Figure 22.44:** Mixtures of color pigments run almost directly toward black. However, this a*b* diagram also shows that mixtures of yellow with black drift into olive green.

Today their main applications are in the decorative, industrial and automotive sectors. Different versions are also available for artists and restorers.

In addition to the so-called silver bronzes based on aluminum, there are also gold bronzes consisting of copper and copper/zinc alloys. Depending on their composition, the latter have such illustrious names as rich gold, rich pale gold or pale gold. In the following, silver bronzes are presented and discussed.

Liquid aluminum is atomized to produce aluminum pigments. This produces potato-shaped particles that look flat-rolled like cornflakes. If the atomization is carried out under inert gas such as helium, round particles are produced that look like "silver dollars". Due to their regular shape, "Silverdollars" reflect the incident light more evenly and strongly than "Cornflakes".

Besides this difference, the aluminum pigments can have different particle sizes. Fine particles have a higher opacity, while coarse particles glitter and shine more (Figure 22.46).

If aluminum pigments are not wetted by the binder, they float up in the wet film, i.e. to the surface. This property is known as "leafing". If the pigments are completely wetted by the binder, they distribute themselves evenly in the binder. This behavior is known as "non-leafing".

Aluminum pigments appear gray to us, although they are lighter in gloss than white. The reason for this lies in the human perception, which does not focus on small

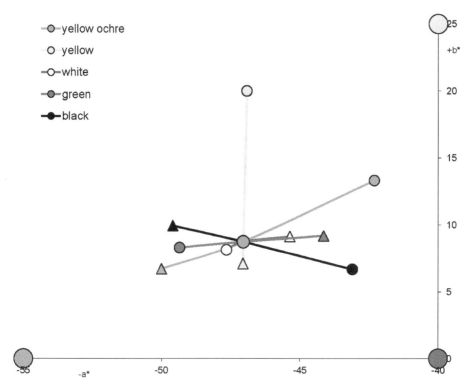

**Figure 22.45:** Starting color (light green circle) is a mixture of yellow ochre, yellow, green, white, and black. The individual mixed colors have been increased (circles) or decreased (triangles) in quantity. The changes are clearly visible in this a*b* diagram.

areas. This is also where an optical property of aluminum pigments comes into play: the optical properties of aluminum pigments are dependent on the angle, the angle of illumination and the angle observer. Since the aluminum pigments have no color, their brightness changes above all. They are brightest close to the angle of gloss – i.e. where the incident light is reflected to the maximum – and are therefore the brightest. The further away from this gloss angle, the lower the brightness (Figure 22.47).

This flop, as the brightness effect is called, can be influenced in its properties by additives. For example, the same aluminum pigment with additive near the gloss angle can be darker than one without additive. This impression turns away from the gloss. Black pigments can also cause this effect (Figure 22.48).

Aluminum pigments can be mixed with color pigments to create colorful metallic effects, as shown in Figure 22.49. However, there are limitations, especially with white and yellow pigments. White pigments are relatively large and can minimize the flop effect. Yellow pigments tend to be greenish.

The best mixing results are achieved with blue and green as well as black pigments, as shown in Figure 22.50. Here, luminous metallic effects can be achieved

**Figure 22.46:** Comparison between an aluminum paint (left) and a chromatic paint.

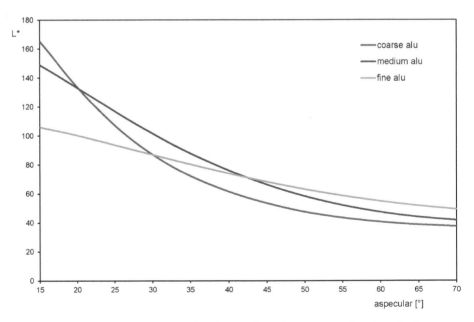

**Figure 22.47:** The closer the distance to the gloss angle is, the more strongly reflect aluminum pigments. Their reflections depend on the particle size.

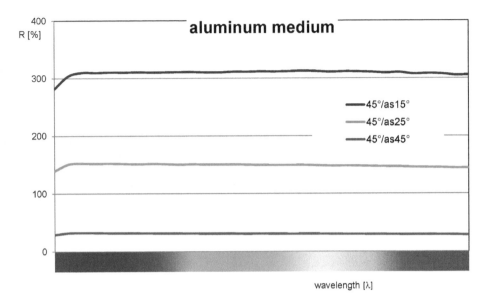

**Figure 22.48:** Aluminum pigments reflect very strongly near the gloss angle. The reflections decrease the further the distance to the gloss angle is.

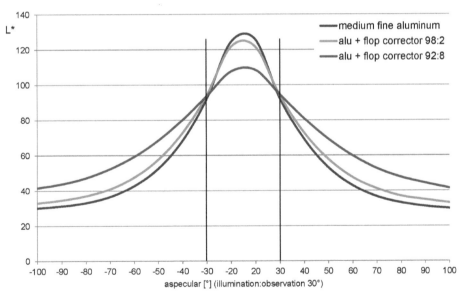

**Figure 22.49:** Flop of an aluminum pigment can be influenced with a flop corrector. This can reduce the reflection of the aluminum pigment close to the gloss and increase it away from the gloss.

which are also popular in the automotive sector. Even mixtures of colored and aluminum pigments mainly only change their brightness when the viewing angle is changed. There is practically no color shift.

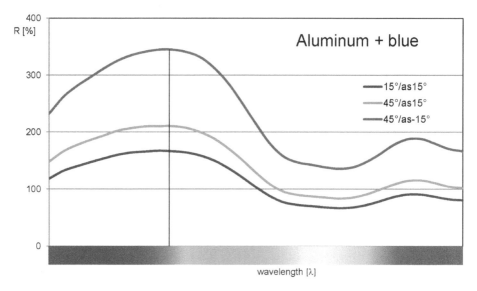

**Figure 22.50:** Even when mixed with color pigments, aluminum pigments do not cause any reflection and color shifts.

## 22.7 Interference pigments

Interference pigments have their model in nature, e.g., butterfly wings, beetle shells and snail shells (Figure 22.51). They are described in detail in chapter Special Effect Pigments. Nowadays, there are different interference pigments that are very popular in car paints, cosmetic products and other decorative objects (Figure 22.52).

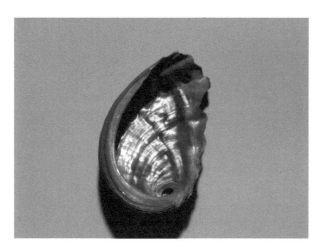

**Figure 22.51:** Nature is showing the way: Iridescent colors.

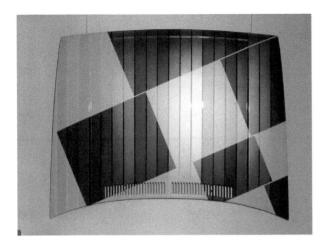

**Figure 22.52:** Three "color states" can be observed on this hood: Above a black background the reflection colors are intensive, above a white background they are clearly weaker. The corresponding transmission colors can be observed in an angle away from gloss above white background.

As their name suggests, their colors and effects are based on the principle of interference [5]. This is where light waves meet and superpose to form a resultant wave of greater, lower, or the same amplitude. Light waves are defined by their length and their maximum. The wavelength are characterized by the wave movement in 1 sec. From 0 over the maximum at 90° back to 0 at 180°, then the wave passes through its minimum at 270°, then back to zero at 360°. Typically, this is what sinusoidal curves look like.

If one wave meets another wave of the same wavelength, the maximum and minimum are amplified. If the two waves had a shift of half a wavelength, the maximum of one wave would meet the minimum of the second wave. Then the waves are extinguished. However, a large wave can also interfere with a small one. Depending on the coincidence, parts of the resulting wave increase or decrease. Pigments reflect in the whole spectral range. Interference phenomena occur for every wavelength. The result is an interference color which is typical for the respective pigment [7].

The interference pigments influence the incident light in such a way that parts are shifted against each other. This causes them to interfere with each other. These pigments consist of a wafer-thin platelet as a carrier material. These platelets are coated with a highly refractive material. If white light hits such a pigment, a part of it is reflected directly on the surface. The remaining part passes through the highly refractive layer by refraction of the light rays. At the boundary layer to the carrier flake, again a part is directly reflected, the other part travels through the carrier flake and the processes are repeated. The part that is reflected at this boundary layer leaves the pigment parallel to the first part. Since the second part travels a longer distance through the highly refractive layer before leaving the pigment, interference occurs. In the process, certain waves are amplified and others are attenuated. These processes depend on the optical path lengthening (path difference), and this depends on the layer thickness of the highly refractive material [8]. The optical principle of light interaction with a particle of an interference pigment is shown in Figure 22.53.

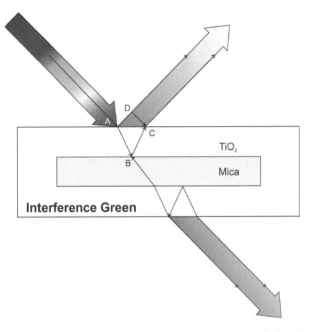

**Figure 22.53:** Interference pigments based on $TiO_2$ coated mica platelets divide the incident light into a reflection and a complementary transmission part.

In most cases, titanium dioxide or iron oxide is used as a highly refractive material to coat the carrier platelets. Natural mica platelets or synthetic materials are used as carrier material. The physical principles of interference are always the same. Vaporized pigments are a special feature: In a high vacuum, different layers are vaporized. Here, too, colors are created due to interference phenomena.

If the light components leave the pigment on the upper side, interference occurs and a typical interference color for this pigment is produced. The remaining light passes through the pigment and also leaves it with interference phenomena. Due to the lack of phase shift – at the transition from the optically thinner to the optically denser medium, but not vice versa – the resulting transmission color is complementary to the reflection color. Figure 22.54 shows the different reflection colors that are visible in the gloss angle on painted metal panels. The dark background color absorbs the transmission colors and is observed off gloss as shown in Figure 22.55.

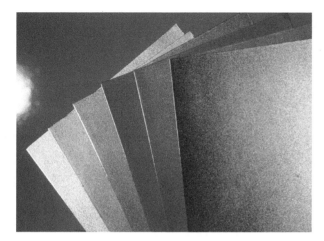

**Figure 22.54:** Different reflection colors are clearly visible in the gloss angle on painted metal panels.

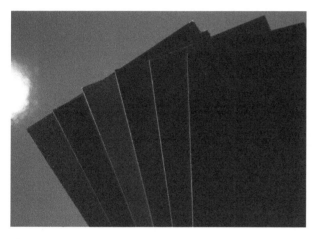

**Figure 22.55:** Away from the gloss angle, the dark background color that absorbs the transmission colors can be seen.

A blue interference pigment appears blue when viewed from above and yellow when viewed through. A green pigment, on the other hand, appears green in the top view and red in the transmission. Figure 22.56 shows that the perceived interference colors are also dependent on the surrounding medium.

If the coating of the carrier platelet (for example mica) is low, silver–white interference pigments are obtained. If the coating is increased, yellow and then red interference pigments can be produced. Here, reflection shifts to longer wavelengths, with a reflection maximum shifting from the UV to the visible spectral range: further increase of the titanium dioxide layer then leads to blue and finally

**Figure 22.56:** The resulting interference colors are also dependent on the surrounding medium. Here, water droplets cause a color shift compared to the "basic color".

to green interference pigments. The dependence of the interference color of $TiO_2$ mica pigments on the thickness of the $TiO_2$ layer is illustrated in Figure 22.57.

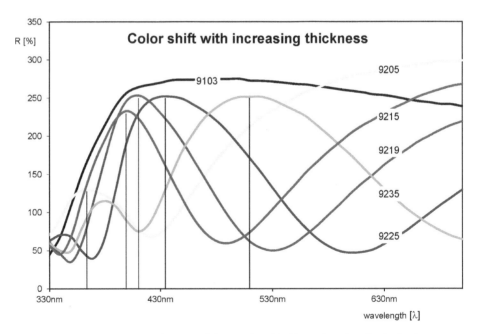

**Figure 22.57:** Different layer thicknesses of the titanium dioxide on the mica platelet lead to different reflection colors. As the thickness of the layer increases, silver–white changes to yellow and then to green via red, purple and blue. A shift of the reflections into longer wavelengths takes place.

The range of interference pigments available varies: titanium dioxide-coated pigments are available in all colors from yellow (gold) to orange, red, purple, blue and green. Iron oxide-coated pigments cover the red range, while so-called combination pigments cover the gold and green ranges [9].

The closer an interference pigment is to the angle of gloss – where the incident light is reflected – the stronger is the reflection color. As with aluminum pigments, their intensity and brightness decrease the further the human eye looks at the pigment from the gloss angle. These pigments have a special feature: The resulting reflection color depends on the angle of the incident light: This determines the optical path extension and thus the reflection color. The range of the color shift due to the angle change can be relatively large and in the extreme case can range from yellow to orange, red, violet and blue to green. Most interference pigments have a range of about 50 nm: Thus, the color of a green interference pigment shifts from yellowish to bluish green when it is illuminated flatter. Examples for these interactions are shown in Figure 22.58–22.61.

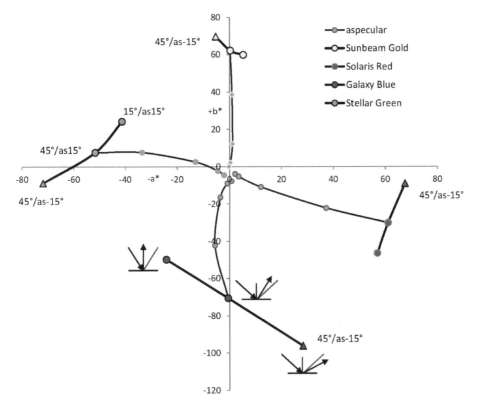

**Figure 22.58:** Reflection color of interference pigments shifts to the shorter wavelength when they are illuminated flatter. A red interference pigment becomes more yellowish, a yellow one greenish, a green one bluish, and a blue one reddish. In the diagram, these color shifts are marked with triangles.

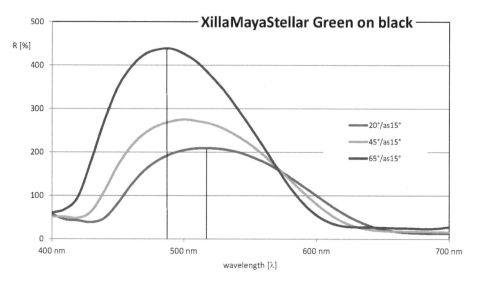

**Figure 22.59:** If interference pigments are illuminated more flatly, their reflection curve shifts to the shorter wavelengths.

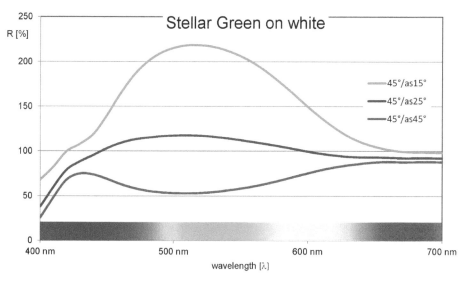

**Figure 22.60:** Above a white background, the reflection color changes to the complementary transmission color.

These shifts are typical for an interference pigment; they always occur in the shorter wavelength spectral range when the pigment is illuminated more flatly. An interference red becomes more yellowish, an interference yellow greenish, an interference green bluish, and an interference blue violet when illuminated at a lower

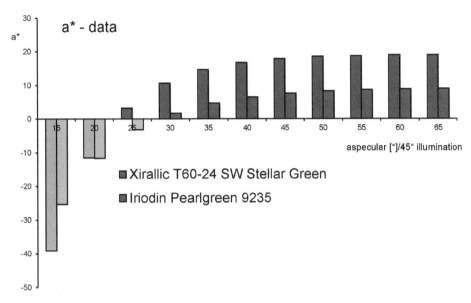

**Figure 22.61:** Change from the reflection color to the complementary transmission color becomes clear with green interference pigments when the a* values are plotted against the difference angles to the gloss (aspecular).

level. This optical property can also be used for their identification and characterization [10].

An interference pigment is characterized by its interference and aspecular line [11]. The interference line results from the measured values at steep, classical and flat illumination while difference angle to gloss (aspecular) keeps the same. With the exception of white interference pigments, there is a color shift that is typical for the respective interference pigment. It always occurs toward shorter wavelengths when illumination is flatter. If the pigment is measured at a fixed illumination angle with different angles of difference from the gloss angle (aspecular), the measured values form the aspecular line. These two lines form an anchor shape that characterizes each interference pigment and can be used to describe it [12]. The interference line always runs counterclockwise in the a*b* chart under flat illumination, as shown in Figures 22.62 and 22.63. The ball in Figures 22.64 and 22.65 shows the strong influence of the angle of illumination on the visible interference effects.

## 22.8 Mixing behavior of interference pigments

Interference pigments are usually transparent and divide the incident light into two parts: The reflection parts give the typical reflection color and the transmission parts give the complementary transmission color. As these pigments divide the

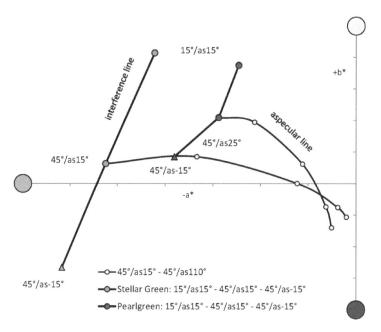

**Figure 22.62:** Two different interference pigments can be clearly distinguished by their interference and aspecular lines.

light only and do not change anything else, they behave like lights, which show an additive mixing behavior [13].

An interference yellow mixes with interference blue to white. With colored pigments one would get green, as is well known. An interference green mixes with an interference red to produce a light yellow. These blends can be extended to achieve any desired intermediate colors [14].

Interesting results are also achieved with colored pigments. In the beginning these were mainly mixed with white interference pigments, see differences in Figure 22.66. These mixed colors already had their color appeal, with colored interference pigments this can be increased considerably. When mixing, similar behavior applies as when mixing colored pigments with white: green and blue colored pigments increase the chroma up to a turning point. Yellow colored pigments are too light to give interference pigments even more color effect. White pigments are too large and, as with aluminum pigments, interfere with the effect. Black, on the other hand, increases the effect while absorbing the transmission color. Since the possibilities of mixtures are so manifold, each user has to carry out his own experiments. Figure 22.67 shows exemplary the manifold design possibilities with interference pigments.

Since most interference pigments are transparent, the background color plays a major role: If transparent interference pigments are applied to a white background,

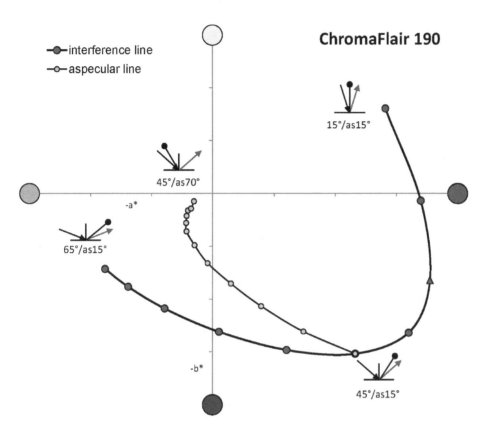

**Figure 22.63:** Pigment in the diagram has the largest color gradient from yellow to red, violet and blue to green.

the reflection color typical for this pigment can be seen near the gloss angle. If one moves away from the gloss angle at the observation point, one looks through the interference pigment and recognizes its transmission color that is reflected by the white background. If a transparent interference pigment is applied over a black background, the reflection color appears much stronger. And the transmission color is absorbed by the black background (Figures 22.54 and 22.55).

Interesting effects can be "conjured up" in this way with different colored backgrounds!

In industrial applications such as car painting, one tries to keep this transparency as low as possible. In most cases this is done by adding aluminum pigments to the car paint. Although aluminum pigments interfere with the actual interference effect, small quantities ensure a better application. It is interesting here that aluminum pigments reflect much more strongly near the gloss angle than white interference pigments.

**Figure 22.64:** Interference effects can be displayed on a ball.

**Figure 22.65:** Color changes on the ball depend on the angle of illumination.

## 22.9 Luminescent pigments

Luminescent pigments are divided into fluorescent and phosphorescent pigments. They have a strong attraction due to their extraordinary luminosity. When luminescent pigments are illuminated by UV-light or even visible light, the incident light is converted into light of longer wavelengths. Especially yellow and red luminescent colors appear very intense. In addition to these colors, blue, green (usually mixtures of blue and yellow luminescent pigments) and orange luminescent pigments are available. If the light conversion of luminescent pigments occurs immediately, they are called fluorescent pigments. If the conversion takes place with a time delay, they are called phosphorescent, but also afterglow pigments. Examples of luminescent pigments under normal light conditions or no illumination are shown in Figures 22.68 and 22.69.

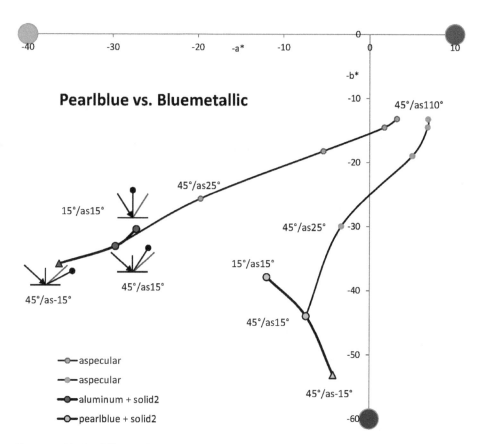

**Figure 22.66:** The difference between mixtures of an aluminum pigment and an interference pigment with a color pigment can be seen in the interference line. In the case of the aluminum pigment, the interference line runs almost as a continuation of the aspecular line. The interference pigment shows the typical anchor shape with the interference line running counterclockwise.

Low lightfastness is a common attribute of all organic fluorescent colorants. It is recommended to apply these pigments in a highly pigmented manner and to cover them with a UV clear coat. Figures 22.70–22.72 show application examples with fluorescent pigments.

## 22.10 And that too!

In addition to color as an optical property, structure, texture and sparkle are also used for description. Together with the color, they give the so-called appearance. These properties influence the human perception of color. For example, if a dashboard of a

**Figure 22.67:** There are almost no limits to the color designs with interference pigments.

**Figure 22.68:** Luminescent pigments are rather inconspicuous under normal lighting.

**Figure 22.69:** Luminescent pigments develop their bright colors in the dark after being illuminated.

**Figure 22.70:** Luminescent colors have a special charm. Compared with normal chromatic colors, their colors are intensified by the conversion of UV light or visible light.

**Figure 22.71:** Organic fluorescent pigments require special application techniques and care because they are not light stable.

vehicle is coarse or fine structured, the human perception changes. Artists have also dealt with textures and processed them in their pictures.

Sparkle is one of the properties used to describe an interference pigment more precisely. Today, the determination of Sparkle is extended to all effect pigments, but this makes little sense. Sparkle's current calculation does not result in physically defined values. The Sparkle values are not valid for the respective pigment and can only be used in direct comparison. Modern measurement methods can better distinguish effects by color measurement.

In all considerations concerning colors, it should always be in the foreground that colors only exist in people's minds. Physical-optical processes that manipulate white sunlight in some way take place in front of human eyes. When rays of light reach the eye, they trigger an optical stimulus there, which is translated into a color in the brain.

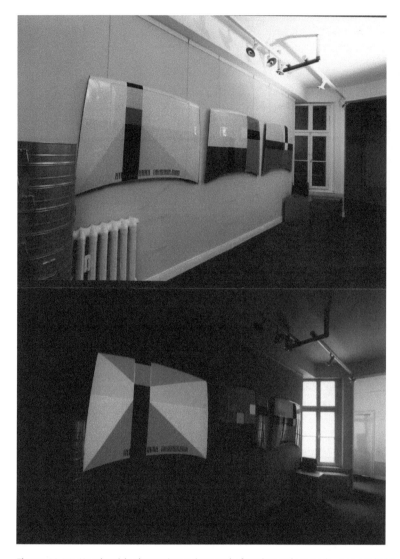

**Figure 22.72:** Hoods with absorption colors and afterglow colors under normal light conditions and in the dark.

## References

1. Grassmann H. Zur Theorie der Farbenmischung. Poggendorf Ann Phys. 1853;vol. 165:69.
2. Goethe JW. Zur Farbenlehre. 1812.
3. Hering E. Zur Lehre vom Lichtsinn. 1878.
4. Völz HG. Ber Bunsen-Ges Phys Chem. 1967;71:326.

5. Hempelmann U. In industrial inorganic pigments. Buxbaum G, Pfaff G, editors. 3rd ed. Weinheim: Wiley-VCH Verlag, 2005:11.
6. Pfaff G. Inorganic pigments. Berlin/Boston: WalterdeGruyterGmbH, 2017:1.
7. Cramer WR. Autolackdesign. Stuttgart: Gentner Verlag, 1987.
8. Cramer WR. Paint Coat Ind. 2001;9:36.
9. Cramer WR. Paint Coat Ind. 2006;9:38.
10. Cramer WR. Color Res Appl. 2002;8:276.
11. Kirchner E, Cramer WR. Color Res Appl. 2012;6:186.
12. Cramer WR. China Coat J. 2019;9:52.
13. Cramer WR. PCI China. 2018;9:34.
14. Cramer WR. China Coat J. 2018;6:64.

# Index

https://doi.org/10.1515/9783110588071-205